AF449150

High Temperature Gas Dynamics

Springer
Berlin
Heidelberg
New York
Hong Kong
London
Milan
Paris
Tokyo

Physics and Astronomy ONLINE LIBRARY

springeronline.com

Tarit K. Bose

High Temperature Gas Dynamics

With 81 Figures, 32 Tables,
and 36 Exercises

 Springer

Dr. Tarit K. Bose
17853 Sherman Way
Reseda, CA 91335-3327
USA
email: tkbose@earthlink.net

Cover picture: A supersonic free plasma jet at moderately low pressures (around 100 Torr).
Courtesy of Joachim V. Heberlein, Minneapolis/MN (USA)

Enlarged, completely revised and updated edition of T.K Bose's *High Temperature Gas Dynamics*
First published in 1979 by THE MACMILLAN COMPANY OF INDIA LIMITED
(c) Tarit Kumar Bose, 1979

Library of Congress Cataloging-in-Publication Data.

Bose, T.K. (Tarit Kumar), 1939– .
High temperature gas dynamics/Tarit K. Bose. p.cm.
Includes bibliographical references and index.
ISBN 3-540-40885-1 (acid-free paper)
1. Gases at high temperatures. 2. High temperature plasmas. 3. Gas dynamics. I. Title.
QC164.5.B67 2004   533'.2–dc22   2003061497

ISBN 3-540-40885-1 Springer-Verlag Berlin Heidelberg New York

Springer-Verlag is a part of Springer Science+Business Media

springeronline.com

© Springer-Verlag Berlin Heidelberg 2004
Printed in Germany

Typesetting: Data prepared by the author using a Springer LATEX macro package
Final processing: Frank Herweg, Leutershausen
Cover design: *design & production* GmbH, Heidelberg

Printed on acid-free paper   55/3141/tr   5 4 3 2 1 0

To My Teacher
Late Professor Dr.-Ing. Fran Bosnjakovic
Former Director and Chair
Institut für Thermodynamik der Luft- und Raumfahrt
Technical University, Stuttgart, Germany

Preface

Since the late fifties, when the race for more and more ambitious projects in Aerospace in many countries started, there has been a necessity for a good book in this subject area for engineering students, which is the motivation for writing this book. There is no doubt that a number of books are already available in the market, but they are generally either too mathematical, or they cater mainly to the interests of students of physics.

This book is an outgrowth of lectures given by the author to the undergraduate and graduate students of aerospace engineering at the Indian Institute of Technology, Madras, India, specializing in the fields of aerodynamics and propulsion. In their later profession these students are increasingly called to tackle real gas problems of hypersonic flight speeds including reentry, high rate of heat flux in nozzles and reentry bodies, and for various exotic and sophisticated high temperature manufacturing processes. It was, therefore, necessary to develop a course of lectures containing the fundamentals of the high temperature gases, the effects of the high temperature on the thermophysical, transport and other properties, the diagnostic techniques, and the preliminaries about the behaviour of the ionized gases in electromagnetic fields. These topics, however, belong to such diversified areas as statistical thermodynamics, kinetic theory of gases, plasma physics, plasma diagnostic techniques, magnetogasdynamics and conventional gasdynamics. Treatment of these topics have been kept to a level at which a student with adequate mathematics and physics background should understand. Therefore at various places in the book the derivation of equations has been done in considerable detail. In this connection mention must be made of the chapter on diagnostic techniques, because some reviewers felt that the optical techniques discussed are too preliminary since very advanced techniques are nowadays used for collection and evaluation of the optical signals. However this author feels the necessity of an approach with prisms and lenses for the understanding of the fundamentals. No single book can possibly cover all the topics handled in the book adequately and there could be differences of opinion about the best way of treating a topic. For example, the so-called the Monte Carlo method may be considered to be the method for radiation gas dynamics, but this author feels that for general engineering students a more deterministic method would be appropriate. A book, including all above topics was, therefore, published several years back and in a limited indian edition, but it is not available now. In the meantime, there has been considerable progress in the subject, and it was thought necessary to re-write the book with the inclusion of much additional material.

Some of the subject areas of immediate interest, which were developed during the last thirty years concern two- and multi-temperature plasmas and these have been described in this book in considerable detail. The topics have been chosen no doubt from the personal interest and areas of research of this author. Study of interaction between the hot gas containing charged particles and the electromagnetic fields, especially the conditions under which Alfvén and other electromagnetic shocks are generated was, personally speaking, very fascinating. I would be happy to discuss any of the topics in this book with readers if I am contacted through e-mail under "tkbose@earthlink.net". While the manuscript of this book was originally written at the time the author was a Professor of Aerospace Engineering at the Indian Institute of Technology, Madras (current name: Chennai), India, he has, however, retired in 1998 and moved to California.

Reproduction from several sources has been done in this book with the permission of the authors and publishers. This has been acknowledged at proper places. Thanks are also due to the publisher, Springer-Verlag, for the excellent job done in publishing this book, especially in helping to convert from *Corel Ventura*, the desktop language used initially to write the manuscript, to LaTeX, which is the preferred Springer typesetting system for book production.

The figures for this book were drawn generally with a computer, but where it was too cumbersome, these were drawn originally by Mr. Karuppaiah of the Indian Institute of Technology Madras. Charts in Appendices are reproduced from my earlier book, which again were reproduced from the book by Bosnjakovic by permission of Verlag Theodor Steinkoff. Similarly the statistical weights and energy levels given in Table 6.2 were published by permission of Springer-Verlag and similar tables given in appendix A from my earlier book. Some other results in Chaps. 8 and 9 were published in my earlier book by permission of Dover Publications. This book was typeset completely on a PC by the author personally. In addition, I would like to thank Professor Heberlein of the High Temperature Lab, Dept. of Mechanical Engineering, University of Minnesota, Minneapolis, USA, for the use of his figure of a supersonic free plasma jet at moderately low pressures.

Finally, I would like to thank my wife, Preetishree, and the three children, Mohua, Mayukh and Manjul, for having put up with me during the writing of this book.

Reseda/CA, USA, October 2003   *Tarit Bose*

Contents

Physical Constants

c = velocity of light in vacuum = 3.0×10^8 ms^{-1}

e = elementary charge = 1.602×10^{-19} As

h = Planck's constant = 6.63×10^{-34} Js

k_B = Boltzmann constant = 1.38×10^{-23} J.K^{-1} = 8.61×10^{-5} eV.K^{-1}

M_e = mass of an electron = 9.108×10^{-31} kg

M_p = Mass of a proton = 1.672×10^{-27} kg = 1836.5 M_e

N_A = Avogadro number = $(6.02544 \pm 0.0004) \times 10^{26}$ molecules.kmole^{-1}

R^* = universal gas constant = $k_B.N_A$ = 8314 J.(kmole.K)$^{-1}$

R_H = Rydberg constant for hydrogen = 10967757.6 ± 3.12 m^{-1}

ϵ_o = dielectric constant in vacuum = 8.8550×10^{-12} As.(Vm)$^{-1}$

μ_o = magnetic permeability in vacuum = 1.25664×10^{-6} Vs.(Am)$^{-1}$

σ = Boltzmann constant of radiation = 5.672×10^{-8} Wm^{-2}K^{-4}

σ = Boltzmann constant of radiation = 3.0×10^8 ms^{-1}

Conversion Factors

$$
\begin{aligned}
1 \text{ Newton (N)} \quad &= 1 \text{ kgm}^2\text{s}^{-1}\\
1 \text{ Joule (J)} \quad &= 1 \text{ Nm} = 1 \text{ kgm}^2\text{s}^{-2}\\
1 \text{ Watt (W)} \quad &= 1 \text{ Js}^{-1} = 1 \text{ Nms}^{-1} = 1 \text{ kgm}^2\text{s}^{-3}\\
&= 1 \text{ Volt(V)} \times \text{Ampere(A)}\\
1 \text{ Gauss (G)} \quad &= 10^{-4} \text{ Vsm}^{-2} = 10^{-4} \text{ Tesla (T)}\\
1 \text{ Ohm } (\Omega) \quad &= 1 \text{ VA}^{-1}\\
1 \text{ kmole} \quad &= 1 \text{ kg-mole}\\
1 \text{ Poise (P)} \quad &= 10^{-1} \text{ kgm}^{-1}\text{s}^{-1}\\
1 \text{ Angstrom(\AA)} \quad &= 10^{-10} \text{ m}\\
1 \text{ bar} \quad &= 10^5 \text{Nm}^{-2} = 10^5 \text{ Pascal (Pa)}\\
1 \text{ atm} \quad &= 1.013 \text{ bar}
\end{aligned}
$$

Energy units (explained later in text) used generally are in J, K, cm^{-1} and eV. These can be converted from one unit to another using the following relations:

$$
\begin{aligned}
1 \text{ electron volt (eV)} \quad &= 1.602 \times 10^{-19} \text{ J} = 11614.4 \text{ K} = 8058.26 \text{ cm}^{-1}\\
1 \text{ J} \quad &= 7.246377 \times 10^{22} \text{ K}\\
1 \text{ cm}^{-1} \quad &= 1.441304 \text{ K} = 1.989 \times 10^{-23} \text{ J}\\
1 \text{ Tesla} \quad &= 1 \text{ Vsm}^{-2}
\end{aligned}
$$

1 Introduction

All in a hot and copper sky
The bloody Sun at noon
Right up above the mast did start
No bigger than the moon.
(From "The Ancient Mariner"
by Samuel Taylor Coleridge)

From ancient times the sun, apparently the largest and the brightest of stars visible to the naked eye, has been both an object of reverence and awe for the common man and a source of inspiration and inquisitiveness. It is also the source of all conventional energy on the Earth. In ancient Egypt, the sun god Ra was the dominating figure among the high gods. In India the sun is glorified in the vedic hymns and there is even a dynasty of sun kings. There are over 200 known quotations relating to the sun in English poetry. In recent times, all over the world, the scientists, faced with the crisis of the conventional energy sources, are working hard to develop an artificial controllable sun in the laboratory by nuclear fusion.

While the development of a practical artificial controllable sun will take some more time, indeed years, there are many technologies in practical life in which the application of high temperature gas is required. For example, it is a well-known fact to thermodynamicists and heat power engineers that the thermal cycle efficiency of a system depends on the ratio of the highest to the lowest temperature in the system. Thus, any consideration for an increase in the thermal efficiency of the system, especially in view of the very severe energy crisis these days, must necessarily mean an increase of this temperature ratio. Now in a thermodynamic cycle the lowest temperature is restricted by the ambient temperature. It is, therefore, evident that any effort to increase the thermal cycle efficiency of a system will mean necessarily that the highest temperature in the system must be pushed higher and higher. Corollaries to this are that ways and means must be found to have better cooling of strongly heated parts of the equipment, better materials, and more radical designs. As an example, in gas turbine engines the maximum possible gas temperature at the time of development of the first Whittle engine was around $750°C$, which has increased to about $1,500°C$ in recent years. It is probable that this temperature may not increase substantially in the future, since the maximum stoichiometric temperature for aviation kerosene with air is around $1,800°C$. Even if a different fuel is used and air is replaced by pure oxygen, an increase in the turbine inlet temperature is very unlikely, despite the best efforts to cool the blades and to develop new materials. It has, therefore, been suggested to combine a conventional gas turbine engine with a previously unconventional magnetogasdynamic system, which, with hardly any moving parts, can operate at much higher temperatures. Furthermore, for aerospace applications like rocket motors or in high-speed wind tunnels, the gas should be heated to as high a temperature as possible before expansion to high speeds through convergent-divergent nozzles. Currently this means densities of particles on the order of 10^{24} per m^3 and temperatures up to about 15,000 K. This temperature limit is increased by several orders of magnitude in atomic bomb fireballs, the solar interior, and in hydrogen bomb explosions.

Since late 1950s, there has been tremendous space travel activity with substantial focus on bringing human passengers back to the earth safely. Man has gone to the moon and returned, and man-made probes have gone to most of the planets around the sun. The technologies associated with space travel have brought with them a betterment of human life here on Earth, for example in meteorology, communication, weather forecasting, television, etc. For these purposes there is a need for long-term positioning of satellites in Earth orbit. For orbit correction, ion and plasma propulsion devices have been in use for a number of years. The industry uses several plasma devices regularly, for example, in plasma cutting, plasma spraying, etc. All of these require an understanding of the behavior of high temperature gases and their interaction with electric and magnetic fields.

It can thus be seen that there are many cases where knowledge of thermo-physical and transport properties of high temperature gases, as well as their gas dynamic behavior in the presence of electric and/or magnetic fields is essential. For example, the properties of flow through a convergent-divergent nozzle, at moderate temperatures, can easily be calculated for known constant value of specific heat ratio $\gamma = C_p/C_v$. For a gas like nitrogen, in which the contribution of the vibrational energy to the total energy at room temperatures is small, $\gamma = 7/5 = 1.4$, but at an elevated temperature, say 3,000 K, there is a full contribution of the vibrational energy and $\gamma = 9/7 = 1.29$. At temperatures around 7,000 K and a pressure around 1 atm, the nitrogen molecules are completely dissociated, and for these a value of $\gamma = 5/3$ is valid. Finally at temperatures above 20,000 K, where the nitrogen atoms at a pressure of 1 atm are completely broken into singly or multiple-charged ions and electrons, because of very large values of the ionization energy, it is of the order of one. Furthermore, in the above temperature range, the mole mass of the nitrogen gas decreases continuously from a value for a nitrogen molecule gas of 28 kg/kmole, to the value for a nitrogen atom gas of 14 kg/kmole, and then further to a very small value for the gas mixture in which the electrons are the main constituents. It can, therefore, be seen that for nozzle calculations for high temperature gases, variation of the specific heat ratio and the mole mass of the gas have to be taken into account, complicating the computation procedure considerably. Similar to the preceding gas properties, other transport properties like the viscosity coefficient, the thermal conductivity coefficient and the diffusion coefficient change their values considerably with temperature, as well as some of the dimensionless transport properties parameters like the Prandtl or Schmidt numbers.

We ask now the question. What is the temperature? While classical thermodynamics fails to give a definite answer on this, it is answered through the concept of equilibrium. According to the zeroth law of thermodynamics, if two bodies brought in contact with each other are in equilibrium, then they both have the same temperature. Further, the second law states that heat will flow from the higher to lower temperature, which is a non-equilibrium condition. Here, again, the temperature itself is not defined.

A slightly different question is about the nature of different energy forms in solids, liquids or gases, and pursuit of an answer has led to the concepts of kinetic and potential energies. Therefore, the gas molecules have energy forms like translational, rotational, vibrational, electronic excitation energy, etc., and any changes in any of these energy forms manifest themselves in the form of radiative electro-magnetic energy. The concept of various energy forms has led also to the question of their distribution and to the definition of temperature in terms of statistical distribution of the energy form under equilibrium conditions. Any flow of mass, momentum or kinetic energy, or electric current is now computed from the first order perturbation of the equilibrium values.

A related question is, therefore, what is non-equilibrium? It can be again of various types. A chemical non-equilibrium is when the molecular particles react with each other to reach a chemical equilibrium state. On the other hand, a thermal non-equilibrium may be when the various energy forms can be described by different temperatures. For example if a strong current is passed through a gas, the electrons may absorb more electro-magnetic energy than the heavy particles, but they may not be able to give up that energy to the heavy particles because of inefficient energy transfer due to collision and, therefore, the electrons and heavy particles may have effectively two different temperatures. Similarly in an expanding gas flowing through a convergent-divergent nozzle or in a gas dynamic shock, the translational and rotational temperature may change quickly, but not the vibrational temperature.

Because of the dissociation and ionization of gases at higher temperatures, there would be areas of variable concentration of different specie, as a result of which the particles tend to diffuse from the regions of higher concentration to those of lower concentration. Thus in electric discharges, the ions and the electrons tend to diffuse out from hotter to cooler regions and recombine there; similar recombinations take place at lower temperatures between the atoms to become molecules. These diffusions and recombinations give rise to energy transport from hotter to cooler regions and may allow energy transfer at least of the same order of magnitude as by pure conduction.

Ionization of gases at high temperatures gives additional properties to gases in the presence of electromagnetic fields. This induced many authors to talk about the ionized gas as *the fourth state of matter*, the first three states being the solid, the liquid and the non-ionized gas state. Another name for the gas mixture consisting of charged particles was used by the American scientist I. Langmuir, who called such a gas mixture a *gas plasma* – a similar name to blood plasma, known to the physicians and the medical students, being only incidental. With the help of electromagnetic fields, the magnitude and the direction of velocity of such high temperature gases can be altered and they can be confined into a space without touching the solid boundaries, increasing the prospects for radical designs of equipment. For the uninitiated, however, the distinguishing feature of an ionized gas is its ability to conduct electricity.

This book starts with a discussion on the rudimentaries of modern quantum mechanics, followed by an introduction to statistical mechanics and the methods to obtain the thermophysical properties. From the concept of energy levels, the principles to determine the radiative properties of high temperature gases are discussed exhaustively, so that the ranges of *optically thin* and *optically thick* radiation are clearly delineated. Further, the collision processes between different particles are discussed, and expressions for the collision frequency and the mean free path are given. It is shown that the charged particles of a collisionless plasma gyrate around the magnetic field lines, and in a collision-dominated plasma the effect of a strong magnetic field is to give tensor properties to the transport properties.

Separate chapters deal with the production of high temperature gases, and with electron emission, if the gas is heated by electrical means. It is shown that there is actually a considerable reduction in the adiabatic flame temperature during combustion, as well as in the temperature behind a strong shock, only if real variations of the gas properties are taken into account. Finally a chapter is devoted to diagnostic techniques for the high temperature gases, followed by a chapter on the gas-dynamic equations and gas-dynamic interactions with the electromagnetic fields, which brings us to the actual title of this book.

2 Introduction to Quantum Mechanics

In 1802 *Dalton* formulated the *law of multiple proportions*, which states that if two elements combine in more than one proportion to form different compounds the masses of one of the elements with identical amounts of the second element are in the ratio of integral numbers. In 1833 *Faraday* found the *law of electrolysis* as a proof for the existence of an electrical elementary quantum of charge. These discoveries supported the postulation of the particle (atom, molecule) theory of matter. On this basis, during the second half of the 19th century, the *mechanical theory of heat* was first formulated by *Clausius*, and was further developed by *Maxwell* and *Boltzmann*. Mechanical explanation of the pressure of a gas in a closed vessel, as well as the phenomenon of linear increase of pressure with temperature was possible in this way. In 1811 *Avogadro's hypothesis* was formulated, which stated that equal volume of different gases, under the same conditions of temperature and pressure, contain equal number of molecules. In 1869 *Mendelejeff* and *Lothar Meyer* developed *Periodic Tables*. They arranged the elements columnwise on the basis of certain chemical properties, and it was found subsequently, that these properties are dependent on the number of electrons on the outermost orbit around the nucleus of the element. Proceeding from *Rayleigh-Jean law of radiation* for large wave lengths, and from *Wien's law* for short wavelengths, *Planck* in 1900 combined these two laws semi-empirically and found the famous law of radiation that bears his name. From his analysis, for the first time, the existence of an elementary quantum of radiation was found. A rigorous explanation of *Planck's law* was, however, left to *Albert Einstein*, who in the 1920s applied the results of a statistic developed by an Indian scientist *Satyendra Nath Bose* to the light particles (photons). In 1905, based on astronomical experiments, Einstein also formulated his *theory of relativity* and gave for the first time a mass-energy equivalence principle. From the alpha particle scattering experiments in 1906–13, *Rutherford* concluded that the mass of the atom should almost be totally concentrated around a very dense nucleus. He further suggested a simple model of an atom consisting of a very dense nucleus around which the electrons move in orbits. The electrons are kept in orbits by a balance between the *centrifugal* and *Coulomb forces*. His theory, however, could not explain how the electrons could stay in orbit without any dipole radiation, which would result in their slowing down. An explanation for this was left to the Danish scientist, *Niels Bohr*, who, by drawing help from another branch of physics, namely, spectroscopy, could explain in 1913 the nature and radius of the electron orbit. This is discussed in detail in the following section.

Although the principles of radiation energy, quantum energy and the energy-mass equivalence were known at the time of *Bohr*, particles and radiation energy were considered as two different natural objects. However, various experiments showed that under certain circumstances light behaved like particles, and particles exhibited the wave nature. Thus in 1924, *De Broglie* formulated his principle of *particle-wave dualism*. Based on this, in 1926, *Heisenberg* and *von Schrödinger* formulated, by two independent methods, the two important theories named after them, and thus founded the basic principles of modern quantum mechanics.

2.1 Line, Band and Continuous Spectra Bohr's Atomic Theory

When the science of spectroscopy was in its infancy, the spectra were divided by their appearance, as given in Fig. 2.1, into *line*, *band* and *continuous spectra* . Before man learnt the origin of these different types of spectra, which have been identified later as due to atomic, molecular and solid body radiation respectively, it was obvious from the line spectra of hydrogen, given schematically in Fig. 2.1a, that there must have been a relation between the wave length or frequency of different lines.

The first series law of spectra in the visible region of hydrogen was found in 1885 by *Balmer*, a school teacher from Basel, and is given by the relation

$$\nu \propto (2^{-2} - n^{-2}), (n = 3, 4, 5, \ldots) \quad . \tag{2.1}$$

Although this equation could not be explained by the *Rutherford model of atomic structure*, it was thought that the orbiting electron around the nucleus must represent a dipole and thus must be capable of radiation. The main difficulty, however, was that this model could not explain the occurrence of a

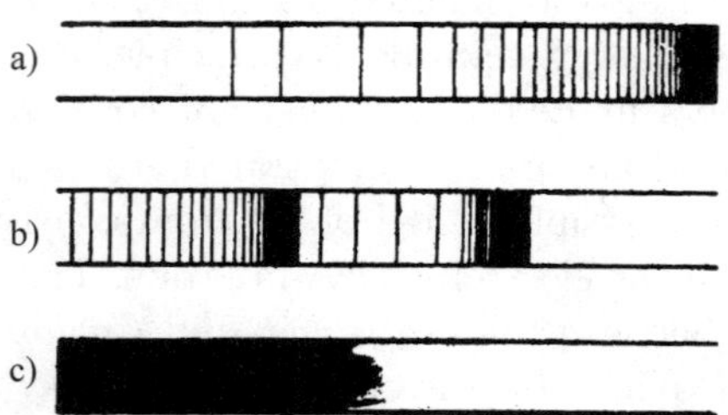

Fig. 2.1. Types of spectra: (**a**) line, (**b**) band and (**c**) continuous spectra

discrete line radiation on account of their being no possibility of having stable orbits. In 1913 *Bohr* gave the solution to this problem by postulating that the electrons travel in certain stationary orbits in which they do not radiate energy, and that in such an orbit the circular line integral of the electron angular momentum should be a multiple of the basic quantum number h. Thus the mathematical formulation of *Bohr's hypothesis* for a circular orbit of a hydrogen atom is

$$\oint M_e w_e \mathrm{d}s = 2\pi M_e w r = nh, (n = 1, 2, 3, 4) \tag{2.2}$$

where w is the orbiting velocity of the electrons. Further, *Bohr* postulated that the atom emits (or absorbs) a quantum of electromagnetic radiation when the electron transits from one orbit to another. The frequency of the radiation is dependent on the energy difference ΔE between the two orbits and is given by

$$\Delta E = h\nu \quad . \tag{2.3}$$

Total energy in an orbit consists of the potential and kinetic energies

$$E = E_{\mathrm{pot}} + E_{\mathrm{kin}} = -C\frac{e^2}{r} + \frac{1}{2}M_e w^2 \tag{2.4}$$

where $C = 1/(4\pi\epsilon_o) = 8.986\times10^9$ Vm(As)$^{-1}$.

The force balance equation (centrifugal force = Coulomb force) in the orbit for hydrogen atom (one proton, one electron) is

$$C\frac{e^2}{r^2} = M_e \frac{w^2}{r} \quad . \tag{2.5}$$

From (2.2) and (2.5), we get two equations for orbital speed and radius,

$$w = r\omega = \frac{2\pi e^2 C}{nh} \tag{2.6}$$

and

$$r = \frac{Ce^2}{M_e w^2} = \frac{n^2 h^2}{4\pi e^2 M_e C} \tag{2.7}$$

where $\omega = w/r$ is the *orbital radian frequency*. Substituting the approximate values for electrons, we get for *hydrogen atom*

$$w = 2.18 \times 10^6/n, \mathrm{ms}^{-1} \tag{2.8}$$

and

$$r = 0.53 \times n^2, \text{Å} \quad . \tag{2.9}$$

Thus the *orbital frequency* for the hydrogen atom is

$$\nu_{orb} = \frac{w}{2\pi r} = \frac{6.563 \times 10^{15}}{n^3}, \mathrm{s}^{-1} \quad . \tag{2.10}$$

In the ground state ($n = 1$) for hydrogen atom, the collision cross section, therefore, is $\pi r^2 = 0.8825$ (Å)^2 and is of the same order of magnitude as it is obtained from the measurement of the transport properties. Substituting the relations for w and r, (2.6, 2.7), into (2.4), the total energy of the orbiting electron is

$$E = -\frac{2\pi^2 e^4 C^2 M_e}{n^2 h^2} \, . \tag{2.11}$$

which, after substitution of relevant values for electrons in the hydrogen atom, gives

$$E = -2.17 \times 10^{-18}/n^2 [\text{J}] = -13.54/n^2 \text{ , eV} \, . \tag{2.12}$$

From (2.12) it is seen that for $n = 1$, E is a negative quantity, but for n going to infinity (ionization!), where the electron is at an infinite distance from the nucleus, E goes to zero. Thus, according to (2.12), the ionized atom has zero energy, whereas the bound atom has a negative energy. However, from convention, it is found convenient to put zero energy at the ground level ($n = 1$). This is done easily by subtracting the ground level energy and (2.11) becomes

$$E = \frac{2\pi^2 e^4 C^2 M_e}{h^2} \left(1 - \frac{1}{n^2} \right) \, . \tag{2.13}$$

Now substituting (2.11) or (2.13) into (2.3), one gets

$$\nu = \frac{\Delta E}{h} = \frac{E'' - E'}{h} = \frac{2\pi^2 e^4 C^2 M_e}{h^3} \left(\frac{1}{n'^2} - \frac{1}{n''^2} \right) \, . \tag{2.14}$$

From the science of spectroscopy, the spectral frequency ν is where the transition from one energy level to the other takes place. If $n'' > n'$, a transition takes place from the higher energy level E'' to lower energy level E' with consequent release of energy in emission. If however, $n'' < n'$, the radiative transfer is due to absorption. While the frequency of the radiative energy is in s^{-1}, from the convention in spectroscopy it is easier to work with the wave number $\bar{\nu} = \nu/c = 1/\lambda$ where λ is the wave length. Thus from (2.14), the wave number for hydrogen is given by the relation

$$\bar{\nu} = \frac{1}{R'_H} \left(\frac{1}{n'^2} - \frac{1}{n''^2} \right) \tag{2.15}$$

where

$$R'_H = \frac{2\pi^2 e^4 C^2 M_e}{h^3 c} = 10973731.2 \pm 0.8 \text{ , m}^{-1} \tag{2.16}$$

is the so-called *Rydberg constant* for hydrogen. There is a small discrepancy, however, between the frequency or wave length of radiation for hydrogen as given by (2.15) and experimental results. This discrepancy can be taken care of by introducing a small correction, which, as it is physically explained, is due to the electron and the nucleus moving around a common axis, instead of the electron only moving around the nucleus. The corrected Rydberg constant is now

$$R_H = R'_H \left[1 + \frac{M_e}{M_{\text{atom}}} \right] = 10967757.6 \pm 3.12 \text{ , m}^{-1} \text{ .} \tag{2.17}$$

From (2.15), it is now possible to calculate exactly the wave length region of the spectra. In case $n' = 1$ and $n'' = n$, we get from (2.15)

$$\bar{\nu} = \lambda^{-1} = R_H \left[1 - \frac{1}{n^2} \right], n = 2, 3, 4, \ldots \tag{2.18}$$

which is in a series form and is called after its discoverer, the *Lymann series* for the hydrogen atom radiation. The wavelengths of the Lymann series are given by the relation

$$\lambda = \frac{1}{R_H \left[1 - \dfrac{1}{n^2} \right]} = \frac{912}{\left[1 - \dfrac{1}{n^2} \right]}, \text{Å, n} = 2,3,4, \ldots \text{ .} \tag{2.19}$$

For $n = 2, 3, 4, \ldots$, the corresponding wavelengths are 1216, 1026, 972.8, $\ldots$, 912 Å. It is seen that all these lines are in the ultraviolet region. It is, therefore, not surprising that this series was discovered later. The earliest series discovered is the *Balmer series* given by the relation

$$\lambda = \frac{1}{R_H \left[\dfrac{1}{2^2} - \dfrac{1}{n^2} \right]}, n = 3, 4, 5, \ldots \tag{2.20}$$

which is in the visible region. Similarly there are other series as follows:

- *Paschen series*: $\bar{\nu} = \lambda^{-1} = R_H[1/3^2 - 1/n^2]$, $n = 4, 5, 6, \ldots$
- *Bracket series*: $\bar{\nu} = \lambda^{-1} = R_H[1/4^2 - 1/n^2]$, $n = 5, 6, 7, \ldots$
- *Pfund series*: $\bar{\nu} = \lambda^{-1} = R_H[1/5^2 - 1/n^2]$, $n = 6, 7, 8, \ldots$

Series with $n' > 5$ have no names, but the wavelengths for each of these series can be determined easily by the procedure given above.

From (2.13) and the definition of the Rydberg constant, we see now, that the energy level is given by the expression

$$E = R_H hc \left[1 - \frac{1}{n^2} \right] \tag{2.21}$$

in which the proportionality constant $R_H hc = 2.18148702 \times 10^{-18} \text{J} = 13.62$ eV.

For $n = 1,2,3,\ldots$ the corresponding energy levels can be computed from (2.21), and these in electron-volt (eV) are 0, 10.21, 12.11, 12.77, $\ldots$, $13.62 \ldots$ eV. These energy levels for the hydrogen atom are shown in Fig. 2.2, with the corresponding values of n. One can show Fig. 2.2 with possible tran-

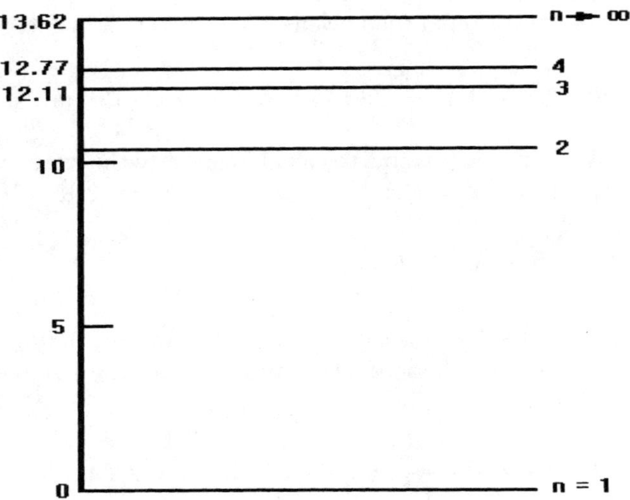

Fig. 2.2. Energy levels for hydrogen atom

sitions, both in emission and absorption. For example, for the Lymann series, all transitions take place from a higher energy level to the ground level, and in absorption from the ground level to a higher energy level.

The unit of the energy level in the present case is shown in J or eV, and these are not the only units currently in existence. We can write energy also in m^{-1} or in cm^{-1}, which is obtained by dividing (2.21) by (hc). In fact, many of the energy level tables given in books of reference are in cm^{-1}, which is the reason to give in this book the conversion factor from cm^{-1} to J or eV. In other books the energy levels are given in K. These can be converted into J or eV by multiplying with the respective conversion factor.

Bohr's atomic model is successful in explaining the spectra of the hydrogen atom in a very lucid manner and with great accuracy. This is reflected in the fact, that the Rydberg constant is obtained theoretically with great accuracy. The model is also valid for all particles which are similar to the hydrogen atom, namely, a single electron around a nucleus like He^+, Li^{++}, ..., etc. In such cases, one can think of an electron around a nucleus of charge $+Ze$, where Z is the charge number. The reader should work out, in analogy to those for hydrogen, the details of calculation to show that the Rydberg constant R'_H in such cases is to be multiplied simply by Z^2 to R'_H. Further it is found that for those atoms which have just one electron in the outermost orbit, which is the case for alkali atoms, the earlier analysis is valid, at least in principle. The failure of the Bohr model for the line spectra of multielectron atoms led Sommerfeld to consider that besides circular orbits, elliptical orbits may also be possible. To describe elliptical orbits more than one quantum number is needed. Hence, n is called the *"principal quantum number"* and can have values $n = 1,2,3,\ldots$. In addition, the following are the other quantum numbers with possible values against each of them:

$- l = angular\ quantum\ number = 1,2,\ldots,(n\text{-}1),n;\ n$ values
$- m = magnetic\ quantum\ number$
    $= -(l-1),-(l-2),\ldots,-2,-1,0,1,2,\ldots,(l-2),(l-1);\ (2l\text{-}1)$ values
$- s = spin\ quantum = \pm\ 1/2;\ 2$ values.

In case we assume, that the electron energy in an atomic structure is dependent on the value of the four quantum numbers n, l, m and s, and assuming, according to *Pauli principle*, that no two electrons in the ground state may have exactly the same energy, it is possible to determine the maximum number of electrons in the orbit (shell) in the ground state. For example, for $n = 1$, l and m can have only 1 value each and s can have only two values. Therefore, the innermost orbit $(n = 1)$ can have a maximum of only two electrons. Similarly for the next higher principal quantum number $(n = 2)$, l can have two values (1 and 2), m can have only 4 values (1 value for $l = 1$ and 3 values for $l = 2$), and s can have two values for each l. Therefore, for the second orbit $(n = 2)$, if $l = 1$, there can be only two values, and if $l = 2$, there can be only six values - a total of 8 values. In a similar manner, we can show that for the next higher principal quantum number $(n = 3)$, there can be a total of 18 values. These results can be generalized by writing $2n^2$ as the total number of electron energy values available in ground state, and has some relevance with the chemical properties of the pure gas.

The *angular quantum numbers* are written, by convention, as $l = $ s, p, d, f, $\ldots$, instead of $l = 1, 2, 3, 4, \ldots$ Even with modification from the single principal quantum number, Bohr's atomic theory can explain and interpret the wave lengths of spectra due to hydrogen or hydrogen-like atoms. However, for many other cases, like the anomalies in *Zeeman effect*, *tunnel effect*, etc., where the particles go through a *potential barrier*, and so on, new theories had to be developed.

2.2 Wave-Particle Dualism and Wave Mechanics

In spite of success with the wave theory of light in providing explanations for phenomena such as interference and refraction, there are many phenomena such as the photoelectric effect, that could not be explained by this theory. The explanation of these phenomena led to the quantum theory of radiation. *Max Planck* in A.D. 1900 postulated that in its interaction with matter, electromagnetic radiation behaved as though it consisted of particles, or quantum of energy, called photons, having an energy given by $E = h\nu$. This showed the dual character of electromagnetic radiation and led *de Broglie*, in 1924, to suggest that a similar dualism might exist for material particles and electrons. From the *mass-energy equivalence* relation of Einstein, $E = Mc^2$, we obtain the relation for momentum of photons as $Mc = h\nu/c = h/\lambda$. Assuming that this equation also applies to material particles and electrons, we have for a particle of mass M moving with velocity w the momentum $p = Mw = h/\lambda$, so that

$$\lambda = h/(Mw) = h/p \; . \tag{2.22}$$

This equation gives the wavelength λ of the hypothetical matter waves associated with the material particles and is the fundamental equation of *de Broglie* theory. Equation (2.22) is now used to work out two specific cases. First, consider a man of mass 60 kg being stationary, which means $w = 0$ and the equivalent wave length $\lambda \to \infty$. If the man walks at the speed of 5 km.h^{-1}, then the equivalent wavelength is 10^{-35} Å. For either of these cases, and also at other velocities in between, the equivalent wave lengths are either too small or too large to detect or to measure. Similarly, while investigating the particle nature of a light of wavelength 5,000 Åwith the speed of light $w = c = 3 \times 10^8$ ms^{-1}, the equivalent mass is $M = 10^{-37}$ kg, which is too small to be measured. Thus it is found that under most test conditions it is possible to recognize either the wave aspect or the particle aspect, but not both at the same time. This raises questions regarding the actual character of the universe bringing these to the realm of philosophical speculations, as is found in the *Upanishads*.

For monochromatic radiation the wavelength may be known, and thus from (2.22), the equivalent momentum can be determined. In 1927 *Heisenberg* put forward the uncertainty principle, which states that the exact simultaneous determination of the position and momentum of a particle is impossible. If Δx is the uncertainty involved in the measurement of the coordinate of a particle [in m] and Δp is the uncertainty in the simultaneous measurement of its momentum [in kgms^{-1}], then

$$\Delta x.\Delta p \approx h \; , \; \text{Js} \; . \tag{2.23}$$

It has been shown clearly from refraction experiments, that the uncertainty is due to the *wave-particle dualism*, and not due to errors in the measurement of certain quantities.

From (2.2) and (2.22), one can write further

$$2\pi M_e wr = nh = nM_e w\lambda \tag{2.24}$$

and thus, the circumference of the orbit is $2\pi r = n\lambda$ ($n = 1,2,3, \ldots$). For $n = 1$, in the ground level, the circumference is equal to a certain wavelength λ. Similarly for an arbitrary n, Bohr's quantum condition is reduced to the condition that there must be an integer number of wavelengths that would just fit into the circular orbit. Such integer number of wavelengths would generate a standing wave instead of a stationary *Bohr orbit*. As a result, while the orbiting electrons move in a path of wave pattern, the orbit, as defined by the Bohr's quantum condition, becomes the path in which an electron has the maximum probability to stay.

This concept has further been exploited by *Schrödinger* by introducing a *wave function*, ψ, which satisfies the general wave equation

$$\frac{1}{c^2}\frac{\partial^2 \psi}{\partial t^2} = \nabla^2 \psi \tag{2.25}$$

where the velocity of propagation is $c = \nu\lambda = h\nu/(Mw)$. Now, if E is the total energy and U the potential energy, then

$$\frac{M}{2}w^2 = E - U \tag{2.26}$$

and therefore,

$$Mw = \sqrt{2M(E-U)} = \frac{h\nu}{c} \ . \tag{2.27}$$

Thus further,

$$c = \frac{h\nu}{\sqrt{2M(E-U)}} \ . \tag{2.28}$$

Let $\psi(x\ ,y,\ z,\ t) = \bar{\psi}(x\ ,y,\ z)\exp^{-2\pi j\nu t}$ in which $j = \sqrt{(-1)}$ and $\bar{\psi}$ is an amplitude function. Thus from (2.25), the final form of *Schrödinger wave equation* is

$$\nabla^2\bar{\psi} + \frac{8\pi^2 M}{h^2}(E-U)\bar{\psi} = 0 \ . \tag{2.29}$$

By comparison with classical problems of wave propagation, it is clear that the solution of $\bar{\psi}$ has the character of an *eigen function*, and is complex. It has further been shown by theoretical physicists that the square of the absolute value obtained by multiplying with its *conjugate complex* $\bar{\psi}^*$ gives the probability of a particle being found at time t, in a volume element at x, y, z formed between x and $x + \mathrm{d}x$, y and $y + \mathrm{d}y$, and z and $z + \mathrm{d}z$. While the general analysis of (2.29) is outside the scope of this book, a few sample results will now be derived.

2.2.1 Rigid Rotors

For this case it is assumed that two equal masses are rotating around a common axis separated by a distance r. Further, let $U = 0$. Replacing M by $2M$, where M is the mass of a single atom, and if φ is the azimuthal coordinate, (2.29) becomes

$$\frac{\mathrm{d}^2\bar{\psi}}{\mathrm{d}\varphi^2} + \frac{16\pi^2 Mr^2}{h^2}E\bar{\psi} = 0 \ . \tag{2.30}$$

Since solutions found are of the type $\bar{\psi}(\varphi) = \bar{\psi}(\varphi + 2\psi)$, condition for periodicity, the general solution sought is of a simple harmonic nature. That is, the trial solution is $\bar{\psi}(\varphi) = \exp^{jJ\varphi}$, where $j = \sqrt{(-1)}$. By substituting the trial solution into the differential equation one gets the particular solution

$$E_J = \frac{h^2}{16\pi^2 M r^2} J^2 = hcBJ^2 = k_B \Theta_r J^2 \tag{2.31}$$

where $J = 0, \pm 1, \pm 2, \ldots$ = *rotational quantum number*, $B = \Theta_r k_B/(hc)$ and Θ_r = *characteristic rotational temperature*.

It may be pointed out that $|J|$ is a measure of the number of nodes in the value of $\bar{\psi}$. Further the characteristic rotational temperature is inversely proportional to Mr^2, which is the *mass moment of inertia.*

For a diatomic molecule, in which two mass, M_1 and M_2 are connected to each other and are at a distance r_1 and r_2 respectively from the *center of gravity*, it is evident, that $r_1 + r_2 = r$, the distance between the atoms, and $M_1 r_1 = M_2 r_2$. Therefore, one can show that the *mass moment of inertia*

$$I = M_1 r_1^2 + M_2 r_2^2 = \mu r^2 \tag{2.32}$$

where $\mu = M_1 M_2/(M_1 + M_2)$ is the *reduced mass.* Thus the *characteristic rotational temperature* and the moment of inertia for a homopolar diatomic molecule are related by the relation

$$\Theta_r = h^2/[32\pi^2 k_B I] \, . \tag{2.33}$$

We discuss now the more general case when we determine for a given body the mass moment of inertia around various axes through the one and the same point (*center of mass*). We find, according to a theorem of mechanics, that there are three mutually perpendicular directions for which the moment of inertia is a maximum or minimum. These directions are called the *principal axes* and the corresponding moment of inertia as the *principal moments of inertia*, I_A, I_B and I_C. Diatomic molecules and multi-atomic molecules, in which two principal moments of inertia are equal and that of the third is zero or very small, are called *linear molecules.* If the three principal moments of inertia are all equal, then the molecule is called the *spherical top molecule,* but if they are all unequal to each other, then the molecule is called the *asymmetric top (asymmetric rotor) molecule.* On the other hand, if at least two of the principal moments of inertia are equal, then it is a *symmetric top molecule (symmetric rotor)* molecule.

In the more elaborate theory of quantum mechanics for a diatomic molecule J^2 in (2.31) is replaced by $J(J + 1)$. Thus the energy of a rigid diatomic rotor is

$$E_J = k_B J(J + 1)\Theta_r \, . \tag{2.34}$$

The frequencies of the spectral lines for transition between pure rotational energy levels can be calculated for rigid diatomic rotor from *Planck's radiation law*

$$\nu = \frac{\Delta E}{h} = \frac{E'' - E'}{h} = \frac{k_B \Theta_r}{h}[J''(J'' + 1) - J'(J' + 1)]$$
$$= \frac{k_B \Theta_r}{h}(J'' - J')(J'' + J' + 1) \, . \tag{2.35}$$

Now the selection rule for such transitions is $\Delta J = \pm 1$. Thus, the equation for the frequencies of the lines due to transitions between pure rotational energy levels in a diatomic rotor is

$$\nu = \frac{2k_B\Theta_r}{h}J''$$ (2.36)

for emission, and

$$\nu = \frac{2k_B\Theta_r}{h}(J'+1)$$ (2.37)

for absorption.

It is clear that these are equidistant lines. It may be pointed out that for homopolar molecules a change in the rotational quantum number does not mean any change in the electronic dipole moment, and as such for these molecules, transitions between pure rotational energy levels may take place by collision without radiation. However, the rotational bands may occur in Raman spectrum, as well as at the time of coupling with the electronic energy levels. In Table 2.1, the characteristic values of Θ_r and some other data for different diatomic molecular gases have been given, which have been determined from spectroscopical data. The electronic and vibrational energy levels are assumed to be in the respective ground state.

From the values of Θ_r given in Table 2.1 and (2.36) and (2.37) , one can estimate the wavelength of radiation of lines for transition between pure rotational energy levels, $\lambda \approx hc/(2\Theta_r k_B) = 7.21\times 10^7/\Theta_r$, where λ is determined in Å. Thus it can be seen that these lines should appear in the wavelength range of 70 to 1.3×10^5 microns (far infrared). Further, the energy difference for these transitions, $\Delta E \approx 2\Theta_r k_B J$ (in eV) is extremely small at small quantum numbers, but increases proportionately to the quantum number J.

In addition, Table 2.2 contains rotational data for various molecules with more than two atoms, in which Θ_A, Θ_B and Θ_C are characteristic temper-

Table 2.1. Characteristic data for different diatomic molecular gases

Gas	Θ_r(K)	Θ_v(K)	x	v_{max}	E_D(eV)	α(K)
H_2	87.6	6333	0.0266	18.8	5.125	4.31e+0
N_2	2.89	3400	0.0061	82.0	11.997	2.69e-2
O_2	2.082	2280	0.0076	65.8	6.457	2.27e-2
Cl_2	0.351	815	0.0290	17.2	0.605	2.86e-2
Br_2	0.117	468	0.0124	40.3	0.812	2.44e-3
I_2	0.054	309	0.0028	178.6	2.375	1.68e-4
OH	27.2	5375	0.0022	22.7	5.259	1.03e+0
CN	2.74	2981	0.0063	79.4	10.185	2.50e-2
CO	2.78	3130	0.0062	80.6	10.867	2.52e-2
NO	2.46	2745	0.0073	68.5	8.094	2.56e-2
HCl	15.26	4310	0.0174	28.7	5.332	1.61e-1

Table 2.2. Characteristic rotational data for multiatomic molecules

Molecule	Rotor type	σ	$\Theta_r(K)$	$\Theta_A(K)$	$\Theta_B(K)$	$\Theta_C(K)$
HCN	(a)	1	0.213	0.213		
CO_2	(a)	2	0.056	0.056		
CS_2	(a)	2	0.016	0.016		
N_2O	(a)	1	0.060	0.060		
C_2H_2	(a)	2	0.170	0.170		
SO_2	(a)	2	0.949	0.949		
NH_3	(b)	3	0.840	1.433	0.908	
BF_3	(b)	3	0.027	0.050	0.024	
C_2H_6	(b)	6	0.102	0.095	0.366	
H_2O_2	(b)	2	0.188	0.118	1.449	
C_2H_4	(b)	4	0.156	0.131	0.701	
CH_4	(c)	12	0.517	0.757		
H_2O	(d)	2	1.527	2.090	4.005	1.337
H_2S	(d)	2	0.750	1.303	1.498	0.681

atures along the principal rotational axis. The characteristic rotational temperature, Θ_r, is obtained from these values according to following formulas:

(a) *Linear molecules* $(\Theta_A = \Theta_B)$: $\Theta_r = \Theta_A$;
(b) *Symmetric top molecules* $(\Theta_B = \Theta_C)$: $\Theta_r = (\Theta_A \Theta_B^2/\pi)^{1/3}$
(c) *Spherical top molecules* $(\Theta_A = \Theta_B = \Theta_C)$: $\Theta_r = \Theta_A/\pi^{1/3}$
(d) *Asymmetric top molecules* $(\Theta_A \neq \Theta_B \neq \Theta_C)$: $\Theta_r = (\Theta_A \Theta_B \Theta_C/\pi)^{1/3}$

In complex molecules, in addition, the rotation of one group of atoms relative to another (for example of the group CH_3 about the bond C-C in ethane) must be taken into consideration. Internal rotation may be hindered as well as free, because a molecule has a force field tending to orient a group of atoms in a definite position where the force, inhibiting rotation is minimum to the position, where it is maximum, and is called the energy or potential barrier.

2.2.2 Harmonic Oscillator

Let there be two particles of equal mass M connected with a spring of spring constant k (Fig. 2.3). If the spring-mass system oscillates symmetric around a mean position, it is called a *harmonic oscilator*. From the theory of vibrations it can be shown that the frequency of oscillation for such a case is given by the relation

$$\nu = \frac{\sqrt{k/M}}{2\pi} . \tag{2.38}$$

Since the force $F = -kx = -dU/dx$, where U is the potential, $U = kx^2/2 = 2\pi^2 M\nu_{osc}x^2$.

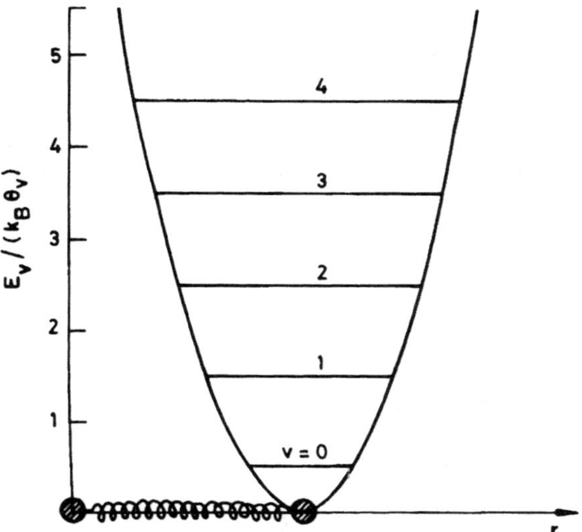

Fig. 2.3. Model and energy levels of a harmonic diatomic oscillator

Thus (2.29) becomes

$$\frac{d^2\bar{\psi}}{dx^2} + \frac{8\pi^2 M}{h^2}(E - 2\pi^2 M\nu_{\mathrm{osc}}^2 x^2)\bar{\psi} = 0 \; . \tag{2.39}$$

With $\xi = 2\pi x \sqrt{M\nu_{\mathrm{osc}}/h}$, $C = 2E/(h\nu_{\mathrm{osc}})$, the *wave equation* becomes

$$\frac{d^2\bar{\psi}}{d\xi^2} + (C - \xi^2)\bar{\psi} = 0 \; . \tag{2.40}$$

Let there be a general trial function $\bar{\psi}(\xi) = H(\xi)\exp^{-\xi^2/2}$. Substituting this into the wave equation, the wave equation becomes

$$\frac{d^2 H}{d\xi^2} - 2\xi\frac{dH}{d\xi} + (C - 1)H = 0 \; . \tag{2.41}$$

This equation is solved with the boundary conditions

$$\xi \to \infty : \frac{d^2 H}{d\xi^2} \to 0, \frac{dH}{d\xi} \to 0 \; , \; H \text{ is finite} \quad . \tag{2.42}$$

This equation is the so-called *Hermite equation*, which has a finite solution only when $(C-1)/2$ is an integer. Assuming $(C-1)/2 = v$, where v is an integer, we get $C = 2v + 1 = 2v + 1 = 2E/(h\nu_{\mathrm{osc}})$ and thus, the allowed vibrational energy levels of the diatomic harmonic oscillator are

$$E_v = h\nu_{\mathrm{osc}}(v + 1/2), v = 0, 1, 2, \ldots \tag{2.43}$$

where v is called the *vibrational quantum number* ($v = 0$, 1, 2, 3, ...). From *Planck's radiation law*, the frequency of radiation is $\nu = |\Delta E|/h = \nu_{\mathrm{osc}}|\Delta v|$ and since from quantum mechanical considerations the most probable selection rule is $|\Delta v| = \pm 1$, the frequency of radiation is equal to the *frequency of oscillation* ($\nu = \nu_{\mathrm{osc}}$) of a single line.

Defining a *characteristic vibrational temperature* $\Theta_v = h\nu/k_B$, (2.43) becomes

$$E_v = k_B \Theta_v (v + 1/2) \; . \tag{2.44}$$

Values of Θ_v for various diatomic gases given in Table 2.1 are applicable when the electron energy is in the ground level ($v = 0$). However, in the vibrational ground state ($v = 0$), the molecule possesses a non-zero vibrational energy $E_o = k_B \Theta_v/2$. Consistent with the convention adopted to measure the vibration energy from the ground state, we subtract this energy, and we get for the vibration energy as

$$E_v = v k_B \Theta_v \; . \tag{2.45}$$

It may now be seen, that for pure vibrational transitions, the energy difference ($|\Delta v| = 1$) is about two to three orders of magnitude larger than for pure rotational transitions. Thus any vibrational transition is almost always accompanied by rotational transitions. For rigid rotors, it has already been pointed out that the spectral frequency in emission is given by the relation $\nu = 2BcJ''$, where $B = \Theta_r k_B/hc$. For a not so rigid rotor, there is a new value of Θ_r or B for every combination of the electron and vibration energy levels. Since the difference between the electronic energy levels are quite large in comparison to the difference between the vibrational energy levels, it may be assumed that the transitions between the vibrational energy levels may take place at the ground electronic energy level. The change in the value of Θ_r for different values of the vibrational quantum number is given approximately by the formula

$$\Delta\Theta_r = (\Theta_r)_{v=0} - \alpha v \tag{2.46}$$

where values of α for various diatomic molecular gases are given in Table 2.1. Thus for a not so rigid rotor with vibrational and rotational transitions, and with $\Delta J = J'' - J'$, the wave number of radiation is

$$\begin{aligned}
\bar{\nu} = \nu/c &= B'' J''(J'' + 1) - B' J'(J'' + 1) + \bar{\nu}_v \\
&= (B'' - B')(J' + 1)J' + B'' \Delta J(2J' + \Delta J + 1) + \bar{\nu}_v
\end{aligned} \tag{2.47}$$

where $\bar{\nu}_v = k_B \Theta_v \Delta\nu/(hc)$.

For the three cases of $\Delta J = -1$, 0 and $+1$, and we get

$$\begin{aligned}
\Delta J = -1 &: \bar{\nu} = (B'' - B')(J' + 1)J' - 2B'' J' + \bar{\nu}_v \\
\Delta J = 0 &: \bar{\nu} = (B'' - B')(J' + 1)J' + \bar{\nu}_v \\
\Delta J = 1 &: \bar{\nu} = (B'' - B')(J' + 1)J' + 2B''(J' + 1) + \bar{\nu}_v \; .
\end{aligned} \tag{2.48}$$

Actually, corresponding to these three cases, there are three branches of spectral lines which together give a band structure. This is further repeated in the hyperfine structure of the spectra in the case of an electronic transition coupled with transitions between vibration–rotational energy levels. It may again be pointed out, as in the case of transitions between pure rotational energy levels, that for homopolar molecules such transitions without an electronic transition do not involve a change in electronic dipole moment. Thus, no electromagnetic wave radiation is possible for these homopolar molecules, except when these transitions are coupled with transitions between electronic energy levels also.

For *heteropolar diatomic molecules* with two atomic masses M_1 and M_2 connected with a spring of spring constant k, the two differential equations are

$$M_1\ddot{x}_1 + k(x_1 - x_2) = 0 \text{ and } M_2\ddot{x}_2 + k(x_2 - x_1) = 0 \ . \qquad (2.49)$$

Thus, $M_1\ddot{x}_1 + M_2\ddot{x}_2 = 0$. Let $x_1 = x_{m1}\cos(\omega t)$ and $x_2 = x_{m2}\cos(\omega t)$. Therefore, $x_{m2} = -M_1 x_{m1}/M_2$ and from the first differential equation, the *frequency of oscillation* is

$$\nu_{\text{osc}} = \frac{\omega}{2\pi} = \frac{1}{2\pi}\sqrt{\frac{k}{\mu}} \qquad (2.50)$$

where the *reduced mass* is $\mu = M_1 M_2/(M_1 + M_2)$.

For further analysis (2.43) and the subsequent text are valid. Figure 2.3 shows the energy levels which are the sum total of the kinetic and potential energies of the vibrating atoms in the molecule, calculated from (2.43), as a function of the vibrational quantum number. The wavelength range of the spectra for the transition between the two vibrational energy levels can easily be computed from the values of the characteristic vibrational temperature in Table 2.1, and this should be between 10^4 to 10^6 Å to which transitions between the rotational energy levels are superimposed as the hyperfine structure of the spectra. While the harmonic oscillator model is adequate as long as the ratio of T/Θ_v is small, the model fails with increasing temperature because the dissociation and the increase in the volume with temperature cannot be explained. Therefore, under these conditions the model of an anharmonic oscillator is studied, which is discussed in the next section.

For multiatomic molecules (number of atoms in the molecule > 2), there can be different characteristic temperature for each degree of freedom of vibration; values of the characteristic vibration temperature for a select number of gases are given in Table 2.3 with the associated *degeneracy* given within the parenthesis.

20   2 Introduction to Quantum Mechanics

Table 2.3. Characteristic vibrational temperature for selected multiatomic gases with degeneracy within parenthesis

Gas	$\Theta_{vk}(\mathrm{K})$											
	k=1	2	3	4	5	6	7	8	9	10	11	12
H_2O	4170	1820	4290									
H_2S	3760	1860	1870									
SO_2	1660	760	1920									
NO_2	1900	930	2340									
CO_2	1930 (2)	960 (2)	3390									
CS_2	950	570 (2)	2190									
N_2O	1850	850 (2)	3200									
O_3	1020 (2)	1500 (2)	2510 (2)									
C_2H_2	4860	2840	4740	880 (2)	1050 (2)							
C_2N_2	3350	1220	3100	730 (2)	330 (2)							
NH_3	3800	1080	3900 (2)	1860 (2)								
BF_3	1280	1010	2160 (2)	690 (2)								
H_2O_2	4130	2070	2030	1250	4910	1970						
CH_4	4200	2200 (2)	4350 (3)	1880 (3)								
C_2H_4	4350	2340	1930	1190	4720	1510	1370	1360	4480	1430	4310	2080
SF_6	1120	930	1390	890	760	520						

2.2.3 Anharmonic Oscillator

While the concept of vibration rests on the assumption of sufficiently small amplitude, actually the amplitudes are by no means infinitesimal small and for accurate calculations higher order terms in the potential energy are required. Energy levels for such an oscillator are modified from (2.43), and are given by the relation

$$E_v = (v + 1/2)h\nu_{\mathrm{osc}} - x(v + 1/2)^2 h\nu_{\mathrm{osc}} \tag{2.51}$$

where x is a material constant and is acting as a correction factor in the harmonic oscillator model. The value of x is determined from the maximum value of $v = v_{\mathrm{max}}$ when dissociation takes place. Since for $v \to v_{\mathrm{max}}$,

$E_v \rightarrow E_D$ = *dissociation energy* and $dE_v/dv = 0$, it is evident that $v_{\max} = (1-x)/(2x)$ and $E_D = h\nu_{\rm osc}/(4x)$. Thus from (2.51)

$$\frac{E_v}{E_D} = \frac{(1+2v)}{1+2v_{\max}}\left[2 - \frac{(1+2v)}{1+2v_{\max}}\right] . \tag{2.52}$$

For a harmonic oscillator $x \rightarrow 0$, $\nu_{\max} \rightarrow \infty$ and $E_D \rightarrow \infty$ but for anharmonic oscillator as $v \rightarrow v_{\max}$ the ratio E_v/E_D goes to one. Therefore use of (2.52) is not meaningful, since there is no dissociation. However, for the purpose of comparison with anharmonic oscillator, E_D is kept the same for both oscillators. Therefore for a harmonic oscillator, the expression taken is $E_v/E_D = (v+1/2)/(4x)$. With a value of $v_{\max}$ around seven, corresponding to a value of x around 0.0667, values of E_v/E_D are plotted in Fig. 2.4 along with energy levels and approximate potential distribution curves for both harmonic and anharmonic oscillators. It is seen that for a given value of the vibrational quantum number, a harmonic oscillator has a higher energy level than an anharmonic one. Further $v_{\max}$ need not have a round figure, and is only an approximation obtained from the energy levels determined spectroscopically. From Table 2.1, it is clear that the above value of $v_{\max} = 7$ is no way near the actual value for a diatomic molecular gas, and the smallest value of $v_{\max}$ is 18. This has the effect that the vibrational energy levels for small values of v are not much different for harmonic and anharmonic vibrational models. From (2.51), one obtains the equation for the wave number of the vibration spectra of an anharmonic oscillator for an energy level jump from quantum

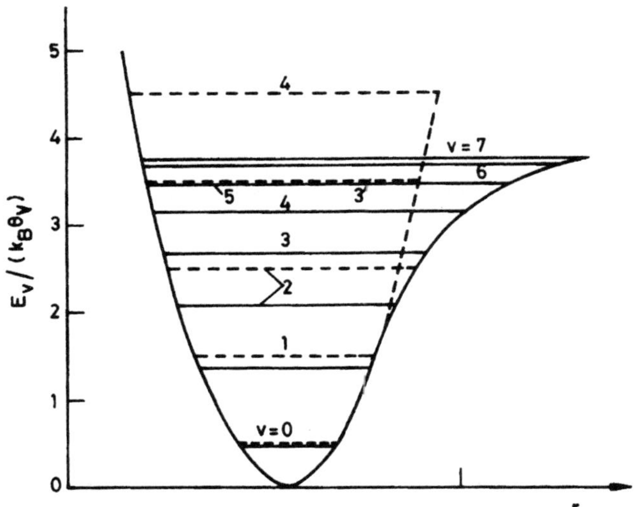

Fig. 2.4. Energy levels of an anharmonic oscillator ($v_{\max} = 7$)

number v'' to v' as

$$\bar{\nu}_v = \frac{\nu_v}{c} = \frac{\Theta_v k_B}{hc}(v'' - v')[1 - x(v'' - v' + 1)] \, . \tag{2.53}$$

In emission, for $\Delta v = v'' - v' = 1$,

$$\bar{\nu}_v = \frac{\Theta_v k_B}{hc}[1 - 2x(v' + 1)] \, . \tag{2.54}$$

Since x is much smaller than one, for small values of v' the spectral frequency of the vibrational lines is dependent only on the characteristic vibrational temperature of the gas. By accurately measuring the frequency of the spectral lines of vibration it is possible to determine Θ_v, x and the dissociation energy. These are given in Table 2.1.

While we have assumed so far, that there is no coupling between the rotational and vibrational energy levels, in actual practice this is not so. In the case of non-rigid rotor, the rotational energy for a diatomic molecule is

$$E_J = J(J+1)k_B\Theta_r - 4k_B\Theta_r(\Theta_r/\Theta_v)^2 J^2(J+1)^2 \tag{2.55}$$

in which Θ_r is calculated at the vibrational ground level, in which for most of the molecular specie the effect of change of Θ_v at high rotational quantum number J is neglected.

For a multiple atomic gas, depending on the symmetry, there can be a number of values of v and x. In addition, there are cross-terms linking different modes of vibration, for which the vibration energy becomes

$$E_v = E_v(v_1, v_2, v_3)$$
$$= k_B \sum_{i=1}^{3} \left[\Theta_{vi} \left(v_i + \frac{1}{2} \right) - \sum_{k=1}^{3} x_{ik} \left(v_i + \frac{1}{2} \right) \left(v_k + \frac{1}{2} \right) \right] \tag{2.56}$$

in which x_{ik} are the *anharmonicity constants*. For some typical molecules these are given in Table 2.4.

Finally, we would discuss, the effect of putting the ground vibrational energy level to zero as per the convention. For the anharmonic oscillator, this results in the relation $v_{\max} = 1/(2x)$, but the relation between E_D and $h\nu_{\mathrm{osc}}$

Table 2.4. Anharmonicity constants for some multi-atomic molecules

Gas	x_{11}	x_{22}	x_{33}	x_{12}	x_{13}	x_{23}
CO_2	−0.3	−1.3	−12.5	5.7	−21.9	−11.0
HCN	52.0	−2.8	−35.5	−4.2	−14.4	−19.5
N_2O	−3.2	−2.2	−13.7	4.7	−12.4	−26.1
H_2O	−43.8	−19.5	−46.3	−20.0	−155.	−19.8

does not change. Equation (2.52) becomes

$$\frac{E_v}{E_D} = \frac{v}{v_{\max}} \left[2 - \frac{v}{v_{\max}} \right] \, . \tag{2.57}$$

While we have discussed at least partially, the energy levels in molecules in the form of rotational, vibrational and electronic energy, there are other energies like translational energy which we did not discuss. Further we have not discussed so far about the distribution of these energies among the molecules. This we would discuss further in the next chapter.

2.3 Exercise

2.3.1 Estimate the spectral wavelengths for He^+ and Li^{++}.

2.3.2 Calculate the wavelengths and energy exchanged in the Balmer series lines of hydrogen atom.

2.3.3 Calculate the spectral wavelengths in rotation and vibration separately for the heteropolar diatomic molecules OH, CN, CO, NO and HCl.

2.3.4 Using the *characteristic vibrational temperature* for various molecules given in Table 2.1 and 2.3, calculate the spectral wave length of vibrational-rotational band and the corresponding energy.

2.3.5 Verify the values of dissociation energy of diatomic molecules given in Table 2.1, when the anharmonicity constant x is given.

3 Introduction to Statistical Mechanics

It is probably clear by now that the gas particles, of which there are several different varieties, have electronic excitation, vibrational, rotational, spin and translational energies. While it is impossible to keep track of the exact total quantum energy an individual particle may possess at any given time, it is possible to determine for a large number of particles the approximate percentage distribution of the energy with suitable auxiliary conditions like the total energy being kept constant. Methods by which the statistics of the possible energy distributions are made do vary no doubt, but the most probable distribution for a large number of particles to be accommodated in a still larger number of energy levels seems to give similar results, as will be shown later.

While trying to obtain the statistics of the energy distributions of the particles, we denote the particles, considered to be balls, with letters like a, b, c, d, ... , so that we can exactly distinguish these from each other, and place these in boxes each having a definite energy. Thus, merely placing one ball in one particular box allows the ball to attain the particular energy. Now let the number of balls (particles) be N and the number of boxes (energy levels or energy states) be g. If the balls are distinguishable from each other (since they are labelled) and they are put into the boxes without any restriction on the number of balls in each box, the distribution is called *Boltzmann statistic*. In case the balls are not distinguishable, then it is *Bose statistic* developed by *Satyendra Nath Bose* of India in the twenties of the twentieth century, and subsequently used by *Einstein* for the statistics of electromagnetic radiating particles (photons). However, if the particles are not distinguishable from each other, and if the number of particles in each box is restricted to a maximum of one for each box, since, according to *Pauli principle* no two particles may have exactly the same energy, then it is the *Fermi statistic*. Obviously in this last case, N is less or equal to g. While the consequence of these different statistical procedures are examined later, in the sections that follow we discuss without proof some possible arrangements, without any auxiliary restrictions on the total energy.

(a) N balls are all distinguishable from each other, and they are placed maximum one in g boxes. Obviously, N is less or equal to g. While the first ball can be placed in any of the g boxes, the second ball can be placed only in

$(g-1)$ boxes. Thus for the two balls there are $g(g-1)$ possibilities, and in a similar fashion for N balls placed maximum one in each box, the *number of possibilities* is

$$W_1 = g(g-1)(g-2)\ldots(g-N+1) = \frac{g!}{(g-N)!} \ . \qquad (3.1)$$

If $g = N$, obviously, $W_1 = g!$ since $0! = 1$.

Example: $N = g = 3$, the balls are numbered a, b, c. Possible arrangements in the three boxes are

$$\begin{array}{|cccccc|}
a & a & b & b & c & c \\
b & c & a & c & a & b \\
c & b & c & a & b & a
\end{array}$$

and $W_1 = 3! = 6$.

(b) Out of N balls N_1 are not distinguishable and a maximum of one ball is placed in each box. Obviously $N_1 \le N \le g$. Thus the number of possibilities is

$$W_2 = \frac{g!}{(g-N)!N_1!} \ . \qquad (3.2)$$

Example: $N = g = 3$, $N_1 = 2$, and the balls are numbered a, a, b. Possible arrangements in these boxes are

$$\begin{array}{|ccc|}
a & a & b \\
a & b & a \\
b & a & a
\end{array}$$

and $W_2 = 3$.

(c) Out of a total of N balls, $N_1, \ldots, N_p$ balls in each group are not distinguishable from each other in a particular group, and are to be arranged in g boxes. Obviously N is smaller than or equal to g and N is smaller than or equal to $(N_1 + N_2 + \ldots + N_2)$. Thus, the number of possibilities is

$$W_3 = \frac{N!}{(g-N)!N_1!N_2!\ldots N_p!} \ . \qquad (3.3)$$

(d) Let there be N balls, which can not be distinguished from each other, which are to be distributed to a number of boxes $(= g)$ without restriction on the number of balls in each box. Note that this is actually the *Bose statistic*. The number of possibilities is

$$W_4 = \frac{(N+g-1)!}{(g-1)!N!} \ . \qquad (3.4)$$

(e) Let there be N balls, which can not be distinguished from each other, and not more than one ball can be placed in a box, number of which is g. This is the *Fermi statistic*. The number of possibilities is

$$W_5 = \frac{g!}{(g-N)!N!} \ .$$ (3.5)

Example: 3 balls not distinguishable from each other are to be placed in five boxes. Thus $N = 3$ and $g = 5$, and from (3.4), $W_4 = 35$ and from (3.5), $W_5 = 10$. The possible distributions are given in Fig. 3.1 in each column, the rows being the different boxes. It can be seen, that the distribution of type A, which can have 3 particles in each box, has 5 possibilities; the distribution B, which can have a maximum of 2 balls in each box, has 20 possibilities; and the distribution C with maximum one ball in each box has 10 possibilities. Thus in *Bose statistic*, A distribution has a probability of 5/35, B distribution has 20/35 and C distribution has 10/35. In case the balls are distinguishable (*Boltzmann statistic*), distribution A remains the same, but in distribution B one must distinguish about which two should be in each box, and in the distribution C again each of the balls in the box has to be distinguished. Thus for the *Boltzmann statistic*, there will be 5 possibilities in the distribution A, and 60 possibilities in the distribution B and 60 in the distribution C, and there is a total of 125 possibilities with probability of 5/125 in distribution A, 60/125 in distribution B and 60/125 in distribution C. The last value for single occupancy in the Boltzmann statistic can be found out from (3.1). Note, that for very large number of particles and energy levels, single occupancy is rather a rule than an exception, and it is quite adequate to use (3.1) for the *Boltzmann statistic*. In case the number of balls is restricted to one in each

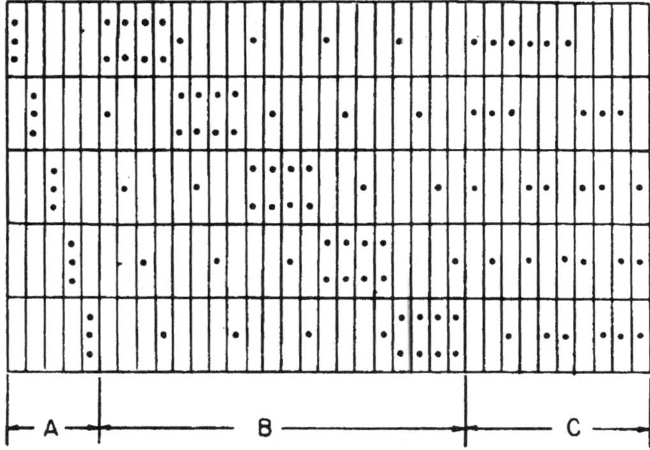

Fig. 3.1. Possible arrangements in Bose Statistics

box and the balls are not distinguishable (*Fermi statistic*), there are only 10 possibilities in C distribution, which can be evaluated from (3.5).

It is, therefore, in order, at least to estimate the number of possibilities for Boltzmann, Bose and Fermi statistic, that we consider (3.1), (3.4), and (3.5), respectively, and the corresponding number of possible distributions are: 60 (distribution C, distinguishable), 35 and 10. Thus the number of possibilities, without considering the limit of the total energy, is largest in the *Boltzmann statistic* (even neglecting the other two distributions for it), followed by the *Bose* and *Fermi statistic*. We are, however, not bothered only about the number of possibilities, but about the most probable distribution for a given total energy. For this purpose let us consider again the case of three balls denoted by a, b and c put in five boxes with energies 0, 1, 2, 3 and 4. Further let the total energy be 4. The possible arrangements are given in Table 3.1. If now the particles are not distinguishable, but more than one particle in each box is allowed (*Boltzmann statistic*), the distribution of type A has 3, B has 6 possibilities and C has 3 possibilities. On the other hand for *Bose statistic*, there are 1 possibility of type A, 1 possibility of type B and 1 possibility of type C, that is a total of 3 possibilities. For the Fermi statistic only 1 possibility of type B is available. Hence, in all the three statistics, the B distribution under given total energy of four is at least one of the probable, if not the most probable. This is further elaborated in the next section.

Table 3.1. Possible arrangements for distinguishable particles with $g = 5$, $N = 3$ and total energy restriction

Energy State	A			B						C		
0	bc	ac	ab	c	b	c	b	a	a	c	b	a
1				b	c	a	a	c	b			
2										ab	ac	bc
3				a	a	b	c	b	c			
4	a	b	c									

3.1 Bose, Boltzmann and Fermi Statistics

We have just given the essential differences among the three statistical procedures, and now we shall examine the most probable distribution of energy levels with a given total energy. For this purpose, we consider only the Bose and Fermi statistics, and we derive the expression for the Boltzmann statistic as a limiting case for the other two. Let the energy levels be grouped in such a way, that the approximate energy levels are in one group. Let there be g_i

energy levels in the group with approximate energy E_i, in which N_i particles are to be distributed. Let g_i be larger than or equal to N_i. Since the total number of possibilities is the product of the number of possibilities in each energy level, one can write from (3.4) and (3.5), that is for
Bose statistic:

$$W = \prod W_i = \prod \frac{(N_i + g_i - 1)!}{(g_i - 1)!N_i!} \tag{3.6}$$

and
Fermi statistic:

$$W = \prod W_i = \prod \frac{g_i!}{(g_i - N_i)!N_i!} \tag{3.7}$$

with auxiliary conditions $N = \sum N_i$, and total energy $E = \sum N_i E_i$

Now the approximate *Sterling formula*

$$\ln x! = x \ln x - x \tag{3.8}$$

is used, which is valid for large value of x, as is shown in Table 3.2.

Table 3.2. Validity of Sterling formula

x	1	2	4	6	10	20
$\ln x$	0	0.693	3.18	6.57	15.35	42.4
$x \ln x - x$	-1.0	-0.614	1.54	4.74	13.03	39.9

By taking the logarithm of (3.6) and (3.7) with the help of (3.8), we get for $g_i > 1$ and $(g_i + N_i) > 1$ the following equations:
Equation (3.6):

$$\ln W = \sum \ln \frac{(g_i + N_i - 1)!}{N_i!(g_i - 1)!} \approx \sum \ln \frac{(g_i + N_i)!}{N_i!g_i!}$$

$$= \sum [(g_i + N_i) \ln(g_i + N_i) - g_i \ln g_i - (g_i + N_i) + (g_i + N_i) - N_i \ln N_i]$$

$$= \sum [(g_i + N_i) \ln(g_i + N_i) - g_i \ln g_i - N_i \ln N_i] \tag{3.9}$$

Equation (3.7):

$$\ln W = \sum [g_i \ln g_i - g_i - N_i \ln N_i + N_i - (g_i - N_i) \ln(g_i - N_i) + g_i - N_i]$$

$$= \sum [g_i \ln g_i - N_i \ln N_i - (g_i - N_i) \ln(g_i - N_i)] . \tag{3.10}$$

From (3.9) and (3.10), the most probable distribution is obtained by making for each of the statistic the derivative of $\ln W$ with respect to N_i equal to zero under the auxiliary conditions, $\partial N/\partial N_i = 0$ and $\partial E/\partial N_i = 0$. As is well known, such problems can be solved by introducing the *Lagrange multiplier*,λ and μ, with which the auxiliary conditions are multiplied and then added to the main equation. Thus, the equation to be satisfied is

$$\frac{\partial \ln W}{\partial N_i} + \lambda \frac{\partial N}{\partial N_i} + \mu \frac{\partial E}{\partial N_i} = 0 \tag{3.11}$$

and one obtains the expression

$$\ln \left(\frac{g_i \pm N_i}{N_i} \right) + \lambda + \mu E_i = 0 \ . \tag{3.12}$$

Herein the upper sign is for *Bose statistic* and the lower sign is for the *Fermi statistic*.

From (3.12) we get the number of particles in the energy state E_i to be

$$N_i = \frac{g_i}{\exp^{-(\lambda + \mu E_i)} \pm (-1)} \ . \tag{3.13}$$

Now for the case that $\mu E_i \ll 0$, a condition that will be examined later, the exponential term in the denominator is much larger than one.

Since

$$N = \sum N_i = \exp^{\lambda} \sum g_i \exp^{\mu E_i} \tag{3.14}$$

we write

$$\frac{N_i}{N} = \frac{g_i \exp^{\mu E_i}}{\sum g_i \exp^{\mu E_i}} = \frac{g_i \exp^{\mu E_i}}{Z} \tag{3.15}$$

where

$$Z = \sum g_i \exp^{\mu E_i} \tag{3.16}$$

is called the *partition function*. After (3.13), two systems, A and A$'$, isolated from each other and from outside, have number of particles with energy E_i as follows:

$$N_i = \frac{g_i}{\exp^{-(\lambda + \mu E_i)} \pm 1} \quad \text{and} \quad N_i' = \frac{g_i'}{\exp^{-(\lambda' + \mu' E_i)} \pm 1} \tag{3.17}$$

with auxiliary conditions

$$\sum N_i = N, \sum N_i' = N', \sum N_i E_i + \sum N_i' E_i' = E \ . \tag{3.18}$$

In case the energy transfer is allowed between the two systems, but not the transfer of particles, there is only one *Lagrange multiplier* common to both the systems, which is also a function of the energy or the temperature common to both the systems. Thus,

$$\mu = \mu' = \mu(T) \ . \tag{3.19}$$

Now the kinetic (translational) energy of a gas particle is given by the relation

$$E = \frac{1}{2}Mw^2 = \frac{M}{2}(w_x^2 + w_y^2 + w_z^2) \ . \tag{3.20}$$

We denote with w the velocity of a single particle with respect to the laboratory coordinate and with v the kinetic speed (associated with temperature, as will be shown later) of a single particle with respect to the mass-average velocity. For the present, we assume that the mass-average velocity is zero. We consider the number of particles, whose kinetic speed in respective coordinate direction is between v_x and $v_x - \mathrm{d}v_x$, v_y and $v_y + \mathrm{d}v_y$, and v_z and $v_z + \mathrm{d}v_z$, denoted as $(v_x, v_x + \mathrm{d}v_x)$, etc. From (3.13), the number of particles in this volume element in the velocity space is proportional to the volume, and is given by the relation

$$\mathrm{d}N = C\frac{N}{Z} \exp^{\mu \frac{M}{2}(v_x^2 + v_y^2 + v_z^2)} \mathrm{d}v_x \mathrm{d}v_y \mathrm{d}v_z \ . \tag{3.21}$$

The proportionality constant C is of the order of Z in magnitude, but it takes care of the dimension. Thus, C has the dimension per unit velocity space volume ($\mathrm{s}^3\mathrm{m}^{-3}$). Similarly, μ has the dimension of the inverse kinetic energy (J^{-1}). By integrating twice over all speeds in y and z directions, we get the relation for the number of particles between $(v_x, v_x + \mathrm{d}v_x)$

$$\mathrm{d}N_x = \frac{NC}{Z} \left[\int\!\!\int_{-\infty}^{\infty} \exp^{\mu \frac{M}{2}(v_x^2 + v_y^2 + v_z^2)/2} \mathrm{d}v_y \mathrm{d}v_z \right] \mathrm{d}v_x \ . \tag{3.22}$$

Now we consider a cubic element and calculate the momentum transfer on one plane surface perpendicular to the x-axis. In time dt, the particles with speed $(v_x, v_x + \mathrm{d}v_x)$ striking the wall are those which were in the volume $Av_x \mathrm{d}t$; A is the cross-section of the wall of the elementary cube perpendicular to the x-axis. Thus the number of particles with speed $(v_x, v_x + \mathrm{d}v_x)$ reaching the particular wall is $Av_x \mathrm{d}t\mathrm{d}N_x$ and the momentum transferred by each particle to the wall is $2Mv_x$. Thus the total momentum given to the wall in time dt is $2Mv_x.Av_x \mathrm{d}t\mathrm{d}N_x$. Since the force applied on this wall is the momentum transfer per unit time, the force $A\mathrm{d}p = 2MAv_x^2 \mathrm{d}N_x$, where p is the pressure. Thus,

$$\mathrm{d}p = 2Mv_x^2 \mathrm{d}N_x = 2Mv_x^2 \frac{NC}{Z} \left[\int\!\!\int_{-\infty}^{\infty} \exp^{\mu M(v_x^2 + v_y^2 + v_z^2)} \mathrm{d}v_y \mathrm{d}v_z \right] \mathrm{d}v_x \ . \tag{3.23}$$

By integration,

$$p = 2M \int_{-\infty}^{\infty} v_x^2 \mathrm{d}N_x$$

$$= \frac{2MNC}{Z} \int_0^{\infty} v_x^2 \left[\int\!\!\int_{-\infty}^{\infty} \exp^{\mu M(v_x^2 + v_y^2 + v_z^2} \mathrm{d}v_y \mathrm{d}v_z) \right] \mathrm{d}v_x$$

$$= \frac{2MNC}{Z} \int_0^{\infty} v_x^2 \exp^{\mu M v_x^2/2} \left[\int\!\!\int_{-\infty}^{\infty} \exp^{\mu M(v_y^2 + v_z^2} \mathrm{d}v_y \mathrm{d}v_z) \right] \mathrm{d}v_x \ . \tag{3.24}$$

Now from *mathematical tables*,

$$\int_{-\infty}^{\infty} \exp^{-(ax^2+2bx+c)} \, \mathrm{d}x = \sqrt{\pi/a}\, \exp^{(b^2-ac)/a} \tag{3.25}$$

for $(a > 0)$

and

$$\int_{0}^{\infty} \exp^{-\beta x^2} \, \mathrm{d}x = \sqrt{-\pi/\beta} \ . \tag{3.26}$$

Thus,

$$\int_{-\infty}^{\infty} \exp^{\mu M v_x^2/2} \, \mathrm{d}v_x = \sqrt{(-2\pi)/(\mu M)} \tag{3.27}$$

and

$$\int_{-\infty}^{\infty} v_x^2 \exp^{\mu M v_x^2/2} \, \mathrm{d}v_x = \int_{-\infty}^{\infty} \left[\frac{\partial}{\partial \left(\dfrac{\mu M}{2} \right)} \exp^{\mu v_x^2/2} \right] dv_x$$

$$= \frac{\partial}{\partial \left(\dfrac{\mu M}{2} \right)} \sqrt{-\frac{2\pi}{\mu M}}$$

$$= \sqrt{\frac{\pi}{2}} \left(-\frac{\mu M}{2} \right)^{1/2}$$

$$= -\frac{1}{\mu M} \int_{-\infty}^{\infty} \exp^{\mu M v_x^2/2} \, \mathrm{d}v_x \ . \tag{3.28}$$

From (3.24), therefore,

$$p = \frac{MNC}{Z} \left[-\frac{1}{\mu M} \int_{-\infty}^{\infty} \exp^{\mu M v_x^2/2} \left(\int\!\int_{-\infty}^{\infty} \exp^{\mu M (v_y^2+v_z^2)/2} \, \mathrm{d}v_y \mathrm{d}v_z \right) \mathrm{d}v_x \right]$$

$$= -\frac{NC}{\mu Z} \int\!\int\!\int_{-\infty}^{\infty} \exp^{\mu M (v_x^2+v_y^2+v_z^2)/2} \, \mathrm{d}v_x \mathrm{d}v_y \mathrm{d}v_z \ . \tag{3.29}$$

Now from (3.21)

$$N = \int \mathrm{d}N = \frac{NC}{Z} \int\!\int\!\int_{-\infty}^{\infty} \exp^{\mu M (v_x^2+v_y^2+v_z^2)/2} \, \mathrm{d}v_x \mathrm{d}v_y \mathrm{d}v_z \tag{3.30}$$

which means that the value of the triple integral is equal to Z/C. Thus from (3.29), $p = -N/\mu$. Putting N equal to the number density n (number of particles per unit volume, m^{-3}) and using the equation of state of a gas,

$$p = -n/\mu = nk_BT \tag{3.31}$$

we get the relation

$$\mu = -1/(k_BT) \ . \tag{3.32}$$

It was shown earlier that for the condition $\mu E_i \ll 0$, Bose and Fermi statistics tend to converge to Boltzmann statistic. This is now equivalent of stating that for $E_i/(k_B T) \gg 0$, which, for all temperatures of interest T, is always true for a finite E_i, Bose and Fermi statistics tend to converge to Boltzmann statistic. Thus the most probable distribution according to the three statistics are:

Bose statistic:

$$N_i = \frac{g_i}{\exp^{E_i/(k_B T) - \lambda} - 1} \tag{3.33}$$

Fermi statistic:

$$N_i = \frac{g_i}{\exp^{E_i/(k_B T) - \lambda} + 1} \tag{3.34}$$

Boltzmann statistic:

$$\frac{N_i}{N} = \frac{g_i \exp^{-E_i/(k_B T)}}{Z} \tag{3.35}$$

where Z is the *partition function* given by the relation

$$Z = \sum g_i \exp^{-E_i/(k_B T)} \tag{3.36}$$

and g_i is called the *statistical weight* (alternatively also called as the *degeneracy*).

Now we write (3.21) in a slightly different form,

$$\mathrm{d}N = \frac{CN}{N} \exp^{-E_i/(k_B T)} \mathrm{d}E . \tag{3.37}$$

Thus for $T \to 0$, the kinetic energy of all particles tend to zero, which is not permissible according to the *Pauli principle* (*Fermi statistic*), since, according to the principle, the particles cannot all have the same energy even for $T \to 0$. Thus, there is a limit on the minimum of the highest energy of particles at 0 K, which is called the *Fermi limiting energy* given by the relation

$$E_F = \frac{h^2 n^{2/3}}{8M} \tag{3.38}$$

where h is the *Planck constant*, n is the number density of particles and M is the mass of the particles.

3.2 Thermodynamic Properties

Starting point of our present discussion is the concept of distribution of energy in a system. Now the number of distribution possibilities of two systems is equal to the product of the possibilities in individual systems, that is, $W = W_1 . W_2$. However from thermodynamics the total entropy in a reversible mixing is obtained by simple addition. It may be pointed out that the entropy is not the only thermodynamic state variable that is additive, and a similar analysis is possible with other thermodynamic properties as well. It is, however, found that the relationship between the *number of possibilities* W and *entropy* S gives meaningful results. A trial function connecting the two

$$S = C \ln W \qquad (3.39)$$

is, therefore, suggested, in which C is a proportionality constant to be determined later. Now from (3.9), (3.10), (3.12) and (3.32), we get

$$\frac{\mathrm{d}(\ln W)}{\mathrm{d}N_i} = \ln \frac{g_i \pm N_i}{N_i} = \sum \left[\frac{E_i}{k_B T} - \lambda \right] \qquad (3.40)$$

and further from (3.39),

$$\mathrm{d}S = C\mathrm{d}(\ln W) = C \sum \ln \frac{(g_i \pm N_i)}{N_i} \mathrm{d}N_i = C \sum \left[\frac{E_i}{k_B T} - \lambda \right] \mathrm{d}N_i \ . \qquad (3.41)$$

Therefore,

$$T\mathrm{d}S = C \sum \left(\frac{E_i}{k_B} - \lambda T \right) \mathrm{d}N_i \ . \qquad (3.42)$$

Since $N = \sum N_i$ and $E = \sum N_i E_i$,

$$\begin{aligned}
T\mathrm{d}S &= C \sum \left[\frac{1}{k_B} \left(\mathrm{d}(N_i E_i) - N_i dE_i \right) - \lambda T \mathrm{d}N_i \right] \\
&= C \left[\frac{1}{k_B} \left\{ \mathrm{d} \left(\sum N_i E_i \right) - \sum N_i dE_i \right) \right\} - \lambda T \mathrm{d} \left(\sum N_i \right) \right] \\
&= \frac{C}{k_B} \mathrm{d}E - \frac{C}{k_B} \sum N_i dE_i - \lambda C T \mathrm{d}N \ .
\end{aligned} \qquad (3.43)$$

The physical meaning of the second and the third terms in the right hand side of (3.43) is as follows:

$$\sum N_i \mathrm{d}E_i \begin{cases} = 0 \ , \text{ inside the system} \\ \neq 0 \ , \text{ work done external to the system} = \mathrm{d}L \ . \end{cases}$$

$$\lambda C T \mathrm{d}N = \xi \mathrm{d}N \begin{cases} = 0 \ , \text{ to conserve particles} \\ \neq 0 \ , \text{ to introduce or remove particles} \ . \end{cases}$$

Thus from (3.43), we can write

$$T\mathrm{d}S = (C/k_B)\mathrm{d}E - (C/k_B)\mathrm{d}L - \xi \mathrm{d}N \qquad (3.44)$$

which can be compared with the energy equation, as it is known from thermodynamics,

$$T\mathrm{d}S = \mathrm{d}E - \mathrm{d}L - \xi \mathrm{d}N \ . \qquad (3.45)$$

It is now evident, that $C = k_B$ and the *trial function*, (3.39), is correct. From (3.9) for Bose statistic, which is essentially similar to the Fermi statistic, one can, therefore, write

$$S = k_B \ln W = k_B \sum [(g_i + N_i) \ln(g_i + N_i) - g_i \ln g_i - N_i \ln N_i]$$

$$= k_B \sum \left[(g_i + N_i) \ln \left(g_i \left(1 + \frac{N_i}{g_i} \right) \right) - g_i \ln g_i - N_i \ln N_i \right]$$

$$= k_B \sum \left[(g_i + N_i) + (g_i + N_i) \ln \left(1 + \frac{N_i}{g_i} \right) - g_i \ln g_i - N_i \ln N_i \right]$$

$$= k_B \sum \left[N_i \ln g_i + (g_i + N_i) \ln \left(1 + \frac{N_i}{g_i} \right) - N_i \ln N_i \right] . \tag{3.46}$$

Noting that for $x \ll 1$,

$$\ln(1 + x) = x - x^2/2 + x^3/3 - x^4/4 + - \ldots \tag{3.47}$$

and thus for $N_i < g_i$

$$S = k_B \ln W = k_B \sum \left[N_i \ln \left(\frac{g_i}{N_i} \right) + (g_i + N_i) \left(\frac{N_i}{g_i} \right) \right]$$

$$= k_B \sum \left[N_i \ln \left(\frac{g_i}{N_i} \right) + N_i + \frac{N_i^2}{g_i} \right]$$

$$\approx k_B \sum \left[N_i \ln \left(\frac{g_i}{N_i} \right) + N_i \right] . \tag{3.48}$$

For large temperatures, that is for $g_i \gg N_i$ and $\lambda k_B T \gg 1$, it has already been shown that both Bose and Fermi statistics merge into the Boltzmann statistic. Under these and also the equilibrium condition, the number density is given by (3.35), which becomes

$$\ln N_i = \ln N + \ln g_i - E_i/(k_B T) - \ln Z . \tag{3.49}$$

Thus,

$$\ln(g_i/N_i) = \ln(Z/N) + E_i/(k_B T) . \tag{3.50}$$

Substituting the above expression into (3.48), we get therefore,

$$S = k_B \sum \left[N_i \ln \left(\frac{Z}{N} \right) + \frac{N_i E_i}{k_B T} + N_i \ln(\exp^1) \right]$$

$$= k_B \sum \left[N_i \ln \left(\frac{Z . \exp^1}{N} \right) + \frac{N_i E_i}{k_B T} \right]$$

$$= \frac{1}{T} \left[E + N k_B T \ln \left(\frac{Z . \exp^1}{N} \right) \right] . \tag{3.51}$$

If N is now put equal to N_A, the *Avogadro number*, the entropy becomes molar entropy (per kmole). Further noting that $N_A k_B = R^*$, *universal gas constant*,

$$S = \frac{1}{T} \left[E + R^* T \ln \left(\frac{Z \exp^1}{N_A} \right) \right] . \tag{3.52}$$

From (3.35) and noting that $N = N_A$, one may write

$$E = \sum N_i E_i = \frac{N_A}{Z} \sum g_i E_i \exp^{-E_i/(k_B T)} \tag{3.53}$$

and also

$$\frac{\mathrm{d}\ln Z}{\mathrm{d}T} = \frac{1}{Z}\frac{\mathrm{d}Z}{\mathrm{d}T} = \frac{1}{Z}\sum \frac{g_i E_i}{k_B T^2}\exp^{-E_i/(k_B T)} = \frac{E}{N_A k_B T^2} = \frac{E}{R^* T^2} \ . \tag{3.54}$$

Thus it is now possible to derive the general expressions for the following thermodynamic variables:

Internal energy:

$$E = R^* T^2 \frac{\partial \ln Z}{\partial T} \tag{3.55}$$

Entropy:

$$S = R^* T \frac{\partial \ln Z}{\partial T} + R^* \ln\left(\frac{Z\exp^1}{N_A}\right) \tag{3.56}$$

Enthalpy:

$$H = R^* T + R^* T^2 \frac{\partial \ln Z}{\partial T} \tag{3.57}$$

Free energy:

$$F = TS - E = R^* T \ln\left(\frac{Z\exp^1}{N_A}\right) \tag{3.58}$$

Free enthalpy:

$$G = TS - H = R^* T \ln\left(\frac{Z}{N_A}\right) \tag{3.59}$$

Specific heats:

$$C_p = \partial H/\partial T \ , \ C_v = \partial E/\partial T \ . \tag{3.60}$$

It may be noted that for the internal energy it is assumed, that for $T \to 0$, $E \to 0$, and it does not contain any reaction or bond energy. Thus the bond or reaction energy at $T \to 0$ is added to the internal energy or enthalpy in (3.55, 3.57 to 3.59). Determination of the thermodynamic properties is now reduced to merely determining the *partition function* Z, which will be determined for different modes of energy in the sections which follow.

3.2.1 Contribution of Translational Energy

While discussing the *wave-particle dualism* it was noted, that the particles can be thought of as equivalent to waves of *wave length* λ. We consider now *standing* waves between two parallel plates at a distance L apart. For the *standing wave* pattern $L = k_x \lambda/2$, where $k_x = 1, 2, 3, \ldots$, and has the character of an *eigen value* (multiplier of a basic wave length) and $(k_x - 1)$

is the number of nodes between the two plates. Therefore, with the help of
(2.22),

$$(Mv)^2 = 2M(Mv^2/2) = 2ME; k_x = 2L/\lambda = 2LMv/h \tag{3.61}$$

and it follows that

$$k_x^2 = 8L^2 ME/h^2 \ . \tag{3.62}$$

It is seen that k_x corresponds to the kinetic energy contributed by the velocity
in x-direction. If

$$k^2 = k_x^2 + k_y^2 + k_z^2 \tag{3.63}$$

then the volume of a shell of sphere of radius k and thickness dk represents
the number of energy state g. Between k_x, k_y, and k_z, there is 1/8-th of a
solid angle 4π. Thus,

$$g(E) = \frac{4\pi}{8} k^2 dk \cong \frac{\pi}{2} k_x^2 dk_x$$

$$= \frac{\pi}{2} \left(\frac{8L^2 ME}{h^2} \right) \frac{L}{h} \sqrt{2M/E} dE = 4\sqrt{2}\pi \frac{M^{3/2}}{h^3} L^3 \sqrt{E} dE \tag{3.64}$$

and the partition function for the translational energy is

$$Z_{\text{trans}} = \sum g_i \exp^{-E_i/(k_B T)} \cong \sum g(E) \exp^{-E_i/(k_B T)}$$

$$= \frac{4\sqrt{2}\pi M^{3/2} L^3}{h^3} \int_0^\infty \sqrt{E} \exp^{-E/(k_B T)} dE \ . \tag{3.65}$$

The integral, as can be seen from mathematical tables, has the value

$$\sqrt{\pi}(k_B T)^{3/2}/2 \ . \tag{3.66}$$

Noting further that L^3 is the volume of the cube, for which substituting the
value of molar volume V^*, we get the relation for the partition function in
the translational mode

$$Z_{\text{trans}} = V^* \left(\frac{2\pi k_B MT}{h^2} \right)^{3/2} \ . \tag{3.67}$$

Now the *equation of state* is $pV^* = R^* T$ and the *mole mass* is $m = M.N_A$.
Since further for translation,

$$\frac{\partial \ln Z}{\partial T} = \frac{3}{2T} \tag{3.68}$$

we get the expressions to determine the contribution of the translational
energy to the following thermodynamic properties:
Internal energy:

$$E_{\text{trans}} = \frac{3}{2} R^* T \ . \tag{3.69}$$

Entropy:

$$S_{\text{trans}} = R^* \left[\frac{5}{2} + \ln \frac{V^*}{N_A} \left(\frac{2\pi M k_B T}{h^2} \right)^{3/2} \right]$$

$$= R^* \left[\frac{5}{2} + \ln \frac{k_B T}{p} \left(\frac{2\pi m k_B T}{h^2 N_A} \right)^{3/2} \right]$$

$$= R^* \left[\frac{5}{2} + \frac{5}{2} \ln T - \ln p + \frac{3}{2} \ln m + \ln \left(k_B \left(\frac{2\pi k_B}{h^2 N_A} \right)^{3/2} \right) \right]$$

$$= R^* \left[-1.15548 + 2.5 \ln T - \ln p + 1.5 \ln m \right] \ . \tag{3.70}$$

where T is in K, p is in bar and m is the mole mass in kg.kmole^{-1}.

Enthalpy:

$$H_{\text{trans}} = \frac{5}{2} R^* T \ . \tag{3.71}$$

Free energy:

$$F_{\text{trans}} = T S_{\text{trans}} - E_{\text{trans}} \ . \tag{3.72}$$

Free enthalpy:

$$G_{\text{trans}} = T S_{\text{trans}} - H_{\text{trans}} \ . \tag{3.73}$$

Specific heats:

$$C_{p,\text{trans}} = (5/2)R^* \ , \ C_{v,\text{trans}} = (3/2)R^* \ . \tag{3.74}$$

Note that for free electrons the contribution of the direction of the electron spin with respect to the direction of motion is taken into account by multiplying the relation for the partition function, (3.67), by a factor 2. Thus, for the entropy of the electrons a term ln2 must be added within the bracket. Further since for electrons, $m = 1/1836.5$, we get the following relation for the molar entropy of the electrons only:

$$S_{\text{trans,el}} = R^* [-11.735758 + 2.5 \ln T - \ln p] \ . \tag{3.75}$$

Herein p is again in *bar* and T is in *K*. It is interesting to note, that as $T \to 0$, the *molar entropy* $S \to -\infty$. However, as $T \to 0$, the pressure p must also tend to zero and the molar entropy becomes indeterminate, when no known substance in gaseous form can exist. It is, therefore, stated that the molar entropy of a substance is zero as $T \to 0$ (*Third law of Thermodynamics*).

3.2.2 Contribution of Rotational and Vibrational Energy Forms

While for the translation of a particle there are three degrees of freedom, for rotation of diatomic molecules there are only two degrees of freedom. We write (in analogy to (3.67)) by replacing the exponent 3/2 in (3.67) for three degrees of freedom of a particle by the exponent $2/2$ for two degrees of freedom of each of the two atoms, and adding them. The relation for the rotational partition function of a diatomic molecule is now

$$Z_r = V_1 \left(\frac{2\pi k_B M T}{h^2} \right)^{2/2} + V_2 \left(\frac{2\pi k_B M T}{h^2} \right)^{2/2} . \qquad (3.76)$$

Further, let r_1 and r_2 be the distance of the two masses from the common axis of rotation. Thus,

$$M_1 r_1 = M_2 r_2 \quad \text{and} \quad r_1 + r_2 = r . \qquad (3.77)$$

Now as per the usual definition of the *mass moment of inertia, $I = M_1^2 r_1^2 + M_2^2 r_2^2 = \mu r^2$,* where $\mu = M_1 M_2/(M_1 + M_2)$ is the *reduced mass*. In addition, analogous to volume V in (3.67) and justified from dimensional considerations, we write the expression for surfaces for the rotational case

$$V_1 = 4\pi r_1^2, V_2 = 4\pi r_2^2 \qquad (3.78)$$

and the *rotational partition function*, is

$$Z_r = \frac{8\pi^2 k_B T}{h^2} \left(M_1 r_1^2 + M_2 r_2^2 \right) = \frac{8\pi^2 k_B I T}{h^2} = \frac{T}{\Theta_r} \qquad (3.79)$$

where Θ_r is the *characteristic rotational temperature* of the gas, values of which are given in Table 2.1 for some diatomic gases. Equation (3.79) is slightly modified to

$$Z_r = \frac{T}{\sigma \Theta_r} \qquad (3.80)$$

where σ is the *symmetry factor* with value equal to two for *homopolar molecules* like H_2, N_2, O_2, etc., and is equal to one for *heteropolar molecules* like HCl, CO, etc. Further for most of the high temperature gases $T > \Theta_r$, and (3.80) is a sufficiently good approximation for the rotational partition function. While determining the contribution of the rotational states to the thermodynamic properties, as well as to determine the contribution of the vibrational and electronic states later, it may be noted that the total partition function is the product of the partition function for each of the different kinds of energy states. Thus all constants in (3.56), (3.58) and (3.59) appearing under the natural logarithm may be considered only once, which has already been done in the case of the translational energy. Further, the difference between the enthalpy, free enthalpy and the specific heat at constant pressure on the one side, and the internal energy, free energy and the specific heat at constant volume on the other need not be considered again, since it has already been done in the case of the translational energy. Under the approximation of $T \gg \Theta_r$ only, and keeping the preceding discussion in mind, the following equations are derived from (3.55–3.60) to determine the contribution of the rotational energy to the thermodynamic properties of the *diatomic gases*:

Internal energy:

$$E_{\text{rot}} = R^* T \ . \tag{3.81}$$

Entropy:

$$S_{\text{rot}} = R^* [1 + \ln(T/\Theta_r)] \ . \tag{3.82}$$

Enthalpy:

$$H_{\text{rot}} = E_{\text{rot}} = R^* T \ . \tag{3.83}$$

Free energy:

$$F_{\text{rot}} = T S_{\text{rot}} - E_{\text{rot}} = R^* T \ln(T/\Theta_r) \ . \tag{3.84}$$

Free enthalpy:

$$G_{\text{rot}} = F_{\text{rot}} = T S_{\text{rot}} - E_{\text{rot}} = R^* T \ln(T/\Theta_r) \ . \tag{3.85}$$

Specific heat:

$$C_{p,\text{rot}} = C_{v,\text{rot}} = R^* \ . \tag{3.86}$$

It has been mentioned already that (3.80), derived analogous to the translational energy by a semi-rigorous method is valid only under the condition $T > \Theta_r$. When this condition is not satisfied, we have to get a more exact relationship, which will now be explained. For this purpose, the rotational energy level for a diatomic molecule, given by (2.34), as $E = k_B J(J + 1)\Theta_r$ is taken. Further, degeneracy for rotation, that is the number of states with identical energy is $g = 2J + 1$. Thus, the *rotational partition function* is

$$Z_r = \frac{1}{\sigma} \sum_{J=0}^{\infty} (2J + 1) \exp^{-\Theta_r J(J+1)/T} \tag{3.87}$$

where σ is equal to one or two depending on whether we are considering a heteropolar molecule or a homopolar molecule, as it has already been mentioned after (3.80).

For $\Theta_r/T \ll 1$, the above summation can be replaced by an integral

$$Z_r = \frac{1}{\sigma} \int_0^{\infty} (2J + 1) \exp^{-\Theta_r J(J+1)/T} dJ = \frac{1}{2\sigma} \int_0^{\infty} \exp^{-z\Theta_r/T} dz \tag{3.88}$$

where $z = J(J + 1)$. One can show easily, that the above integral leads to (3.80).

While evaluation of the above expressions for Z_r and the temperature derivative of $\ln Z_r$ is quite straight forward, it requires numerical evaluation of very time consuming exponential functions. For most of the cases, one can, therefore, use a simpler formula

$$Z_r = \frac{T}{\sigma \Theta_r} \left[1 + \frac{\Theta_r}{T} + \frac{1}{15} \left(\frac{\Theta_r}{T} \right)^2 + \ldots \right] \ . \tag{3.89}$$

For any of the above formulations determination of the thermophysical properties like enthalpy, entropy and specific heats is quite straight forward by using a digital computer to add the terms. For this purpose, let us define the following identifiers:

$$D_1 = \sum_{J=0}^{\infty} (2J+1)\exp^{-J(J+1)\Theta_r/T} \ .$$
(3.90)

$$D_2 = \sum_{J=0}^{\infty} J(J+1)(2J+1)\exp^{-J(J+1)\Theta_r/T} \ .$$
(3.91)

$$D_3 = \sum_{J=0}^{\infty} J^2(J+1)^2(2J+1)\exp^{-J(J+1)\Theta_r/T} \ .$$
(3.92)

We can now write for the contribution of the rotational energy in a general case as follows:

Internal energy:
$$E_{\rm rot} = R^*\Theta_r(D_2/D_1) \ .$$
(3.93)

Entropy:
$$S_{\rm rot} = \frac{E_{rot}}{T} + R^*\ln\left(\frac{D_1}{\sigma}\right) \ .$$
(3.94)

Enthalpy:
$$H_{\rm rot} = E_{\rm rot} = R^*\Theta_r(D_2/D_1) \ .$$
(3.95)

Free energy:
$$F_{\rm rot} = TS_{\rm rot} - E_{\rm rot} = R^*T\ln\left(\frac{D_1}{\sigma}\right) \ .$$
(3.96)

Free enthalpy:
$$G_{\rm rot} = F_{\rm rot} = R^*T\ln\left(\frac{D_1}{\sigma}\right) \ .$$
(3.97)

Specific heats:

$$C_{p,\rm rot} = C_{v,\rm rot} = R^*\left(\frac{\Theta_r}{T}\right)^2\left[\left(\frac{D_3}{D_1}\right) - \left(\frac{D_2}{D_1}\right)^2\right] \ .$$
(3.98)

Results of calculation of $H_{\rm rot}$, $S_{\rm rot}/R^*$ and C_p/R^* for $T > \Theta_r$, (3.81, 3.82, 3.86), designated as analytical, and for more general case, (3.93, 3.94) and (3.98), designated as exact numerical, are now shown in Fig. 3.2. It is seen, that as T/Θ_r goes to zero, all the values for the general case go to zero, but for the case of $T > \Theta_r$, enthalpy changes linearly with temperature, entropy becomes even negative, and the specific heat remains a constant at the high temperature value. It is seen further, that both sets of results merge to each other, when $(T/\Theta_r) > 4$.

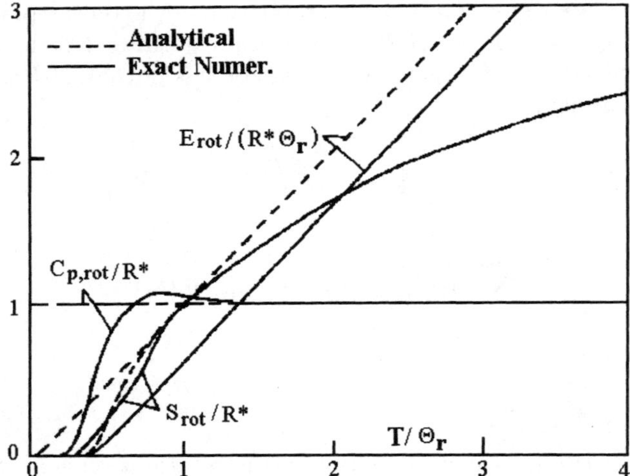

Fig. 3.2. Non-dimensional enthalpy, entropy and specific heat for rotation of diatomic molecules

In molecules with more than two atoms, the characteristic rotational temperature is dependent on those on the principal axes, and the matter has been discussed at some length in Sect. 2.2.1. Of the four types of molecules discussed there, namely (a) the *linear molecules*, (b) the *symmetric top molecules*, (c) the *spherical top molecules*, and (d) the *asymmetric top molecules*, the molecules belonging to the first group are those, which have symmetrical atoms like O-C-O or H-O-H, and for calculation of the thermophysical properties for these, the methodology for diatomic molecules can be used with σ equal to 2. For other molecules the characteristic rotational temperature Θ_r and the symmetry factor σ given in Table 2.2 are to be taken and substituted in (3.55–3.60), as discussed below. In addition, in complex molecules, the rotation of one group of atoms relative to another (for example of the group CH_3 about the bond C-C in ethane) must be taken into consideration. Internal rotation may be hindered as well as free, because a molecule has a force field tending to orient a group of atoms in a definite position, where the force inhibiting the rotation is minimum, to the position, where it is maximum, and is called the *energy* or *potential barrier*.

Now for multiatomic molecules, the rotational energy and the partition function for rotation are dependent on the detailed molecular structure and, as such, can be discussed under the following headings: (a) *Diatomic* and *linear polyatomic molecules*, (b) *Rigid symmetric top molecules*, (c) *Spherical top molecules*, and (d) *Asymmetrical top molecules*. These are discussed one by one in the following:

(a) *Diatomic and linear molecules*: The partition function is given by the relation

$$Z_r = \frac{1}{\sigma} \sum_{J=0}^{\infty} (2J+1) \exp^{-\Theta_r J(J+1)/T}$$

(3.99)

where σ is the *symmetry factor*, and the above relation has been discussed extensively already.

(b) *Rigid symmetrical top molecules* like NH_3, C_2H_4, etc.: Since the mass moment of inertia for various principal axes can be different, the partition function is given by the relation

$$Z_r = \frac{1}{\sigma} \sum_{J=0}^{\infty} \sum_{K=-J}^{J} g_J \exp^{-[\Theta_B J(J+1)+(\Theta_A-\Theta_B)K^2]/T}$$

(3.100)

where

$K = 0,1,2, \ldots ,J$
$g_J = (2J+1)$, if $K = 0$ and $g_J = 2(2J+1)$, if $K > 0$
$\Theta_A = h^2/(8\pi^2 I_A k_B)$ and $\Theta_B = h^2/(8\pi^2 I_B k_B); (\Theta_A < \Theta_B \approx \Theta_C)$
I_A, I_B = *mass moment of inertia* on the two principal axes.

For $T \gg \Theta_B$, the above equation is reduced to

$$Z_r = \frac{1}{\sigma} \left[\frac{\pi T^3}{\Theta_B^2 \Theta_A} \right]^{1/2} = \frac{1}{\sigma} \left(\frac{T}{\Theta_r} \right)^{3/2}$$

(3.101)

where

$$\Theta_r = \left[\frac{\Theta_B^2 \Theta_A}{\pi} \right]^{1/3}.$$

(3.102)

(c) *Spherical top molecules*, like CH_4: Herein $\Theta_A = \Theta_B$, otherwise the relation for the symmetric top molecules is used.

(d) *Asymmetric top molecules*, like H_2O, H_2S, etc.: For $T \gg (\Theta_A, \Theta_B, \Theta_C)$, the relation for the partition function and the characteristic rotational temperature is

$$Z_r = \frac{1}{\sigma} \left[\frac{\pi T^3}{\Theta_A \Theta_B \Theta_C} \right]^{1/2} = \frac{1}{\sigma} \left(\frac{T}{\Theta_r} \right)^{3/2}$$

(3.103)

where

$$\Theta_r = \left[\frac{\Theta_A \Theta_B \Theta_C}{\pi} \right]^{1/3}.$$

(3.104)

From (3.99), (3.101) and (3.103), one can, therefore, write the general expression for rotation of diatomic or multiatomic molecules at higher temperatures as

$$Z_r = \frac{1}{\sigma} \left(\frac{T}{\Theta_r} \right)^n$$

(3.105)

where $n = 1$ for diatomic molecules or rotor type (a) in Table 2.2, or else $n = 3/2$.

By a straight forward application of (3.55–3.60), we can now derive the following relations for the thermodynamic properties:

Internal energy and *enthalpy*:

$$E_{\text{rot}} = nR^*T \ . \tag{3.106}$$

Entropy:

$$S_{\text{rot}} = R^* \left[n + \ln \left((T/\Theta_r)^n / \sigma \right) \right] \ . \tag{3.107}$$

Free energy and *free enthalpy*:

$$F_{\text{rot}} = G_{\text{rot}} = R^* \ln \left[(T/\Theta_r)^n / \sigma \right] \ . \tag{3.108}$$

Specific heats:

$$C_{p,\text{rot}} = C_{v,\text{rot}} = nR^* \ . \tag{3.109}$$

The expression for the partition function for a diatomic molecule as a harmonic oscillator can be determined from (2.44) and (3.36) as

$$Z_{v,\text{harm}} = \sum_{v=0}^{\infty} \exp^{-(v+1/2)\Theta_v/T} = \exp^{-\Theta_v/(2T)} \sum_{v=0}^{\infty} \exp^{-v\Theta_v/T}$$

$$= \exp^{-\Theta_v/(2T)} \left[1 + \exp^{-\Theta_v/T} + \exp^{-2\Theta_v/T} + \ldots \right] \ . \tag{3.110}$$

For Θ_v/T around and larger than one the succeeding exponential terms in the series become negligible with respect to the preceding term. For this condition only, and since for a small value of x,

$$1 + x + x^2 + x^3 + \ldots \approx \frac{1}{1-x} \tag{3.111}$$

the term inside the parenthesis can be substituted by

$$\left(1 - \exp^{-\Theta_v/T} \right)^{-1} \tag{3.112}$$

and thus for a harmonic oscillator under the condition $\Theta_v/T \gg 1$,

$$Z_{v,\text{harm}} = \exp^{-\Theta_v/(2T)} \left(1 - \exp^{-\Theta_v/T} \right)^{-1} = \frac{1}{2\sinh\left[\Theta_v/(2T)\right]} \ . \tag{3.113}$$

Note that since (3.113) is valid strictly only for $T \ll \Theta_v$, that is, when most of the particles are in the vibration energy levels for which the vibration quantum number v is small, this is also the main assumption for the validity of the harmonic oscillator model. However, in many cases, even when this condition is not met, (3.113), because of its very convenient form, is preferred to the anharmonic oscillator model for quick determination of the contribution of the vibrational energy. Further derivation of relations for the contribution of the vibration energy to the thermodynamic properties is a straight forward application of (3.55–3.60), although the limiting values of these at 0K is to be taken into account, and we derive first the following relations.

Internal energy and *enthalpy*:

$$E_{\text{vib,harm}} = H_{\text{vib,harm}} = \frac{R^*\Theta_v}{2}\coth\left[\frac{\Theta_v}{2T}\right] . \tag{3.114}$$

Entropy:

$$S_{\text{vib,harm}} = \frac{R^*\Theta_v}{2T}\coth\left(\frac{\Theta_v}{2T}\right) + R^*\ln\left[\frac{1}{2\sinh[\Theta_v/(2T)]}\right] . \tag{3.115}$$

Free energy and *free enthalpy*:

$$F_{\text{vib,harm}} = G_{\text{vib,harm}} = TS_{\text{vib,harm}} - E_{\text{vib,harm}}$$

$$= R^*T\ln\left[\frac{1}{2\sinh[\Theta_v/(2T)]}\right] . \tag{3.116}$$

Specific heats:

$$C_{p,\text{vib,harm}} = C_{v,\text{vib,harm}} = R^*\left(\frac{\Theta_v}{2T}\right)^2\frac{1}{\sinh^2[\Theta_v/(2T)]} . \tag{3.117}$$

Limiting values of the thermodynamic properties for harmonic oscillator are: for $T \to 0$: $E/(R^*\Theta_v) \to 0.5$, $(S/R^*) \to 0$ and $C_p/R^* \to 0$, and for $T \to \infty$: $C_p/R^* \to 1$. Thus, all properties except the internal energy for a harmonic oscillator model is zero as $T \to 0$. For the internal energy, this limiting energy at $T \to 0$ has to be subtracted from (3.114), since the vibration energy even at the lowest vibration quantum number $(v = 0)$ is not zero. Thus, the equation for determination of the internal energy and enthalpy, substituting (3.113), is

$$E_{\text{vib,harm}} = H_{\text{vib,harm}} = \frac{R^*\Theta_v}{2}\left[\coth\left(\frac{\Theta_v}{2T}\right) - 1\right] . \tag{3.118}$$

While in the above analysis, internal energy at 0K was deducted afterwards, an alternative method, consistent with the convention that the vibration energy in the ground state is zero, is now discussed. For this purpose, the partition function for harmonic oscillator is written from (3.110) by dropping $(1/2)$ in the exponential to get

$$Z_{v,\text{harm}} = \sum_{v=0}^{\infty}\exp^{-v\Theta_v/T}$$

$$= 1 + \exp^{-\Theta_v/T} + \exp^{-2\Theta_v/T} + \ldots = \frac{1}{1 - \exp^{-\Theta_v/T}} \tag{3.119}$$

which gives further the following relations for the thermophysical properties of the harmonic diatomic oscillator:

Internal energy and *enthalpy*:

$$E_{\text{vib,harm}} = H_{\text{vib,harm}} = \frac{R^*\Theta_v}{\exp^{\Theta_v/T} - 1} \ . \tag{3.120}$$

Entropy:

$$S_{\text{vib,harm}} = R^* \left[\frac{\Theta_v}{T} \frac{1}{\exp^{\Theta_v/T} - 1} + \ln \frac{1}{1 - \exp^{-\Theta_v/T}} \right] \ . \tag{3.121}$$

Free energy and *free enthalpy*:

$$F_{\text{vib,harm}} = G_{\text{vib,harm}} = R^* T \ln \frac{1}{1 - \exp^{-\Theta_v/T}} \ . \tag{3.122}$$

Specific heats:

$$C_{p,\text{vib,harm}} = C_{v,\text{vib,harm}} = R^* \left(\frac{\Theta_v}{T}\right)^2 \frac{\exp^{\Theta_v/T}}{(\exp^{\Theta_v/T} - 1)^2} \ . \tag{3.123}$$

Numerical results of $E_{\text{vib,harm}}/(R^*\Theta_v)$ and $S_{\text{vib,harm}}/R^*$, calculated with the help of two different sets of equations above and also by taking more number of terms in calculation of the partition function of a harmonic oscillator between T/Θ_v from 0.1 to 3 have no significant difference between the methods, and the results of calculation of the thermodynamic properties for a harmonic oscillator has been shown in Fig. 3.3. However for $T/\Theta_v > 1$, the thermophysical properties results calculated for an anharmonic oscillator is larger than those calculated from the harmonic oscillator model (typically, for $T/\Theta_v = 3.0$, the internal energy and entropy calculated by anharmonic oscillator model is slightly larger than those calculated by the harmonic oscillator model depending on the value of the *anharmonicity constant x*). We would discuss, therefore, calculation of the thermodynamic properties by the *anharmonic oscillator* model briefly. The starting point of calculation is the definition of the partition function for the anharmonic oscillator

$$D_1 = Z_{v,\text{anh.}} = \sum_{v=0}^{v_{\max}} \exp^{-v(1-xv)\Theta_v/T} \tag{3.124}$$

and two other identifiers (equivalent of temperature derivatives)

$$D_2 = \sum_{v=0}^{v_{\max}} v(1 - xv) \exp^{-v(1-xv)\Theta_v/T} \tag{3.125}$$

and

$$D_3 = \sum_{v=0}^{v_{\max}} v^2(1 - xv)^2 \exp^{-v(1-xv)\Theta_v/T} \ . \tag{3.126}$$

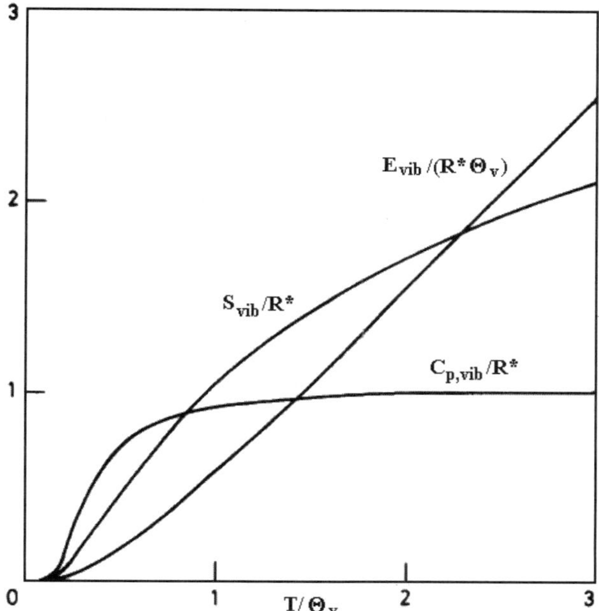

Fig. 3.3. Non-dimensional enthalpy, entropy and specific Heat for harmonic di-atomic oscillators

These identifiers are similar in structure as those given in (3.90–3.92) for rotors, and the thermophysical properties relations are also very similar to those given for rotors in (3.93–3.98). Therefore, the individual thermodynamic properties for an anharmonic oscillator (and for a harmonic oscillator by putting $x = 0$ and v_{max} going to infinity) can be calculated from the following relations:

Internal energy and *enthalpy*:

$$E_{\mathrm{vib}} = H_{\mathrm{vib}} = R^*\Theta_v(D_2/D_1) \ . \tag{3.127}$$

Entropy:

$$S_{\mathrm{vib}} = (E_{\mathrm{vib}}/T) + R^* \ln(D_1) \ . \tag{3.128}$$

Specific heats:

$$C_{p,\mathrm{vib}} = C_{v,\mathrm{vib}} = R^*(\Theta_v/T)^2[(D_3/D_1) - (D_2/D_1)^2] \ . \tag{3.129}$$

For multi-atomic molecules with degeneracy in energy or different modes of vibration in different directions, the partition function of vibration for the entire molecule is obtained as a product of partition function in different modes of vibration (for harmonic oscillator, as a first approximation), and is given by the relation

$$Z_v = \prod_k Z_{v,k} = \prod_k \left[\sum_{v=0}^{\infty} \exp^{-v\Theta_{v,k}/T} \right]^{d_k}$$

$$= \prod_k \frac{1}{\left[1 - \exp^{-\Theta_{v,k}/T} \right]^{d_k}} \tag{3.130}$$

where the characteristic oscillation temperature for different modes of operation, $\Theta_{v,k}$, and *degeneracy*, d_k, for various gases are given in Table 2.3. Accordingly, the following equations can be used for the calculation of the thermophysical properties (harmonic oscillation):

Internal energy and *enthalpy*:

$$E_{\text{vib}} = H_{\text{vib}} = \sum H_{\text{vib},k} = R^* \sum_k \frac{\Theta_{v,k} d_k}{\left(\exp^{\Theta_{v,k}/T} -1 \right)} . \tag{3.131}$$

Entropy:

$$S_{\text{vib}} = \sum_k \left[\frac{H_{\text{vib},k}}{T} + R^* d_k \ln \frac{1}{1 - \exp^{-\Theta_{v,k}/T}} \right] . \tag{3.132}$$

Specific heats:

$$C_{p,\text{vib}} = C_{v,\text{vib}} = R^* \sum_k d_k \left(\frac{\Theta_{v,k}}{T} \right)^2 \frac{\exp^{\Theta_{v,k}/T}}{\left(\exp^{\Theta_{v,k}/T} -1 \right)^2} . \tag{3.133}$$

For multi-atomic anharmonic oscillator model, (3.131–3.133) may be suitably modified as per the procedure discussed already for a diatomic molecule.

It has already been pointed out that generally in a transition between the vibrational energy levels there is always a simultaneous transition between the rotational energy levels, although the reverse is not possible (since energy requirement for vibrational transitions are much larger than for the rotational transitions). We would now discuss the interaction of rotation and oscillation in a molecular structure, which is further accentuated due to expansion of the distance between the two atoms in a diatomic molecule, and also in the change in the mass moment of inertia. Now, the characteristic rotational temperature changes linearly with the vibrational quantum number v. If $\Theta_{r,o}$ is the characteristic rotational temperature with the vibrational energy in the ground state ($v = 0$), then at another v the characteristic rotational temperature is given by the relation

$$\Theta_r = \Theta_{ro} - \alpha(v + 1/2) \tag{3.134}$$

and the rotational energy of a single molecule in the *rotational quantum level* J is

$$E_J = J(J + 1)k_B\Theta_r - 4k_B\Theta_r(\Theta_r/\Theta_v)^2 J^2(J + 1)^2 . \tag{3.135}$$

in which the second term in the right hand side is the non-rigid rotor term. Value of α is given in Table 2.1. For the vibration energy, the anharmonic oscillator model with the anharmonicity function x (given in Table 2.1) can be used. For a multiple atomic gas, depending on the symmetry, there can be a number of values of v, x and α. However, for most of the molecular specie, the effect of the non-rigid term and α on the rotational energy is small enough to be neglected, and these need not be considered for evaluation of thermodynamic properties.

3.2.3 Contribution of Electronic Energy

Since for most of the rotational-vibrational transitions, the electrons are still in the ground state, the electronic energy contribution to the thermodynamic properties need not be considered at moderate temperatures. At higher temperatures, however, these need be considered, for which the partition function is given for Boltzmann statistic by (3.36), in which E_i in each of the energy level of a particular ionization state is calculated from the ground level of that particular ionization state. Tables of g_i, E_i and ionization energy for various gases are readily available in books such as *Atomic Energy Levels* by *Moore*[17] and published by the *National Bureau of Standards* of USA. Some of these recalculated and used by this author for various gases are given in Table in Appendix-I. For this purpose, please note that the roman numeral after the name of the species minus one is the ionization level of the gas. For example, the roman numeral 'I' refers to neutral state, 'II' means singly-charged ion, etc. Further, as the principal quantum number $n \to \infty$ the corresponding statistical weight $g \to \infty$, and the energy given is the ionization energy in cm^{-1}. In actual calculation of the partition function and the thermodynamic properties, it is, therefore, necessary, to have a cut-off condition of energy, so that the partition function does not become infinity. From the several formulas available for this purpose, the following by *Unsoeld* [25] is easy to apply:

$$E_i \leq E_{\text{limit}} = I_i - (3e^2/\epsilon_o)(q_{\text{eff}} n_e)^{1/3} \qquad (3.136)$$

in which

$E_{\text{limit}} = $ *cut-off energy*;
$I_i = $ *ionization potential* from the ground level of the particular ionization state;
$q_{\text{eff}} = $ *effective charge number of electrons* in the outermost shell;
$n_e = $ *electron number density*, m^{-3}; and
$e^2/\epsilon_o = 2.9 \times 10^{-26}$, Jm.

Once again, the contribution of the electronic excitation to the thermodynamic properties can be calculated, as it has been done for other modes of energy. Noting the identifiers,

$$D_1 = \sum g_i \exp^{-E_i/(k_B T)} \tag{3.137}$$

$$D_2 = \sum g_i E_i \exp^{-E_i/(k_B T)} \tag{3.138}$$

and

$$D_2 = \sum g_i E_i^2 \exp^{-E_i/(k_B T)} \tag{3.139}$$

formulas for calculation of the contribution to the various thermodynamic properties are as follows:

Internal energy or *enthalpy*:

$$E_{\mathrm{el}} = H_{\mathrm{el}} = (R^*/k_B)(D_2/D_1) \ . \tag{3.140}$$

Entropy:

$$S_{\mathrm{el}} = (E_{\mathrm{el}}/T) + R^* \ln(D_1) \ . \tag{3.141}$$

Specific heats:

$$C_{p,\mathrm{el}} = C_{v,\mathrm{el}} = [R^*/(k_B/T)^2][(D_3/D_1) - (D_2/D_1)^2] \ . \tag{3.142}$$

The total values of the thermodynamic properties are, of course, obtained by adding individual contributions from different modes of energy. For the internal energy or enthalpy, one has to add further the reaction energy or reaction enthalpy at 0 K. The actual results of the enthalpy and the entropy at 1 bar for different gases and temperature up to 15,000 K, reproduced from *Bosnjakovic*[3], are given in the two tables in Appendix B and C.

3.2.4 Sample Calculations

Results of calculation for molar enthalpy (in MJ.kmole^{-1}), entropy (in MJ.kmole^{-1}K^{-1}) and specific heat at constant pressure (in MJ.kmole^{-1}K^{-1}) for H_2 and CO at two different temperatures, 298.15 and 1000 K, are given in tabular form in Table 3.3 in order to familiarize the reader about the procedure involved. At the top of the Table, values are given for the characteristic rotational temperature, Θ_r, the symmetry constant,σ, the characteristic vibrational temperature, Θ_v, the anharmonicity coefficient, x, and the heat of formation at 0 K,ΔH_o(MJ.kmole^{-1}). Further below are the results of calculation of molar enthalpy, entropy and specific heat at constant pressure for translation (3.70), (3.71) and (3.74), rotation (3.82), (3.83) and (3.86), harmonic oscillator model (3.114), (3.116) and (3.117), anharmonic oscillator model (3.124), (3.125) and (3.126), and the sum of all the above (total values for both harmonic and anharmonic models). It is recommended, that the reader should test the results of exact calculation for a rotor (3.93–3.98) with those under ideal rotor model (3.81–3.86).

Table 3.3. Calculation of molar enthalpy, entropy and specific heat at constant pressure for CO and H_2 molecules

Gas	H_2		CO	
Θ_r(K)	87.6		2.78	
σ	2.0		1.0	
Θ_v(K)	6333.0		3130.0	
x	0.0266		0.0061	
ΔH_o(MJ/kmole)	0.0		-113.88	
Temp. (K)	298.15	1000	298.15	1000
Htrans(MJ/kmole)	6.20	20.78	6.20	20.78
Strans(MJ/kmoleK)	0.11756	0.14271	0.15037	0.17533
Cp,trans(MJ/kmoleK)	0.021	0.021	0.021	0.021
Hrot(MJ/kmole)	2.23	8.07	2.47	8.25
Srot(MJ/kmoleK)	0.01272	0.02279	0.04718	0.05718
Cp,rot(MJ/kmoleK)	0.008	0.008	0.008	0.008
Hhar(MJ/kmole)	0.00	0.09	0.00	1.19
Shar(MJ/kmoleK)	0.00	0.00011	0.00	0.00156
Cp,har(MJ/kmoleK)	0.00	0.001	0.00	0.004
Hanh(MJ/kmole)	0.00	0.11	0.00	1.21
Sanh(MJ/kmoleK)	0.00	0.00013	0.00	0.00159
Cp,anh(MJ/kmoleK)	0.00	0.001	0.00	0.004
Incl. harm.:				
Htot(MJ/kmole)	8.43	28.95	-105.21	-83.65
Stot(MJ/kmoleK)	0.13029	0.16562	0.19756	0.23427
Cp,tot(MJ/kmoleK)	0.029	0.030	0.029	0.033
Incl. anharm.:				
Htot(MJ/kmole)	8.43	28.96	-105.21	-83.63
Stot(MJ/kmoleK)	0.13029	0.16564	0.19756	0.23430
Cp,tot(MJ/kmoleK)	0.029	0.030	0.029	0.033

Table 3.4 contains also the result of calculation of the enthalpy, the entropy and the specific heat for CO_2 by taking the rotor data from Tables 2.1 and 2.2 and the harmonic oscillator data from Table 2.3 and using (3.106–3.109) and (3.131–3.133). Heat of formation for CO_2 is -393.23 $MJ.kmole^{-1}$.

Table 3.4. Calculation of molar enthalpy, entropy and specific heat at constant pressure for CO_2

Temp.(K)	298.15	500	1000
Htrans(MJ/kmole)	6.20	10.39	20.78
Strans(MJ/kmoleK)	0.15601	0.16676	0.18166
Cp,trans(MJ/kmoleK)	0.021	0.021	0.021
Hrot(MJ/kmole)	2.48	4.16	8.31
Srot(MJ/kmoleK)	0.07389	0.07818	0.08395
Cp,rot(MJ/kmoleK)	0.008	0.008	0.008
Hvib,har(MJ/kmole)	0.36	1.75	8.66
Svib,har(MJ/kmoleK)	0.00155	0.00500	0.01426
Cp,vib,har(MJ/kmoleK)	0.004	0.009	0.017
Htot(MJ/kmole)	-384.20	-376.93	-355.47
Stot(MJ/kmoleK)	0.23145	0.24994	0.27937
Cp,tot(MJ/kmoleK)	0.033	0.038	0.046

3.3 Distribution of Energy Levels

In the previous section, statistical methods have been developed to compute the thermophysical properties without any regard to the actual energy distribution which is to be discussed in the present section. In the translational mode, the energy of individual particles is $E = Mv^2/2$. If E is not discrete but continuous, then the number of particles with velocities between v and $v + \mathrm{d}v$ is the number of particles in the volume $4\pi v^2 \mathrm{d}v$. Thus from (3.21) and (3.32),

$$\mathrm{d}N = \frac{CN}{Z} \exp^{-Mv^2/(2k_B T)} 4\pi v^2 \mathrm{d}v \tag{3.143}$$

where C is a constant, which accounts for the dimension and its value will be determined. Now, integrating (3.143) and noting that the integral

$$\int_0^\infty \exp^{-Mv^2/(2k_B T)} v^2 \mathrm{d}v = \frac{\sqrt{2}}{M^{3/2}} \int_0^\infty \sqrt{E} \exp^{-E/(k_B T)} \mathrm{d}E$$

$$= \sqrt{\frac{\pi}{2}} \left(\frac{k_B T}{M} \right)^{3/2} \tag{3.144}$$

we get

$$N = \int \mathrm{d}N = \frac{4\pi CN}{Z} \int_0^\infty \exp^{-Mv^2/(2k_B T)} v^2 \mathrm{d}v = \frac{CN}{Z} \left(\frac{2\pi k_B T}{M} \right)^{3/2} . \tag{3.145}$$

Thus,

$$Z/C = \left(\frac{2\pi k_B T}{M} \right)^{3/2} \tag{3.146}$$

and (3.143) becomes

$$dN = 4\pi N v^2 \left(\frac{M}{2\pi k_B T}\right)^{3/2} \exp^{-Mv^2/(2k_B T)} dv \qquad (3.147)$$

where the *distribution function*

$$f(v) = 4\pi v^2 \left(\frac{M}{2\pi k_B T}\right)^{3/2} \exp^{-Mv^2/(2k_B T)} \qquad (3.148)$$

gives the fraction of total number of particles with *kinetic speed* v and speed interval dv.

Let us now define a non-dimensional kinetic speed

$$v' = v/\sqrt{2k_B T/M} \ . \qquad (3.149)$$

Thus from (3.148), we write

$$dN = N f(v)dv = N f(v')dv' \qquad (3.150)$$

where

$$f(v') = \left(\frac{4}{\sqrt{\pi}}\right) v'^2 \exp^{-v'^2} \ . \qquad (3.151)$$

Now $f(v')\to 0$, if $v' \to 0$ and $v' \to \infty$, and the maximum value is obtained by differentiation with respect to v' and putting it equal to zero; the maximum is at $v' = 1$. Equation (3.150) is integrated numerically between 0 and 3 in steps of $\Delta v' = 0.2$, and is plotted in Fig. 3.4 along with $f(v')$. It can be

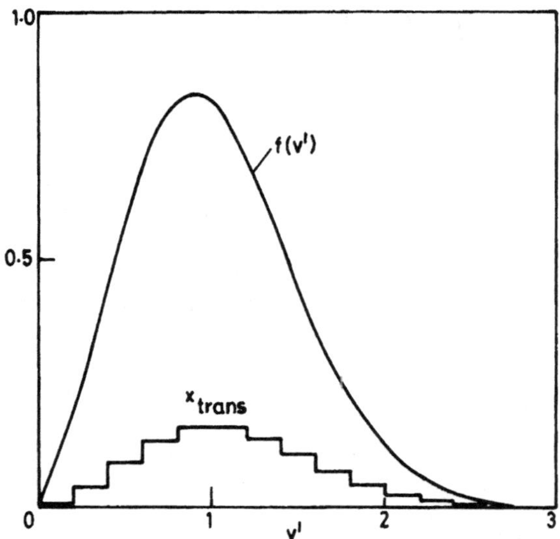

Fig. 3.4. Mole fraction distribution for translational energy at different reduced speeds

seen that 95 percent of all the particles have speed in the range between 0 and 2 with the most probable value is at $v' = 1$. One can, of course, determine the average of any other property, φ, from the integral relation

$$\bar{\varphi} = \int \varphi f(v') dv' = \int \varphi dN \ . \tag{3.152}$$

In this connection, it is worthwhile to consider the integral

$$\int_0^\infty x^\alpha \exp^{-\beta x^2} dx \ . \tag{3.153}$$

By putting $\beta x^2 = y$, $2\beta\, x\, dx = dy$, the above integral can be written in terms of the Gamma function as

$$\frac{1}{2\beta^{(\alpha+1)/2}} \Gamma\left(\frac{\alpha+1}{2}\right) \ . \tag{3.154}$$

Now note the following Gamma function results:

$$\Gamma(0) = \infty; \Gamma\left(\frac{1}{2}\right) = \sqrt{\pi}; \Gamma(1) = 1$$

$$\Gamma\left(\frac{3}{2}\right) = \frac{\sqrt{\pi}}{2}; \Gamma(2) = 1; \Gamma\left(\frac{5}{2}\right) = \frac{3\sqrt{\pi}}{4} \ .$$

By putting $\varphi = 1$, we get the normalizing relation

$$\int_0^\infty f(v') dv' = \frac{2}{\sqrt{\pi}} \Gamma\left(\frac{3}{2}\right) = 1 \ . \tag{3.155}$$

The average non-dimensional kinetic speed of the molecule is obtained by putting $\varphi = v'$ and integrating to get

$$\bar{v}' = \int_0^\infty v' f(v') dv' = \frac{4}{\sqrt{\pi}} \int_0^\infty v'^3 \exp^{-v'^2} dv' = \frac{2}{\sqrt{\pi}} \ . \tag{3.156}$$

Similarly, the mean of the square of the non-dimensional speed is obtained by substituting $\varphi = v'^2$ and integrating to get

$$(\bar{v}'^2) = \int_0^\infty v'^2 f(v') dv' = \frac{4}{\sqrt{\pi}} v'^4 \exp^{-v'^2} dv' = \frac{3}{2} \ . \tag{3.157}$$

From the speed distribution function, it is, therefore, possible to obtain the following three important speeds:

1. *Most probable speed* $v' = v'_p = 1$, that is

$$v_p = \sqrt{2k_B T/M} \ . \tag{3.158}$$

2. *Average speed* $\bar{v}' = (2/\sqrt{\pi})v_p' = 1.129v_p'$, that is

$$\bar{v} = \sqrt{8k_BT/(\pi M)} \ . \tag{3.159}$$

3. *Root mean square of the speed*, (since $\bar{v}'^2 = (3/2)v_p'^2$), that is

$$\sqrt{(\bar{v})^2} = \sqrt{3k_BT/M} \ . \tag{3.160}$$

For comparison, the *isentropic sonic speed* in a gas is

$$a = \sqrt{\gamma R^* T/m} = \sqrt{\gamma k_B T/M} = v_p\sqrt{\gamma/2} \tag{3.161}$$

where m is the mole mass of the gas, γ is the specific heat ratio and R^* is the universal gas constant. In general the isentropic sonic speed is slightly smaller than v_p, but it is of the same order of magnitude. Thus a disturbance propagates in a gas with almost the most probable speed.

While (3.150) gives the mole fraction of particles, which are found in the kinetic speed range between v and $v + dv$, we would now derive similar relations for rotational and vibrational energies. These relations are obtained from the Boltzmann statistic, (3.35). Thus, for the *rotational energy level distribution*, the mole fraction at any rotational quantum number J is

$$x_J = \frac{N_J}{N} = \frac{(2J+1)}{Z_r}\exp^{-J(J+1)\Theta_r/T} = \frac{\Theta_r}{T}(2J+1)\exp^{-J(J+1)\Theta_r/T} \ . \tag{3.162}$$

The expression, plotted in Fig. 3.5, has a maximum, which can be easily obtained by differentiating (3.162) with respect to J and putting it equal to zero to get the maximum at $J = (\sqrt{2T/\Theta_r} - 1)/2$.

Similarly for an anharmonic oscillator model, the mole fraction at any vibration quantum number v is given by the relation

$$x_v = \frac{\exp^{-v(1-xv)\Theta_v/T}}{\sum \exp^{-v(1-xv)\Theta_v/T}} \tag{3.163}$$

which is plotted in Fig. 3.6 for an anharmonicity coefficient, $x = 0.01$.

The physical meaning of the pressure in terms of the mean kinetic speed can now be given also in a simplified manner as follows. Let us consider a cube of side 'a' containing N particles having a mean kinetic speed v. Therefore, $N = na^3$, where n is the number density $[\mathrm{m}^{-3}]$ of the particles. Now the momentum transfer per collision while reflecting from one wall is $2Mv$, and in time $t = 2a/v$, $(1/3)$ of N particles are going to fall on each wall. Therefore, the momentum transfer per second is $pa^2 = (N/3)(2Mv/t) = na^2Mv^2/3$ and noting $Mv^2 = 3k_BT$, we get the equation of state again as $p = nk_BT$.

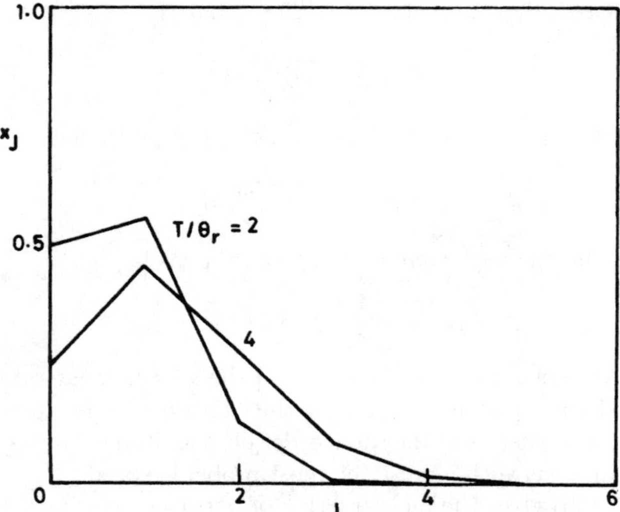

Fig. 3.5. Mole fraction distribution for different T/Θ_r

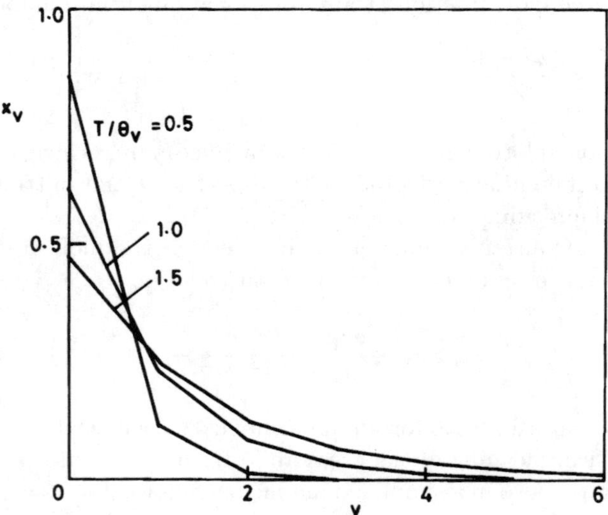

Fig. 3.6. Mole fraction distribution for different T/Θ_v

3.4 Exercise

3.4.1 For various gases estimate the value of molar enthalpy, entropy and specific heat at different temperatures and pressures and compare these with those given in available literatures.

3.4.2 Calculate the non-dimensional values of enthalpy, entropy, and specific heat as a function of the ratio of the temperature and the characteristic temperature separately for diatomic rotor and harmonic oscillator.

3.4.3 Calculate the energy distribution of energy in rotor and oscillator (harmonic and anharmonic) modes as a function of the ratio of the temperature to the characteristic temperature.

3.4.4 Calculate the specific heat ratio of the diatomic molecules in the temperature range 100 to 5,000 K and show how values deviate from the ideal gas values.

3.4.5 For various diatomic molecules given in Table 2.1, estimate the temperature where the contribution of vibration in C_p is less than one percent of the total.

3.4.6 Determine the population distribution with temperature of the first four energy levels of hydrogen atom by Boltzmann distribution.

4 Radiative Properties of High Temperature Gases

4.1 Basic Concepts and Laws

We consider emission of radiation from a surface element $\mathrm{d}A$ in a hemispherical space with radius of the hemisphere being r (Fig 4.1). Further we consider a surface element element $\mathrm{d}A' = r^2 \sin\psi \, \mathrm{d}\psi \, \mathrm{d}\theta$ on this hemisphere, and thus a solid angle $\mathrm{d}\Omega = \mathrm{d}A'/r^2 = \sin\psi \, \mathrm{d}\psi \, \mathrm{d}\theta$. In addition any radiation quantity emanating normal to the surface element $\mathrm{d}A'$ has components in three directions:

$$l_x = \sin\psi \sin\theta, \, l_y = \sin\psi \cos\theta, \, l_z = \cos\psi \ . \tag{4.1}$$

Obviously if l is in the direction in which the radiative energy propagates with its three angles $\{l_x, l_y \text{ and } l_z\}$, then the normal unit vector $\mathbf{n}$ $\{i, j, k\}$ is given by the relation ([110]):

$$\mathbf{n} = i\cos(l_x) + j\cos(l_y) + k\cos(l_z) \ . \tag{4.2}$$

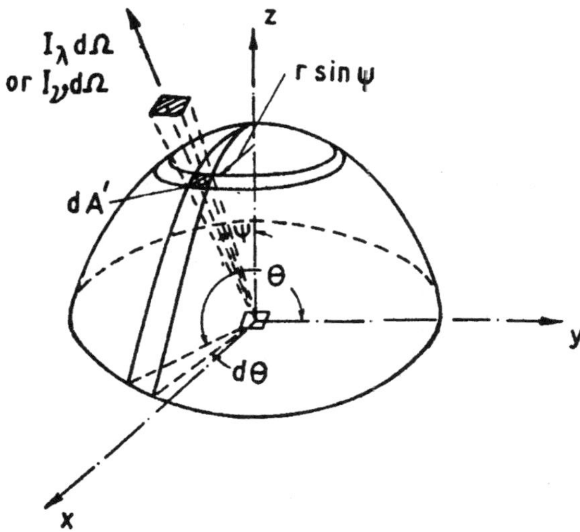

Fig. 4.1. Determining solid angles in spherical coordinates

Radiation in all wavelengths $\lambda = 0$ to ∞ (or all frequencies, $\nu = c/\lambda = 0$ to ∞, $c = $ *speed of light*), or in a particular wavelength λ (or a particular frequency ν) may be considered. Further, we define some of the other radiative quantities as follows.

Total or *integrated radiation* $\dot{Q}$ [in W] is the total radiation in all wavelengths (frequencies), in all directions from the surface of a body, per unit time out of a hemisphere.

Emissive power or *integral density of radiation* $\dot{E}$ [in Wm^{-2}] is now the energy at all wavelengths, in all directions, per unit wave length, per unit surface area of the body, and is related to the *spectral intensity of radiation*

$$B_\lambda = \mathrm{d}\dot{E}/\mathrm{d}\lambda = \dot{E}_\lambda = \mathrm{d}^2\dot{Q}/(\mathrm{d}A\mathrm{d}\lambda), \mathrm{Wm}^{-3} . \tag{4.3}$$

Similarly, in terms of frequency, we write

$$B_\nu = \mathrm{d}\dot{E}/\mathrm{d}\nu = \dot{E}_\nu = \mathrm{d}^2\dot{Q}/(\mathrm{d}A\mathrm{d}\nu), \mathrm{Jm}^{-2} . \tag{4.4}$$

Noting that $B_\lambda \mathrm{d}\lambda = B_\nu \mathrm{d}\nu$, one can write

$$B_\lambda = B_\lambda(c/\lambda^2) = B_\nu \nu^2/c . \tag{4.5}$$

Radiation of energy per unit wavelength, unit surface area and unit solid angle (steradian) of the body in one direction is the *angular spectral intensity of radiation*

$$I_\lambda = \mathrm{d}B_\lambda/\mathrm{d}\Omega = \mathrm{d}^2\dot{E}/(\mathrm{d}\lambda\mathrm{d}\Omega) = \mathrm{d}^2\dot{Q}/(\mathrm{d}A\mathrm{d}\lambda\mathrm{d}\Omega), \mathrm{Wm}^{-3}\mathrm{sterad.}^{-1} . \tag{4.6}$$

Similarly in terms of the spectral frequency the angular intensity of radiation is

$$I_\nu = \mathrm{d}B_\nu/\mathrm{d}\Omega = \mathrm{d}^2\dot{E}/(\mathrm{d}\nu\mathrm{d}\Omega) = \mathrm{d}^2\dot{Q}/(\mathrm{d}A\mathrm{d}\nu\mathrm{d}\Omega), \mathrm{J.m}^{-2}.\mathrm{sterad.}^{-1} . \tag{4.7}$$

Noting (4.5), we get

$$I_\lambda = I_\nu(c/\lambda^2) = I_\nu(\nu^2/c) . \tag{4.8}$$

Integrating (4.6) over all wavelengths

$$I = \int I_\lambda \mathrm{d}\lambda = \mathrm{d}\dot{E}/\mathrm{d}\Omega \tag{4.9}$$

we get the *total angular intensity of radiation* [in Wm^{-2}sterad.$^{-1}$] which is the radiation of energy at all wave lengths per unit surface area of the body and solid angle. Further integrating (4.3) over all wave lengths, we get the relation for the *total intensity of radiation*

$$B = \int B_\lambda \mathrm{d}\lambda = \dot{E}, \mathrm{Wm}^{-2} . \tag{4.10}$$

Now the energy radiated through the hemispherical surface element is $\mathrm{d}\dot{E}_\nu = I_\nu \mathrm{d}\nu\mathrm{d}\Omega$ and the flux of photons through this surface element is $\mathrm{d}\dot{n}_f$. The *photon-flux* (number of photons per unit area, m^{-2}) is given by the relation $d\dot{n}_f = (h\nu)^{-1}d\dot{E}_f$ [in $m^{-2}s^{-1}$], whose three components in the three positive coordinate directions can be obtained by multiplying it with the cosines in the three directions, (4.1). Since the energy per photon is $h\nu$ and the momentum in the j-th coordinate direction is $(h\nu/c)l_j$ [see also Sect. 4.2], the *differential fluxes of radiant spectral energy* and *momentum* are given by the relation

$$d\dot{E}_{j\nu} = I_\nu l_j d\Omega = I_\nu l_j \sin\psi d\psi d\theta \tag{4.11}$$

and

$$d\tau_{ij\nu} = (I_\nu/c)l_i l_j d\Omega = (I_\nu/c)l_i l_j \sin\psi d\psi d\theta \ . \tag{4.12}$$

We would now evaluate (4.11, 4.12) for the special case of an isotropic radiation, in which I_ν or I_λ are independent of the direction of propagation. Assuming j-th coordinate direction is normal to the surface

$$dB_\nu = d\dot{E}_\nu = I_\nu \cos\psi d\Omega = (1/2)I_\nu \sin(2\psi)d\psi d\theta \tag{4.13}$$

which states that the component of the intensity of radiation in the direction of the normal to the surface is proportional to the cosine of the angle ψ formed between the direction of radiation and the normal to the surface of the body (*Lambert's cosine law*). Thus for *isotropic radiation*, the relation between B_ν and I_ν functions can be obtained by integrating over all directions out of a hemisphere [in Jm^{-2}]

$$B_\nu = \frac{1}{2}\int_{\psi=\pi/2}^{0}\int_{\theta=0}^{2\pi} I_\nu \sin(2\psi)d\psi d\theta = \pi I_\nu, Jm^{-2} \ . \tag{4.14}$$

Similarly

$$B_\lambda = \pi I_\lambda, Wm^{-2} \ . \tag{4.15}$$

Further for *isotropic radiation*,

$$\tau_{ij\nu} = \int (I_\nu/c)l_i l_j \sin\psi d\psi d\theta = \frac{I_\nu}{c}\begin{cases} 2\pi/3 & \text{for i=j} \\ 0 & \text{for i} \neq \text{j} \ . \end{cases}$$

Radiative pressure applied to a fully reflecting surface is twice the momentum flux in that direction. Thus in the i-th direction, the *spectral radiative pressure* is given by the average of the sum of all momentum-flux with $i = j$, and leads to the relation

$$p_\nu = \frac{2}{3}\sum_{i=1}^{3}\tau_{ij\nu} = \frac{4\pi}{3}\left(\frac{I_\nu}{c}\right) \ . \tag{4.16}$$

We now examine, for the isotropic radiation, the spectral energy of radiation in transit from a small sphere to a larger sphere, but small enough to consider the value of the intensity of radiation to be uniform. It is also assumed that all radiative energy emanating from the smaller sphere is absorbed on the surface of the larger sphere. The flux of photons is then $4\pi I_\nu/(h\nu)$ and the energy flux is $4\pi I_\nu = u_\nu/c$, where u_ν is the *spectral internal energy of radiation.* Thus for *isotropic radiation,*

$$u_\nu = \frac{4\pi}{c} I_\nu = 3p_\nu \ . \tag{4.17}$$

A very special case of isotropic radiation is the radiation inside a cavity, which we would consider now for its importance, in which the cavity wall at a given temperature is emitting and absorbing the radiative energy in such a manner that an *equilibrium radiation,* also called *black body radiation* exists in the cavity. It is easy to verify that neither I_ν nor u_ν nor p_ν depend on the size of the two spheres.

By combining the empirical relations for radiation, known as *Wien's law* and *Rayleigh-Jean's law* (to be discussed later), *Max Planck* developed a semi-empirical law (which was subsequently named after him) giving a relation for the *spectral intensity of equilibrium radiation* (denoted by an asterisk as superscript)

$$B_\lambda^* = \frac{C_1/\lambda^5}{\exp^{C_2/(\lambda T)} - 1}, \mathrm{Wm}^{-3} \tag{4.18}$$

where λ is the *wavelength* [in m], and C_1 and C_2 are constants whose values have subsequently been found to be

$$C_1 = 2\pi hc^2 = 3.7483 \times 10^{-16}, \mathrm{Wm}^2 \tag{4.19}$$

and

$$C_2 = hc/k_B = 0.014388, \mathrm{mK} \ . \tag{4.20}$$

Results of B_λ^* against wavelength and temperature as a parameter are shown in Fig. 4.2. While *Planck's law of radiation* was good enough to investigate a number of problems, its rigorous theoretical derivation was done by *Albert Einstein* who applied the statistics developed earlier by *Satyendra Nath Bose,* which we would follow. We would now discuss the method to calculate the distribution of photons for an equilibrium radiation. As a consequence of the fact that the total energy of radiation, for a given temperature, is fixed but not the number of photons, only one *Lagrange undetermined multiplier* is used as constraints in the statistical equations discussed in Chap. 3. We write now, therefore, from (3.12),

$$\ln\left(\frac{g_i + N_i}{N_i}\right) = \frac{E_i}{k_B T} \tag{4.21}$$

which can be rewritten as

$$N_i = \frac{g_i}{\exp^{E_i/(k_B T)}} \ . \tag{4.22}$$

Herein $N_i = N_\nu d\nu$ is the number of photons in the volume element in energy space with average energy $h\nu$. Thus $N_\nu = N_i/d\nu$ is the change in the number of photons with respect to ν, the energy of all photons with associated

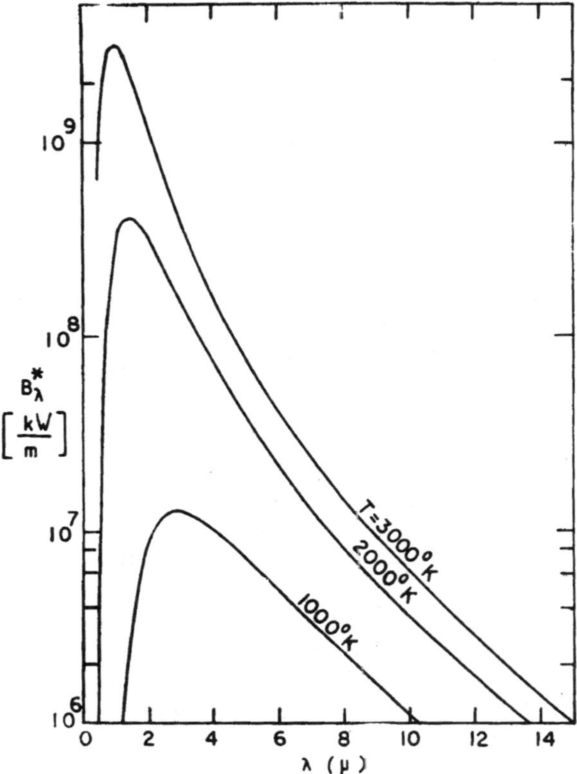

Fig. 4.2. Spectral intensity of a black body radiation

frequency of radiation ν is $N_\nu h\nu$ and the associated energy density in the volume V is

$$u_\nu^* = N_\nu h\nu/V \qquad (4.23)$$

where u_ν^* is the *spectral radiative energy* for equilibrium radiation.

Now for the photons from a discussion regarding the translational energy of particles, (3.64), $g_i = (\pi/2)k^2\mathrm{d}k$. With $k = 2L/\lambda$, $\lambda = c/\nu = $ wave length of radiation, and volume $V = L^3$, we can then write

$$g_i = 4\pi V(\nu^2/c^3)\mathrm{d}\nu \ . \qquad (4.24)$$

Note that for an *unpolarized radiation*, the value of g_i in this relation has to be multiplied by two. Since,

$$N_i \cong N_\nu \mathrm{d}\nu = \frac{4\pi V}{h\nu c}I_\nu^* \mathrm{d}\nu = \frac{u_\nu^* V}{h\nu}\mathrm{d}\nu \qquad (4.25)$$

we write further,

$$\frac{4\pi V}{h\nu c}I_\nu^* d\nu = \frac{1}{\exp^{h\nu/(k_B T)}-1}8\pi V\frac{\nu^2}{c^3}d\nu \tag{4.26}$$

which follows

$$I_\nu^* = \frac{2h\nu^3}{c^2}\frac{1}{\exp^{h\nu/(k_B T)}-1}, \mathrm{Jm}^{-2}.\mathrm{sterad.}^{-1}\;. \tag{4.27}$$

From (4.5), (4.8) and (4.15) for *isotropic* and *equilibrium radiation*, we can also write

$$I_\lambda^* = \frac{2hc^2}{\lambda^5}\frac{1}{\exp^{hc/(\lambda k_B T)}-1}, \mathrm{Wm}^{-3}.\mathrm{sterad.}^{-1} \tag{4.28}$$

$$B_\nu^* = \frac{2\pi h\nu^3}{c^2}\frac{1}{\exp^{h\nu/(k_B T)}-1}, \mathrm{Jm}^{-2} \tag{4.29}$$

and

$$B_\lambda^* = \frac{2\pi hc^2}{\lambda^5}\frac{1}{\exp^{hc/(\lambda k_B T)}-1}, \mathrm{Wm}^{-3}\;. \tag{4.30}$$

By comparing (4.18) with (4.30), the values of constants C_1 and C_2 can be obtained, which are given in (4.19, 4.20). In this process we completed derivation of *Planck's Radiation law* by rigorous statistical method. Result of calculation of B_λ^* has been plotted in Fig. 4.2 and a discussion will follow (4.42).

Now, with the help of Eqs. (4.16) and (4.17), the relations for the *spectral radiative pressure* and the *spectral radiative energy* for equilibrium radiation are

$$p_\nu^* = \frac{4\pi}{3c}I_\nu^* = \frac{8\pi h\nu^3}{3c^3}\frac{1}{\exp^{h\nu/(k_B T)}-1}, \mathrm{Nsm}^{-2} \tag{4.31}$$

and

$$u_\nu^* = 3p_\nu^* = \frac{8\pi h\nu^3}{3c^3}\frac{1}{exp^{h\nu/(k_B T)}-1}, \mathrm{Jsm}^{-3}\;. \tag{4.32}$$

Total quantities of equilibrium radiation can now be obtained by integrating (4.26–4.32) over all frequencies. Thus the *total intensity of equilibrium radiation* is

$$I^* = \int I_\nu^* d\nu = \int_0^\infty \frac{2h\nu^3}{c^2}\frac{d\nu}{\exp^{h\nu/(k_B T)}-1}$$
$$= \frac{2k_B^4 T^4}{h^3 c^2}\int_0^\infty \frac{[h\nu/(k_B T)]^3}{\exp^{h\nu/(k_B T)}-1}d\left(\frac{h\nu}{k_B T}\right)\;.$$

Let $x = h\nu/(k_B T)$. Further, noting for $x < 1$, that,

$$1 + x + x^2 + x^3 + \ldots = \frac{1}{1-x} \tag{4.33}$$

and also

$$\frac{1}{1-\exp^{-x}} = 1 + \sum_{n=1}^{\infty} \exp^{-nx}; \quad \int_0^{\infty} x^3 \exp^{-nx} \, \mathrm{d}x = \frac{6}{n^4} \, . \tag{4.34}$$

Thus,

$$\int_0^{\infty} \frac{x^3}{\exp^x - 1} \mathrm{d}x = \int_0^{\infty} \frac{x^3 \exp^{-x}}{1 - \exp^{-x}} \mathrm{d}x = \int_0^{\infty} x^3 \exp^{-x} \left[1 + \sum_{n=1}^{\infty} \exp^{-nx} \right] \mathrm{d}x$$

$$= \int_0^{\infty} \sum_{n=1}^{\infty} \left(\frac{x^3}{\exp^{nx}} \right) \mathrm{d}x = 6 \sum_{n=1}^{\infty} \left(\frac{1}{n^4} \right) = \frac{\pi^4}{15} \, . \tag{4.35}$$

and we get the following equations for the total quantities of equilibrium radiation.

Total angular intensity of equilibrium radiation:

$$I^* = \frac{2\pi^4 k_B^4 T^4}{15 h^3 c^2} = \frac{\sigma}{\pi} T^4 \, , \, \mathrm{Wm}^{-2}.\mathrm{sterad.}^{-1} \, . \tag{4.36}$$

Total intensity of equilibrium radiation:

$$B^* = \pi I^* = \sigma T^4 \, , \, \mathrm{Wm}^{-2} \, . \tag{4.37}$$

Total equilibrium pressure:

$$p^* = \frac{4\pi}{3c} I^* = \frac{4\sigma}{3c} T^4 = 2.51986 \times 10^{-16} \, , \, \mathrm{Nm}^{-2} \, . \tag{4.38}$$

Total equilibrium radiative internal energy (= radiant energy density):

$$u^* = 3p^* = \frac{4\sigma}{c} T^4 \, , \, \mathrm{Jm}^{-3} \tag{4.39}$$

where

$$\sigma = \frac{2\pi^5 k_B^4}{15 h^3 c^2} = 5.6697 \times 10^{-8} \, , \, \mathrm{Wm}^{-2}\mathrm{K}^{-4} \tag{4.40}$$

is the total radiation constant obtained experimentally by *Stefan* and later substantiated analytically by *Boltzmann, σ,* is usually called the *"Boltzmann constant of radiation"*. Equation (4.37) is the analytical formulation of the *Stefan-Boltzmann's law,* and was formulated chronologically prior to the *Planck's law of radiation,* (4.18). From (4.38), it is found that the radiation pressure of a black body radiation of a gas in a cavity at 100,000 K is around 0.252 bar. Thus, if the total pressure measured at the wall at this gas temperature is 0.252 bar, then the cavity is filled up entirely by the radiative energy as particles. The wall temperature of the cavity has to be also at 100,000 K. In practice, however, the wall temperature of the cavity is at a much lower temperature and the gas may radiate freely without much absorption, so that the radiative pressure is negligible. Thus for most of the practical cases radiative pressure may be considered equal to zero, although the radiative energy flux (intensity of radiation) may be substantial.

We would now discuss the method to calculate the entropy for an equilibrium radiation. For the general case of equilibrium gas radiation, we can use the statistical methods discussed in Chap. 3 with the exception that here we have the auxiliary condition for prescribed total energy but no restriction on the number of photons. We write now, therefore, from (3.12) with Lagrange undetermined multiplier λ put equal to zero for Bose statistics, as

$$\ln\left(\frac{g_i + N_i}{N_i}\right) = \frac{E_i}{k_B T} \tag{4.41}$$

and thus from (3.43), for Bose statistics,

$$dS = k_B d(\ln W) = \sum \ln\left(\frac{g_i + N_i}{N_i}\right) dN_i = \sum \frac{E_i}{k_B T} dN_i$$

$$= \frac{1}{T} \sum E_i dN_i = \frac{1}{T}\left[d\sum(E_i N_i) - \sum N_i dE_i\right] .$$

The first term within the bracket is the change in the internal energy of radiation du^* and the second term vanishes, since no work is done externally by radiation. The concept of entropy here is, however, the same as in classical thermodynamics, that is the radiative heat flux can take place always from the region of higher temperature to lower temperature. Thus when two bodies are in relation to each other, the body at higher temperature gives to the body at lower temperature the heat by radiation, and only in the equilibrium case, when the temperature of the two bodies is common, the radiative heat flux is that for a black body radiation. For the equilibrium case, denoted by an asterisk as superscript, the expression $TdS^* = du^* = (16\sigma/c)T^3 dT$ is obtained and thus for equilibrium radiation the *entropy of radiation* is given by the relation

$$S^* = \frac{16}{3}\frac{\sigma}{c}T^3 , \; \mathrm{Jm^{-3}K^{-1}} . \tag{4.42}$$

We would now discuss some special cases of the *Planck's radiation law*, which is given in equation form by (4.18) or (4.30), for which the values of B_λ^* [in kW.m^{-3}] is plotted against λ [in μ] for different temperatures (Fig. 4.2). It can be seen that the numerator in these equations decreases with increasing λ due to λ^{-5} term , but the inverse of the denominator increases with λ. The product of the two can, therefore, be expected to have a maximum. By differentiating B_λ^* with respect to the wavelength and equating to zero gives the maximum value of B_λ^* at the wavelength designated as $\lambda_{\max}$. The latter is given by the relation

$$\lambda_{\max}T = 2898[\mu.K] \tag{4.43}$$

with the associated maximum emissive power

$$B_{\lambda_{\max}}^* = 1.307 \times 10^{-8}T^5 , \; \mathrm{kW.m^{-3}} . \tag{4.44}$$

Equation (4.43) shows that the wavelength at which the maximum emissive power is radiated shifts to shorter wave lengths at higher temperatures (*Wien's displacement law*). This law was formulated by *Wien* in 1898 by applying *Doppler principle* (shifting in the frequency or wavelength of the radiative energy of a moving source or receiver) to the adiabatic compression of radiation in a perfectly reflecting enclosure, and he deduced that the wave length of each constituent of the radiation should be shortened in proportion to the rise of temperature produced by the compression. Thus at temperatures about the room temperature, the maximum emissive power is very much in the far infrared, in which only the rotational energy level transitions take place (for a multi-atomic molecule). As the temperature of a body is raised, the maximum emissive power shifts to the near infrared, then visible and further to the ultraviolet region at a very high temperature. This is observed, when a metal like iron is heated, when it gives first invisible radiation, then it becomes red hot, and then finally white hot. We can find also from (4.43), that the sun's maximum radiative power is at about 4830 Å, corresponding to the disk temperature of the sun at 6,000 K. It is interesting to note that the maximum sensitivity of the human eye coincides somewhat with the maximum radiative power of the sun, whereas for many animals or birds the maximum sensitivity may be in the infrared, which is why many animals can see in the darkness when the human eye can not.

Assuming *Wien's displacement law* it may be conjectured that the form of the curve representing the distribution of radiative intensity should be the same for different temperatures with the maximum displaced in proportion to the absolute temperature, with the total area increasing proportional to the fourth power of the absolute temperature. This is now shown with the help of the *Planck's law*. Planck's radiation law has two particular solutions, the one for large wavelengths is called the *Rayleigh-Jean's law*, and the other for short wavelengths is the *Wien's law*. By expanding the exponential term in (4.18)

$$\exp^{C_2/(\lambda T)} = 1 + \left(\frac{C_2}{\lambda T}\right) + \frac{1}{2!}\left(\frac{C_2}{\lambda T}\right)^2 + \dots \tag{4.45}$$

in series and terminating the series only with two terms for $\lambda T \gg C_2$, we get the expression

$$B_\lambda^* = C_1 T/(C_2 \lambda^4) \tag{4.46}$$

which expresses the *Rayleigh-Jean's law*. According to this, although B_λ^* increases proportional to the temperature, but it tends to 0 as $\lambda \to \infty$ and one had to find a method to calculate it for very large wave lengths.

In the second extreme case of $\lambda T \ll C_2$, one in the denominator can be neglected compared to the exponential term to express the *Wien's law*:

$$B_\lambda^* = (C_1/\lambda^5)\exp^{-C_2/(\lambda T)} \tag{4.47}$$

from which again the Wien's displacement $\lambda_{\max}.T = C_2/5 = 2877.6\ \mu\text{K}$ can be derived.

Although the above two laws, in addition to the Planck's radiation law, provide expressions for the intensity of radiation for a black body, it should be mentioned, that gases, in general, radiate at moderate pressures only selectively in line or band radiations, which may not be considered even approximately as a grey or black body radiation. Even the gas continuum radiation has features which may not be approximated in a similar manner by the black-body radiation.

4.2 Gas Radiation and Equation of Energy Transfer

While the gases radiate in general very selectively at various frequencies, it is possible to have a radiative energy balance at the radiating frequency taking into account the emission and the absorption of the energy by the gas. In addition to the absorption and the emission, the *"radiant particles"* (photons) may be scattered by gas molecules, which is equivalent of the change of direction by elastic collision between particles. However, it is found for most situations in gas dynamics, that the scattering is unimportant, and hence in the following section consideration of scattering is neglected.

Equation of radiative transfer can be explained more easily, if a one-dimensional model is used. In such a case, it is a continuity equation for the number density, n_R, of the part of the photons emanating through the end surface element having a direction of propagation normal to the surface element in the frequency range between ν and $\nu + d\nu$ with the associated energy $h\nu$. Thus there is a flux $\dot{n}_f = cn_R$ through the surface element. Taken in a cartesian coordinate system with a cubical volume element $dV = dxdydz$, the number of photons in dV is $n_R dxdydz$ and its time rate of change is $\partial n_R/\partial t)dxdydz$ [in s^{-1}], due to the rate of change of the number of photons through the surfaces of this cubical volume as well as a result of changes due to emission and absorption, and to a lesser extent due to scattering, which is neglected in the present treatment of the subject. The contribution of convection is proportional to the divergence of the number flux

$$-\frac{\partial}{\partial x_j}(l_j\dot{n}_f)dxdydz = -cl_j\frac{\partial n_R}{\partial x_j}dxdydz \qquad (4.48)$$

where l_j is again the cosine of the angle of the direction of flux of photons with the normal to the respective surface of the volume element. If we add to it a term

$$\left(\frac{\partial n_R}{\partial t}\right)_{e,a} dxdydz \qquad (4.49)$$

for the net rate of change of the number of photons due to emission and absorption, then the continuity equation for the number of photons becomes

$$\frac{\partial n_R}{\partial t} + cl_j\frac{\partial n_R}{\partial x_j} = \left(\frac{\partial n_R}{\partial t}\right)_{e,a}. \qquad (4.50)$$

This equation can now be written in terms of the *angular spectral intensity of radiation* I_ν. Since

$$\int I_\nu \mathrm{d}\nu \mathrm{d}\Omega = h\nu c n_R \tag{4.51}$$

the integration in the left hand side of the equation is done over the frequency range in which the energy of the photons $h\nu$ is assumed to be constant. Further, the right hand side of (4.50) can be written in terms of properties of the gas through the definition of coefficients of emission and absorption. Thus the *mass emission coefficient* j_ν [in $\mathrm{Jm}^{-3}\mathrm{sterad.}^{-1}$] is defined in such a manner that the radiant energy emitted per unit time and volume is given by the relation

$$\dot{e}_e = j_\nu \mathrm{d}\nu \mathrm{d}\Omega = h\nu(\partial n_R/\partial t)_e \ , \ \mathrm{Wm}^{-3} \ . \tag{4.52}$$

The rate of absorption of the radiant energy by a gas is found to be proportional to the *angular spectral intensity* I_ν. Accordingly, the *absorption coefficient* κ_ν [in m^{-1}] is defined such that the radiant energy absorbed by the gas per unit volume and time is given by the relation

$$\dot{e}_a = -\kappa_\nu I_\nu \mathrm{d}\nu \mathrm{d}\Omega = h\nu(\partial n_R/\partial t)_a \ , \ \mathrm{Wm}^{-3} \ . \tag{4.53}$$

Substituting (4.51 – 4.53) into (4.50), the *equation of radiative transfer*, in absence of scattering, becomes [in $\mathrm{Jm}^{-3}\mathrm{sterad.}^{-1}$]

$$\frac{1}{c}\frac{\partial I_\nu}{\partial t} + l_j \frac{\partial I_\nu}{\partial x_j} = (j_\nu - \kappa_\nu I_\nu) \ , \ \mathrm{Jm}^{-3}\mathrm{sterad.}^{-1} \ . \tag{4.54}$$

One general remark about the diffuse reflection and transmission (scattering) may now be mentioned. If a parallel beam of radiation is incident on a plane-parallel radiation path in a specified direction, it is required to find the angular distribution of the intensities diffusely reflected from the surface and diffusely transmitted below the surface. It can, in principle, be considered in terms of a scattering function and a transmission function. It is, therefore, necessary to distinguish between the reduced induced radiation which penetrates without having suffered any scattering or absorption process, and the diffuse radiation field which has arisen in consequence of one or other of scattering processes (*Chandrasekhar* [4]).

It is also found convenient to use a term called the *source function* S_ν given by the relation

$$S_\nu = j_\nu/\kappa_\nu \ , \ \mathrm{Jm}^{-2}\mathrm{sterad.}^{-1} \tag{4.55}$$

and (4.54) can also be written in the following form:

$$\frac{1}{c}\frac{\partial I_\nu}{\partial t} + l_j \frac{\partial I_\nu}{\partial x_j} = \kappa_\nu(S_\nu - I_\nu) \ . \tag{4.56}$$

For the solution of (4.54) or (4.56), it is necessary to know the values of j_ν and κ_ν or S_ν and κ_ν, determination of which becomes somewhat easier from quantum mechanical considerations applied to special cases of chemical and radiative equilibria. In the case of chemical equilibrium, the molefraction of particles in different energy levels can be determined from statistical considerations, whereas in the case of radiative equilibrium the emitted and the absorbed energies equalize.

As the general case of radiative non-equilibrium, transitions between a higher energy level m and a lower energy level n are considered, in which an energy $h\nu$ is involved. In case we are talking about a line radiation, let it be assumed for simplicity that both I_ν and $h\nu$ are approximately constant over the line width. Even if we are talking about a continuum radiation, let us assume these two to be constant over the small radiation frequency interval $\Delta\nu$. Thus, integrating (4.54) over the line width of a particular line or wave length range being considered, and letting

$$\bar{j} = \int j_\nu \mathrm{d}\nu \quad \text{and} \quad \bar{\kappa} = \int \kappa_\nu \mathrm{d}\nu \; . \tag{4.57}$$

the right hand side of (4.54) is evaluated. Note that $\bar{j}$ has the dimension $\mathrm{Wm^{-3}sterad.^{-1}}$ and $\bar{\kappa}$ has the dimension $\mathrm{m^{-1}s^{-1}}$.

For *spontaneous emission* from the energy level m to the energy level n, the number of transitions is proportional to the number of particles in energy level m. The proportionality constant is A_{mn} $[\mathrm{s^{-1}}]$ and the energy released per unit time and solid angle is

$$\dot{e}_s = \bar{j}_s = \frac{1}{4\pi} A_{mn} n_m h\nu \; , \; \mathrm{Wm^{-3}sterad.^{-1}} \; . \tag{4.58}$$

In addition to the spontaneous emission, there is also the *induced emission* in the direction along the incident radiation, for which the proportionality constant B_{mn} $[\mathrm{m^2 J^{-1}s^{-1}}]$ is used. Thus for the induced emission

$$\dot{e}_i = \bar{j}_i = B_{mn} n_m I_\nu h\nu \; , \; \mathrm{Wm^{-3}sterad.^{-1}} \; . \tag{4.59}$$

Similarly, the energy absorbed per unit time, volume and solid angle is given by the relation

$$\dot{e}_a = \bar{\kappa} I_\nu = B_{nm} n_n I_\nu h\nu \; , \; \mathrm{Wm^{-3}sterad.^{-1}} \tag{4.60}$$

where B_{nm} is the proportionality constant for absorption $[\mathrm{m^2 (J.s)^{-1}}]$. The constants A_{mn}, B_{mn} and B_{nm} are called *Einstein's coefficients for probability of transition*. At this point it would be in order to mention an important application of the induced emission in *lasers*. In the simplest solid state lasers the electrons around the nucleus of a solid state crystal are brought to a higher energy level by subjecting the crystal to an external radiation. This is followed initially by spontaneous emission in all directions. However, if the solid state crystal is placed between two highly reflecting parallel mirrors, one of which may be nearly hundred percent reflecting and the other just about one percent less reflecting. The spontaneous emitted radiation in the direction of the axis of the two mirrors is reflected back and forth and it causes the induced emission in the direction of the radiation intensity to grow as soon as it hits one of the excited atoms. Thus very quickly, a very strong radiation intensity in the direction of the above mentioned axis is built up. In gas lasers a mixture of two gases with similar relevant energy levels is used in which one of the gases has a metastable energy state. The gases are mixed in such a proportion that in one the electrons are brought to a higher energy level in an electric discharge followed by the transfer of energy by collision to the metastable energy level of the second gas, in which a spontaneous emission is comparatively rare. This induced emission is built up in this case also in a similar manner to that in the case of a solid state laser. While (4.58–4.60) are also valid if there is no chemical or radiative equilibrium, validity of these equilibrium conditions mean that

$$I_\nu = I_\nu^* \ , \ \dot{e}_s + \dot{e}_i = \dot{e}_a \quad \text{and} \quad (\bar{j} - \bar{\kappa} I_\nu) = 0 \tag{4.61}$$

$$\text{(for radiative equilibrium)}$$

and

$$\frac{n_n^*}{n_m^*} = \frac{g_n}{g_m} \exp^{-(E_n - E_m)/(k_B T)} = \frac{g_n}{g_m} \exp^{h\nu/(k_B T)} \ . \tag{4.62}$$

$$\text{(for chemical equilibrium)}$$

Herein g is the statistical weight and the asterisk as superscript denotes the equilibrium. Thus from (4.58–4.60) and (4.61, 4.62), and in comparison with (4.27), we get the relation

$$I_\nu^* = \frac{A_{mn}/(4\pi B_{mn})}{\dfrac{g_n}{g_m} \dfrac{B_{nm}}{B_{mn}} \exp^{h\nu/(k_B T)} - 1} = \frac{2h\nu^3}{c^2} \frac{1}{\exp^{h\nu(k_B T)} - 1} \ . \tag{4.63}$$

By comparing the terms, we can, therefore, write

$$\frac{A_{mn}}{B_{mn}} = \frac{8\pi h\nu^3}{c^2} \ ; \ \frac{g_n}{g_m} \frac{B_{nm}}{B_{mn}} = 1 \ . \tag{4.64}$$

Now the *source function* is determined only for chemical equilibrium, and the restriction of the radiative equilibrium will only be considered to make certain conclusions. Noting that $\bar{j} = (\dot{e}_s + \dot{e}_i)$ and $\kappa I_\nu = \dot{e}_a$, the source function is given by the relation

$$\begin{aligned}
S_\nu &= \frac{\bar{j}}{\bar{\kappa}} = \frac{1}{B_{nm} n_n^* h\nu} \left[\frac{1}{4\pi} A_{mn} n_m^* h\nu + n_M^* B_{mn} I_\nu h\nu \right] \\
&= \left[\frac{2h\nu^3}{c^2} + I_\nu \right] \exp^{-h\nu/(k_B T)} \\
&= \left[I_\nu^* \left(\exp^{h\nu/(k_B T)} - 1 \right) + I_\nu \right] \exp^{-h\nu/(k_B T)} \\
&= I_\nu^* + (I_\nu - I_\nu^*) \exp^{-h\nu/(k_B T)} \ .
\end{aligned} \tag{4.65}$$

Equation (4.65) has been derived under the condition that the chemical equilibrium is satisfied. For the *radiative equilibrium*, $I_\nu = I_\nu^*$, and it follows that

$$S_\nu = I_\nu^* = \bar{j}/\bar{\kappa} \ . \tag{4.66}$$

Equation (4.66) is the mathematical formulation of the *Kirchhoff's law of radiation*, the physical explanation of which is given in the following paragraph. Further, it can be seen from (4.65) that for I/I_ν^* less or equal to one, the ratio S_ν/I_ν^* is also less or equal to one. However, for the limit of $S_\nu/I_\nu^* \to 0$, that is when the intensity of radiation is very much smaller than the black body radiation,

$$\lim_{I_\nu \to 0} (S_\nu/I_\nu^*) = 1 - \exp^{-h\nu/(k_B T)} \tag{4.67}$$

which has values between 0 at $\nu = 0$ and 1 at $\nu \to \infty$. Physically, when a body is irradiated like a black body from the surrounding medium, the condition $I_\nu/I_\nu^* = 1$ means that the temperature of the body is the temperature of the radiating energy flux of the surrounding medium and the body emits the same quantity of radiant energy as it absorbs. On the other hand, $I_\nu \to 0$ means the case of a freely emitting gas without absorption. In most of the books dealing with radiation, Kirchoff's law is described physically in the following manner. Let there be two bodies, one black and the other "*not black*" at temperature T within a cavity whose wall is at the same temperature. The second body being "*not black*" absorbs certain percentage of the incident energy, but the first body absorbs everything. In such a case both bodies must emit as much energy as they absorb. Unless this is the case, there will be a temperature difference between the two bodies, which violates the *second law of thermodynamics*. Thus Kirchoff's law of radiation states: When a body emits a quantity of radiative energy at a particular temperature and wave length (frequency), it absorbs also the same energy at the same temperature and the same wavelength. Now for a chemical equilibrium but radiative non-equilibrium,

$$
\begin{aligned}
(\bar{j} - \bar{\kappa}I_\nu) &= n_m^* \left[\frac{A_{mn}}{4\pi} + B_{mn}I_\nu\right] h\nu - B_{nm}I_\nu n_n^* h\nu \\[2mm]
&= n_m^* \left[\frac{2h\nu^3}{c^2} + I_\nu\right] B_{mn}h\nu - \frac{g_m}{g_n} B_{mn}n_n^* I_\nu h\nu \\[2mm]
&= n_n^* B_{mn}h\nu \left[\frac{n_m^*}{n_n^*}\left(\frac{2h\nu^3}{c^2} + I_\nu\right) - \frac{g_m}{g_n}I_\nu\right] \\[2mm]
&= n_n^* B_{nm}h\nu \left[\frac{g_n}{g_m}\frac{n_m^*}{n_n^*}\left(\frac{2h\nu^3}{c^2} + I_\nu\right) - I_\nu\right] \ .
\end{aligned}
$$

Applying (4.63), the above equation can further be reduced to

$$\bar{j} - \bar{\kappa}I_\nu = \bar{\kappa}\left[\exp^{-h\nu/(k_B T)} I_\nu^*\left(\exp^{h\nu/(k_B T)} -1\right) + I_\nu \exp^{-h\nu/(k_B T)} - I_\nu\right] \tag{4.68}$$

in which the individual terms can be designated as follows:

- *Spontaneous emission:* $\bar{j}_s = \bar{\kappa}\left(1 - \exp^{-h\nu/(k_B T)}\right) I_\nu^* = \bar{\kappa}' I_\nu^*$
- *Induced emission:* $\bar{j}_i = I_\nu \exp^{-h\nu/(k_B T)}$
- *Absorption:* κI_ν

where

$$\bar{\kappa}' = \bar{\kappa}\left[1 - \exp^{-h\nu/(k_B T)}\right] \tag{4.69}$$

is the *apparent absorptions coefficient*. We have, therefore, a method to combine the spontaneous volumetric energy coefficient with the corresponding absorptions coefficient.

After combining the various terms in (4.68), we get

$$\bar{j} - \bar{\kappa} I_\nu = \bar{\kappa}'(I_\nu^* - I_\nu) \ . \tag{4.70}$$

It can be verified easily, that the first term within the bracket in (4.70) is the spontaneous emission term and the second term is a combination of the induced emission and the absorption. Therefore, we may interpret $\bar{\kappa} I_\nu$ as the true absorption term, whereas $\bar{\kappa}' I_\nu$ as the apparent absorption term. By dimensional analysis it can be seen that $\bar{\kappa}$ and $\bar{\kappa}'$ are in $[\mathrm{m}^{-1}]$. While the above discussion concerns an entire spectral line by integrating (4.54) across the line width, a general equation need not be restricted to an individual line, but should be equally valid for a continuum radiation. For this purpose, (4.54) becomes

$$\frac{1}{c}\frac{\partial I_\nu}{\partial t} + l_j \frac{\partial I_\nu}{\partial x_j} = \kappa_\nu'(I_\nu^* - I_\nu) \tag{4.71}$$

where

$$\kappa_\nu' = \kappa_\nu\left(1 - \exp^{-h\nu/(k_B T)}\right) \ . \tag{4.72}$$

As before in (4.70), the first term in the right hand side of (4.71) is due to spontaneous emission and the second term is due to the combined induced emission and absorption.

A formal solution of (4.70) is simple if it is a one-dimensional case and if the time derivative term is neglected. The latter is allowed in case the ratio of a characteristic distance and time is smaller than the velocity of light. Since this is found to be the case for almost all problems in gas dynamics, it is concluded that omission of the time derivative term is allowed in most cases. Further, we would measure the coordinate from the point of the observer in the direction opposite to the direction of propagation of the radiation. Thus (4.70) is reduced to

$$\frac{\partial I_\nu}{\partial r} = \kappa_\nu'(I_\nu^* - I_\nu) \ . \tag{4.73}$$

Now $r = 0$ is the point at which the observer is receiving radiation and let, at $r = R$, I_ν has a known boundary value $I_\nu(R)$. Equation (4.73) is an equation of the type

$$\frac{dy}{dx} + P(x)y = Q(x) \tag{4.74}$$

which has a general solution

$$y = K \exp^{-\int P dx} + \int \exp^{\int P dx} Q dx \tag{4.75}$$

where K is an arbitrary constant. The particular solution of (4.73), with the prescribed boundary condition, is

$$I_\nu(0) = I_\nu(R)\exp^{-\int_0^R \kappa'_\nu \, dr} + \int_0^R \kappa'_\nu I_\nu^*(r)\exp^{-\int_0^r \kappa'_\nu dr} \, dr \ . \tag{4.76}$$

Equation (4.76) gives the angular spectral intensity of radiation at the point of the observer, which consists of two parts:

(1) the contribution of the boundary located at a distance R in the direction opposite to that of propagation of the radiation, attenuated by a factor

$$\exp^{-\int_0^R \kappa'_\nu \, dr} \tag{4.77}$$

to account for absorption and induced emission in the intervening gas; and

(2) the contribution of the spontaneous emission from the gas volume element at a varying distance r, each elementary contribution attenuated by a factor

$$\exp^{-\int_0^r \kappa'_\nu \, dr} \tag{4.78}$$

and the whole summed over all the elements between the observer and the boundary.

Since the second term in the right hand side of (4.76) gives the contribution of the spontaneous emission, we investigate this first. Let $I_\nu(R) = 0$ and the first term in the right hand side of (4.76) is put equal to zero. Further, let the *spectral volumetric radiative energy release*(by spontaneous emission) be $\dot{e}_\nu = j_s = \kappa'_\nu I_\nu^*$ [Jm^{-3}]. The *emissivity* is defined by the relation $\epsilon_\nu = I_\nu/I_\nu^*$. It may be noted that if $\epsilon_\nu \ll 1$, spontaneous emission is dominating, and if $\epsilon_\nu = 1$, it is the black-body radiation dominating. Now let $r^* = r/R$, and from (4.76), we get the relation for emissivity coefficient

$$\epsilon_\nu(r^* = 0) = \frac{I_\nu(0)}{I_\nu^*(0)} = \int_0^1 \frac{\dot{e}_\nu R}{I_\nu^*(0)} \exp^{-\int_0^{r^*} \kappa'_\nu R dr^*} dr^* = 1 - \exp^{-\tau_\nu} \tag{4.79}$$

where

$$\tau_\nu = \int_0^R \kappa'_\nu dr \tag{4.80}$$

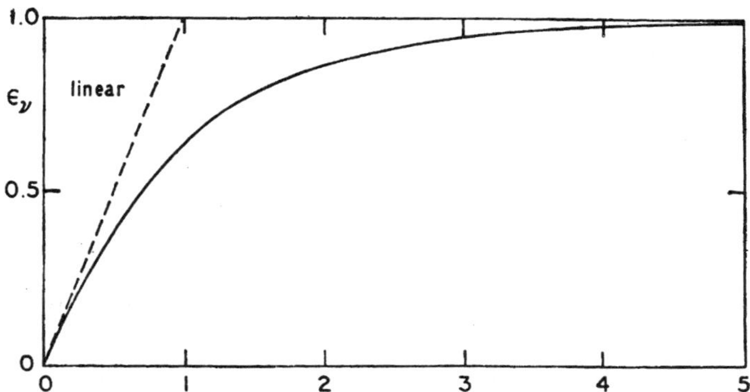

Fig. 4.3. Results of emissivity coefficient as a function of the optical length

is called the *optical length*, which is non-dimensional and is equal to $(\kappa'_\nu R)$ for a constant absorption coefficient. Results of ϵ_ν versus $(\kappa'_\nu R)$ is plotted in Fig. 4.3. It can be seen, that for $(\kappa'_\nu R) < 0.25$, there is a linear relationship between ϵ_ν and $(\kappa'_\nu R)$, where it is possible to assume that the radiation is completely dominated by the spontaneous emission, and the radiating gas column is said to be *optically thin*. On the other hand for $(\kappa'_\nu R) > 5$, $\epsilon_\nu = 1$ and $I_\nu = I^*_\nu$, and the gas column is said to be *optically thick*. Since for free-free and free-bound radiation, which are important for radiative energy transfer in gas dynamics rather than the bound-bound (line) radiation, κ'_ν is proportional to

$$p^m \exp -h\nu/(k_B T) \, . \tag{4.81}$$

It is evident that for large pressures and small frequencies (large wave lengths), the effect of absorptions coefficient may be quite significant.

We can now obtain the total emitted energy for "*optical thin*" gas by integrating emitted energy over all frequencies to get

$$j = \int_0^\infty j_\nu \mathrm{d}\nu = \int_0^\infty \kappa'_\nu I^*_\nu \mathrm{d}\nu = \kappa' \int_0^\infty I^*_\nu \mathrm{d}\nu = \kappa' I^* = \frac{\sigma \kappa'}{\pi} T^4$$
$$[\mathrm{Wm}^{-3}\mathrm{sterad.}^{-1}] \, . \tag{4.82}$$

Herein, κ' is the integrated absorption coefficient given by the equation

$$\kappa' = \frac{\int_0^\infty \kappa'_\nu I^*_\nu \mathrm{d}\nu}{\int_0^\infty I^*_\nu \mathrm{d}\nu} = \frac{1}{\sigma T^4} \int_0^\infty \kappa'_\nu B^*_\nu \mathrm{d}\nu \, , \ \mathrm{m}^{-1} \tag{4.83}$$

where there is no initial radiation. Thus all variables in (4.83) refer to gas. On the other hand if a pencil of black-body radiation emanating from a surface is intervened by a gas volume, then the absorption of the gas is given by

$$\dot{e}_a = \int_0^\infty \kappa_\nu I^*_\nu \tag{4.84}$$

and we get for the *overall gas absorptivity* in comparison to the original black-body surface radiation of the gas column as

$$a = \frac{\pi}{\sigma T_4} \int_0^\infty \kappa_\nu I_\nu^* d\nu = \frac{1}{\sigma T^4} \int_0^\infty \kappa_\nu B_\nu^* d\nu \ , \ \mathrm{m}^{-1} \ . \tag{4.85}$$

It can be seen that between (4.83) and (4.85) the difference is only that while in the former induced emission has been taken into account, the latter deals with regular absorption. Obviously, the absorptivity a from above (4.85) may replace integrated absorption coefficient κ_ν' from (4.82) if induced emission is neglected.

While the above definition of absorptivity requires integration over wave length from 0 to ∞, the method can be extended for wavelength (or frequency) intervals specific for individual gases and frequency band. Starting point of this discussion is the one-dimensional steady radiative transfer equation (4.73), which we integrate in the frequency range ν_1 and ν_2 to get

$$\frac{\mathrm{d}}{\mathrm{d}r} \int_{\nu_1}^{\nu_2} I_\nu d\nu = \int_{\nu_1}^{\nu_2} \kappa_\nu' (I_\nu^* - I_\nu) d\nu \ . \tag{4.86}$$

We write the expressions for total intensity within the range as

$$I_{\Delta\nu} = \int_{\nu_1}^{\nu_2} I_\nu d\nu, I_{\Delta\nu}^* = \int_{\nu_1}^{\nu_2} I_\nu^* d\nu \tag{4.87}$$

and also for the absorptivity within the same range as

$$a_{\Delta\nu} = \frac{1}{I_{\Delta\nu}^*} \int_{\nu_1}^{\nu_2} \kappa_\nu' I_\nu^* d\nu \ . \tag{4.88}$$

We may formally replace κ_ν' in (4.86) by $a_{\Delta\nu}$, although this is not strictly valid when such replacement is done also for the second term in the right hand side of (4.86). Thus we write

$$\frac{\mathrm{d}}{\mathrm{d}r} I_{\Delta\nu} = a_{\Delta\nu} \left(I_{\Delta\nu}^* - I_{\Delta\nu} \right) \ . \tag{4.89}$$

Now the equivalent solution equation of (4.76) is

$$I_{\Delta\nu}(0) = I_{\Delta\nu}(R) \exp^{-\int_0^R a_{\Delta\nu}} \mathrm{d}r + \int_0^R a_{\Delta\nu} I_{\Delta\nu}^*(r) \exp^{-\int_0^r a_{\Delta\nu} \mathrm{d}r} \mathrm{d}r \ . \tag{4.90}$$

In homogeneous gas $a_{\Delta\nu}$ and $a_{\Delta\nu} I_{\Delta\nu}^*$ can be taken out of the integrals, and we write

$$I_{\Delta\nu}(0) = I_{\Delta\nu}(R) \exp^{-a_{\Delta\nu} R} + I_{\Delta\nu}^* \left(1 - \exp^{-a_{\Delta\nu} R} \right) \tag{4.91}$$

from which we get the relation for emissivity as

$$\epsilon_{\Delta\nu} = \frac{I_{\Delta\nu}(0) - I_{\Delta\nu}(R)}{I_{\Delta\nu}^*(0) - I_{\Delta\nu}(R)} = 1 - \exp^{-a_{\Delta\nu} R} \tag{4.92}$$

or if there is no incoming radiative flux at the boundary, $I_{\Delta\nu}(R) = 0$, then

$$\epsilon_{\Delta\nu} = \frac{I_{\Delta\nu}(0)}{I^*_{\Delta\nu}} = 1 - \exp^{-a_{\Delta\nu}R} . \qquad (4.93)$$

We note that

$$I^*_{\Delta\nu} = \int_{\nu_1}^{\nu_2} I^*_{\nu}\,\mathrm{d}\nu = \int_0^{\nu_2} I^*_{\nu}\,\mathrm{d}\nu - \int_0^{\nu_1} I^*_{\nu}\,\mathrm{d}\nu = I^*(F_2 - F_1) . \qquad (4.94)$$

where

$$F = \frac{1}{I^*} \int_0^{\nu} I^*_{\nu}\,\mathrm{d}\nu = \frac{\pi}{\sigma T^4} \frac{2h}{c} \int_0^{\nu} \frac{\nu^2}{\exp^{h\nu/(k_B T)} - 1}\,\mathrm{d}\nu \qquad (4.95)$$

is the fraction of the total emissive power that is emitted in a given frequency band. Let $\zeta = h\nu/(k_B T)$, and we get from (4.95 [19])

$$F = \frac{15}{\pi^4} \int_0^{\nu} \frac{\zeta^3}{\exp^{\zeta} - 1}$$

$$= 1 - \frac{15}{\pi^4} \sum_{n=1}^{\infty} \frac{\exp^{-n\zeta}}{n} \left(\zeta^3 + \frac{3\zeta^2}{n} + \frac{6\zeta}{n^2} + \frac{6}{n^3} + \cdots \right) . \qquad (4.96)$$

For a gas it is customary to measure experimentally the total absorptivity by measuring the *overall emissivity coefficient* of a gas column as

$$\epsilon = 1 - \exp^{-aL} \qquad (4.97)$$

where L is the optical length. If on the other hand, two end-on arcs of length L (seen only in the axial direction) are used in tandem so that the intensity of the individual arcs are I_1 and I_2, respectively and I_S is the sum of intensity of both the arcs, it is now possible to show that

$$\exp^{-aL} = \frac{I_S - I_1}{I_2} . \qquad (4.98)$$

As a complete measure necessary for the calculation of the radiative exchange between an isothermal gas mass and its surroundings or the radiative flux in non-isothermal gas system requires κ_{ν} or κ_{λ} as a function of frequency (or wavelength), temperature and pressure of all gas components in the radiative mixture. Experimental measurement of κ_{ν} is complicated by the limit of resolution of spectrometers, and have been made only for few gases at very narrow range of conditions, from which absorptivity may be evaluated from (4.92). Further discussion of total emissivities or absorptivities must await consideration of various factors influencing which includes broadening of spectral lines and bands, and are now discussed.

(a) *Natural Line Broadening*: An estimate of the width of a line can be done with the help of the *Heisenberg uncertainty principle*, which states that the product of uncertainty in frequency and time is of the order of $1/(2\pi)$. In terms of line broadening in wave length, the equation is

$$\Delta\lambda \approx \lambda^2/(2\pi c \Delta t) \tag{4.99}$$

where c is the speed of light. Now the average life time of an excited electron is of the order of 10^{-6} to 10^{-8} seconds and taking a wavelength of 5,000 Å, the width of the line is between 10^{-4} and 10^{-2} Å.

(b) *Collision Broadening of Lines*: The absorption coefficient in a line due to collision broadening is expressed in terms of the integrated absorption area of the line $S = \int \kappa_\nu d\nu$ [m^{-1}s^{-1}], and the *half-width* of the line b_c at the line mid-frequency ν_o is given by the relation

$$\kappa_\nu = \frac{S}{\pi} \frac{b_c}{(\nu - \nu_o)^2 + b_c^2} \ . \tag{4.100}$$

In the above relation b_c depends on the frequency of collision between the particles and has the same unit as the frequency. It is given approximately by the relation

$$\frac{b_c}{b_{co}} = \frac{p}{p_o} \left(\frac{T_o}{T}\right)^{1/2} \tag{4.101}$$

where the subscript o refers to reference values. Further for a particular gas in a mixture p is the sum of the partial and the total pressure.

(c) *Doppler Broadening*: The expression for line absorption coefficient is again given in terms of the integrated line absorption area S and the line half width b_D by the relation

$$\kappa_\nu = \frac{S}{b_D}\sqrt{(\ln 2)/\pi}\, \exp^{-[(\nu-\nu_o)^2 \ln 2/b_D^2]} \tag{4.102}$$

where

$$b_D = \nu_o\sqrt{2\pi k_B T \ln 2/(Mc^2)} \ . \tag{4.103}$$

Both b_c and b_D have the unit of frequency. Estimate of these show that the collision broadening is important at high pressures and low temperatures, like the conditions which prevail in furnaces, combustion engines, process equipment, etc. On the other hand, the Doppler broadening is of importance at low pressures and high temperatures, like in plasmas.

(d) *Band Emission*: The complexity of evaluating theoretically the bands is such that these are obtained experimentally. For this purpose, the emissivity of homogeneous combustible gas is usually calculated by means of diagrams, for example by *Hottel* and *Sarofim*[11]. The measurements, however, need be performed only at one temperature since a full measure of the effect of temperature is given by the shift in the distribution between the different energy levels. For *rotation-vibration energy level* the population of the higher energy levels increases at the expense of the lower levels and, within each vibration band, the higher energy levels gain in importance. Further, the different vibrational levels interact with radiative energy at slightly different frequencies. Therefore, both the intensity and the band width increase at higher temperatures. Various models have been proposed to circumvent the problem of formulating relation for κ_ν as a function of frequency. Some of these models, discussed in detail by *Hottel* and *Sarofim*[11] are *Shack model*, *Elsasser model* and *Meyer-Goody model*. In addition, if two or more bands of emitting specie overlap, then the emission or absorption by the mixture will be smaller than the individual contribution. While the emissivity diagrams are available for homogeneous gases, there is need to evaluate the emissivity for a general non-homogeneous gas.

From the above discussion it is evident, that the absorptivity of a pure gas (not mixture) is a function of the temperature, and at least a simple polynomial function of pressure. Therefore, we write formally for the absorptivity, as defined in (4.85), as

$$a \equiv (b + cT + dT^2 + eT^3)(p/p_o)^m \qquad (4.104)$$

where $p_o = 1$ bar is a reference pressure. For industrial applications of gas radiation, where combustion-product radiation is dominant, experimentally determined total emissivities of carbon-dioxide. water vapor and carbon monoxide have been available since 1930s. In view of the complexity of the theory and the associated uncertainty in the calculated values of gas emissivities, the absorptivity measurements are, in many respects, the most reliable source of information and *Hottel* and *Sarofim*[11] have given emissitivity coefficient for many gases in graphical form as a function of gas temperature and $(p.L)$, where p is the pressure (in "atm" for *Hottel* and *Sarofim*, but in "bar" here) and L is the *optical length*. The coefficients b to e and m in (4.104) are now obtained as follows: Since the experimental data is given as a function of T and $(p.L)$, we put $L = 1$ m. Further we take different guessed values of m, and for each m, we get b to e by least square fit. Further m is varied to get the best fit with the emissivity values given in the entire range. The above coefficients (m and b to e), are given for various gases in Table 4.1, which are applicable in the temperature range 300 to 3,000 K for pure gases. These are now used to compute the integrated absorptions coefficient for different gases at 1 bar, which have been plotted in Fig. 4.4 as a function of temperature. In a gas mixture the individual gas absorption coefficient can be found by computing these for partial pressures of individual gases and added together. This is quite alright, so long the bands for individual gases do not overlap with each other. This is, however, not true if both CO_2 and H_2O are present, since both have overlapping spectral bands in regions 2.7μ and 15μ wavelengths. Simplified emissivity chart for CO_2-H_2O mixture to show reduction in the value of emissivity have been given by *Hottel* and *Sarofim* as a function of $(p_{CO_2} + p_{H_2O})L$ as a parameter and it has been shown further that these change very little for the ratio (p_{H_2O}/p_{CO_2}) in the range 1 to 2. These

Table 4.1. Calculated coefficients of (4.104)

Gas	m	b	c	d	e
CO_2	0.5	0.1467	1.4618e-4	-9.4592e-8	1.1721e-11
H_2O	0.5	0.5865	-9.1839e-5	2.0066e-9	2.7445e-12
SO_2	0.6	0.2810	3.2002e-3	-3.5875e-6	9.5845e-10
CO	0.4	-0.08269	4.6555e-3	-2.8393e-6	2.4235e-10
NH_3	0.6	1.8529	7.6847e-3	-1.6103e-5	6.4920e-09
HCl	0.6	-0.01224	-1.5444e-4	7.9664e-8	-1.3246e-11
NO_2	0.5	-0.0224	8.1142e-4	-5.7278e-7	1.1917e-10
CH_4	0.5	-0.16794	5.1001e-3	-3.6002e-6	7.4901e-10
CW*	0.8	0.02574	-1.8289e-5	3.2276e-8	-9.7576e-12

CW* = gas mixture containing CO_2 and H_2O

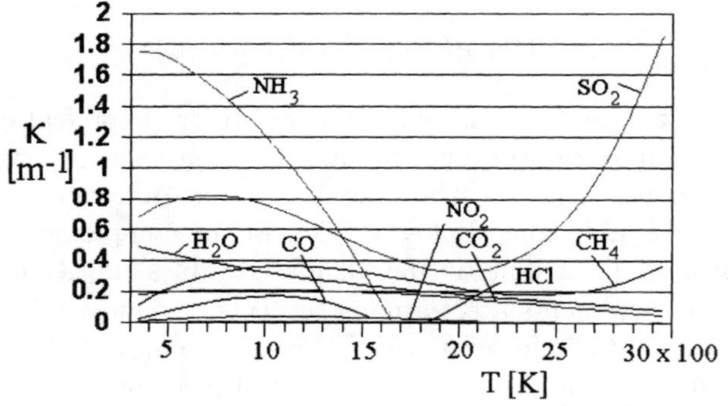

Fig. 4.4. Mean absorption coefficient for some gases

have been used to recompute for both CO_2 and H_2O present to compute
the coefficients as given in Table 4.1, shown as gas CW, and the maximum
value of $\Delta\kappa$ (to be subtracted from the sum of absorptions coefficient for CO_2
and H_2O has been plotted in Fig. 4.5 along with the absorption coefficients
for pure CO_2 and H_2O again. However, we need a relationship to take into
account the effect of (p_{H_2O}/p_{CO_2}). Although such a relationship about contribution to the absorption coefficient is not available, we use an expression
by *Leckner* [77] regarding change in the emissivity. Using the same factor as
his for emissivity, but normalizing, we write the reduction factor to $\Delta\kappa$ as

$$118.867 \left(\frac{\zeta}{10.7 + 101\zeta} - \frac{\zeta^{10.4}}{111.7} \right), \zeta = \frac{p_{H_2O}}{p_{H_2O} + p_{CO_2}} . \tag{4.105}$$

However, for a gas column radiating optically thick, one can consider the
situation also, in which the absorption is significant and I_ν differs only by
a small amount from I_ν^*. The discussion will enable us to estimate the radiative heat-flux when there is locally radiative equilibrium, but the local

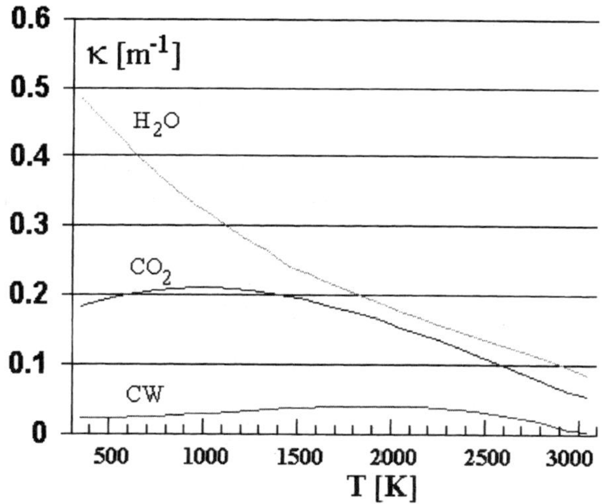

Fig. 4.5. Absorption coefficient for CO_2, H_2O and Maximum overlapping Value

temperature is different at different places. For simplicity, the discussion is limited to the consideration of a point in the gas column that is sufficiently at a distance from any boundary ($R \to \infty$) in terms of the optical length, so that the contribution to I_ν from $I_\nu(R)$ is negligible. Taking the definition of the optical length, (4.80), into account, the formal solution of (4.76) then becomes

$$I_\nu(0) = \int_0^\infty \kappa_\nu' I_\nu^*(r) \exp^{-\int_0^r \kappa_\nu' dr} dr = \int_0^\infty I_\nu^*(\tau_\nu) \exp^{-\tau_\nu} d\tau_\nu \ . \qquad (4.106)$$

Assuming the value of $I_\nu^*(0)$ at the point of the observer deviating only slightly from the value $I_\nu^*(\tau_\nu)$, one can write $I_\nu^*(\tau_\nu)$ in a series form by *Taylor expansion* as

$$I_\nu^*(\tau_\nu) = I_\nu^*(0) + \left(\frac{\partial I_\nu^*}{\partial \tau_\nu}\right)_0 \tau_\nu + \frac{1}{2!}\left(\frac{\partial^2 I_\nu^*}{\delta \tau_\nu^2}\right)_0 \tau_\nu^2 + \ldots \qquad (4.107)$$

and terminate the series after the first two terms only. Substituting the above expression into (4.106) we get

$$I_\nu(0) = \int_0^\infty \left[I_\nu^*(0) + \left(\frac{\partial I_\nu^*}{\partial \tau_\nu}\right)_0 \tau_\nu\right] \exp^{-\tau_\nu} d\tau_\nu$$

$$= I_\nu^*(0)\left[\int_0^\infty \exp^{-\tau_\nu} d\tau_\nu + \frac{1}{I_\nu^*(0)}\left(\frac{\partial I_\nu^*}{\partial \tau_\nu}\right)_0 \int_0^\infty \tau_\nu \exp^{-\tau_\nu} d\tau_\nu\right]$$

$$= I_\nu^*(0)\left[1 + \frac{1}{I_\nu^*(0)}\left(\frac{\partial I_\nu^*}{\partial \tau_\nu}\right)_0\right] \ . \qquad (4.108)$$

Under the assumption that

$$\frac{1}{I_\nu^*(0)} \left(\frac{\partial I_\nu^*}{\partial \tau_\nu}\right)_0 \ll 1 \tag{4.109}$$

we may write $I_\nu(0) \approx I_\nu^*(0)$, and the radiating gas column is approximately optically thick. Equation (4.108) is rewritten into a slightly different form

$$I_\nu = I_\nu^* \left[1 + \frac{1}{I_\nu^*} \left(\frac{\partial I_\nu^*}{\partial T}\right)_0 \left(\frac{\partial T}{\partial \tau_\nu}\right)\right] . \tag{4.110}$$

Introducing a radiant heat flux vector in the x_i coordinate direction

$$\mathbf{q}_i^R = \int_0^\infty \int_0^{4\pi} I_\nu l_i \mathrm{d}\Omega \mathrm{d}\nu , \ \mathrm{Wm}^{-3} \tag{4.111}$$

where l_i is the cosine of the angle of the direction of radiation with the normal to the respective surface. Substituting (4.108) into (4.111) and integrating over the entire solid angle, the first term when integrated is equal to zero and the second term yields

$$\mathbf{q}_i^R = \int_0^\infty \frac{1}{\kappa_\nu'} \left(\frac{\partial I_\nu^*}{\partial T}\right) \left[\int_0^{4\pi} l_i \left(\frac{\partial T}{\partial r}\right) \mathrm{d}\Omega\right] \mathrm{d}\nu . \tag{4.112}$$

Since,

$$\int_0^{4\pi} l_i \left(\frac{\partial T}{\partial r}\right) \mathrm{d}\Omega = -\frac{4\pi}{3} \frac{\partial T}{\partial x_i} \tag{4.113}$$

the *radiative heat flux* becomes

$$\mathbf{q}_i^R = -\frac{4\pi}{3} \frac{\partial T}{\partial x_i} \int_0^\infty \frac{1}{\kappa_\nu'} \frac{\partial I_\nu^*}{\partial T} \mathrm{d}\nu . \tag{4.114}$$

A mean absorption coefficient κ_R called *Rosseland mean absorption coefficient* is now defined by the relation

$$\frac{1}{\kappa_R} \int_0^\infty \frac{\partial I_\nu^*}{\partial T} \mathrm{d}\nu = \frac{4\sigma T^3}{\pi \kappa_R} = \int_0^\infty \frac{1}{\kappa_\nu'} \left(\frac{\partial I_\nu^*}{\partial T}\right) \mathrm{d}\nu \tag{4.115}$$

and the *radiant heat flux* becomes

$$\mathbf{q}_i^R = -\frac{16}{3} \frac{\sigma T^3}{\kappa_R} \frac{\partial T}{\partial x_i} , \ \mathrm{Wm}^{-2} . \tag{4.116}$$

The above equation is similar to the heat conduction equation, which states that the conductive heat flux is proportional to the negative of the temperature gradient. We can write similar for radiative heat transfer. Therefore, for the case of a nearly *optically thick radiation, a radiative thermal conductivity coefficient* (valid only for local radiative equilibrium) can be defined by the relation

$$k_R = \frac{16}{3}\frac{\sigma T^3}{\kappa_R} \ , \ \mathrm{Wm^{-1}K^{-1}} \tag{4.117}$$

where

$$\kappa_R = \frac{2\pi h^2}{4\sigma T^5 c^2 k_B}\int_0^\infty \frac{1}{\kappa'_\nu}\frac{\exp^{h\nu/(k_B T)}}{\left(\exp^{h\nu/(k_B T)}-1\right)^2}\mathrm{d}\nu \ .$$

We have discussed in the present section the method by which we can estimate the radiative energy transfer, provided the absorption coefficient are available from experiments or otherwise. In the following section, we would discuss further how these can be used to estimate the radiative heat flux for ionized gases.

Finally, let us discuss the computational aspects of the radiative transfer equation. While in some typical books on radiative transfer, for example by *Siegel* and *Howell* [19], may discuss various methods of radiative transfer including stochastic methods, there is generally very little about the calculation of the radiative transfer for a highly emitting-absorbing gas. For the present discussion, therefore, we make two simplified assumptions: (1) only the integrated radiative flux is considered, since consideration of the spectral radiative flux is only going to increase computation time many folds; and (2) the total intensity of radiation (radiosity) is considered. We consider further only in one dimension in the cartesian coordinate system, which can no doubt be extended to two or three-dimensional cases fairly easily. When we consider in one coordinate direction, for example in $+x$ direction, let the radiosity in that direction be designated as B^+, and similarly in $-x$ direction let it be B^-. For each of these we write from (4.71) the equation of radiative transfer as

$$\frac{1}{c}\frac{\partial B}{\partial t} + \frac{\partial B}{\partial x} = \kappa'\left(B^* - B\right) \ . \tag{4.118}$$

Now we consider a wall on which B^- radiation is falling and B^+ is emanating. If the wall emissivity is ϵ_w at the wall temperature T_w, then obviously at the wall,

$$B^+ = B^-(1 - \epsilon_w) + \sigma\epsilon_w T_w^4 \ . \tag{4.119}$$

Thus the energy flux being absorbed at the wall is

$$q_w = B^- - B^+ = B^-\epsilon_w - \sigma\epsilon_w T_w^4 \ . \tag{4.120}$$

Similarly for the three-dimensional case. one can solve six B-components in three positive and three negative coordinate directions provided the expressions or data to compute the absorptivity be available. The total radiative energy source (or sink) can, of course, be computed from the sum of the right hand side of the radiative energy transfer in individual directions and this can then act as the radiative energy source (or sink) term in the gas dynamic energy equation.

4.3 Radiative Characteristics for Ionized Gases

As already pointed out, the energy of particles consists of discrete energy levels due to rotation, vibration, electronic excitation, etc., and the continuous energy distribution due to translation. Transition from one discrete energy level to another discrete higher (excited) energy level can, for example, be of the type

$$A + e_{\text{fast}} \to A^* + e_{\text{slow}} \tag{4.121}$$

in which an atom in the ground state or lower energy level is excited by collision with a fast electron which becomes slower after the collision. A reverse process may be somewhat less probable than by radiation.

Reactions for *photo-excitation* and *photo-deexcitation* including the effect of the radiation energy can be as

$$A + h\nu \leftrightarrow A^* \ . \tag{4.122}$$

The reactions by *photoionization* and *photo-recombination* are similar:

$$A + h\nu \leftrightarrow A^+ + e + \frac{1}{2} M_e v_e^2 \tag{4.123}$$

except that, in this case, there is no discrete quantum energy jump giving rise to a continuous free-bound radiation. Ionization, however, is also possible by collision

$$A + X_{\text{fast}} \to A^+ + e + X_{\text{slow}} \ . \tag{4.124}$$

In the reverse reaction, at the time of collision of the ion and the electron, the collision of a slow third partner (*three-body recombination*) is necessary to take away the excess energy, failing which, the ion and the electron have to separate again, except when photo-ionization takes place. For free-free transfer, in which the emission is due to the retardation of the charge particle (electron), which in german is called *Bremsstrahlung* (*radiation due to retardation*), the *radiative energy* $h\nu$ is given by the relation

$$h\nu = \frac{1}{2} M_e \Delta(v_e^2) \ . \tag{4.125}$$

While the continuum spectra can appear when at least one of the energy levels of transition is not discrete, there can also be such spectra for transitions between two normally discrete energy levels when these become quite wide due to internal disturbances or external causes. In addition to these real continuum spectra, there are also *"not real" continuous spectra*, which are found continuous because of the limited resolution capability between neighboring lines in a spectrograph.

In the case of a transition between a discrete energy level E_n and one which the electrons are ionized, the electron possesses ionization energy E_i and also a certain kinetic speed v; the energy balance gives the relation

$$\frac{1}{2} M_e v_e^2 + (E_i - E_n) = h\nu \tag{4.126}$$

the transition corresponds to transition between elliptic and hyperbolic paths. The minimum frequency of radiation ν_n will be for $v_e = 0$, for which $\nu_n = (E_i - E_n)/h$, and the frequency of radiation extends from ν_n to ∞. On the other hand, in the free state of the electron, the percentage distribution of particles in some velocity interval is

$$f(v_e) \cong \exp^{-M_e v_e^2/(2k_B T_e)} \cong \exp^{-h\nu/(k_B T_e)} \tag{4.127}$$

and the intensity of radiation from ν_n falls exponentially with frequency for each of the transitions between the bound energy level and the free state. Around the ionization energy the energy levels are very close to each other, and the individual peaks in intensities merge together to give frequency-independent *spectral volumetric radiation energy*. Different energy levels and transitions, and schematic distribution of the spectral volumetric radiative energy are shown in Fig. 4.6a,b, where $I_i - \Delta I$ is the *effective ionization energy* (= *ionization energy* I_i – *"lowering of the ionization energy"* ΔI). The transition above the effective ionization energy is due to free-free transition of the electrons, while the transition between free electron energy levels up to the minimum energy level E_l has a fairly frequency-independent spectral volumetric radiative energy distribution, which is due to merging of transitions to energy levels very near each other. Transitions from the free electron energy levels to the levels between 0 and E_l have a saw-tooth like character, as shown in Fig. 4.4b.

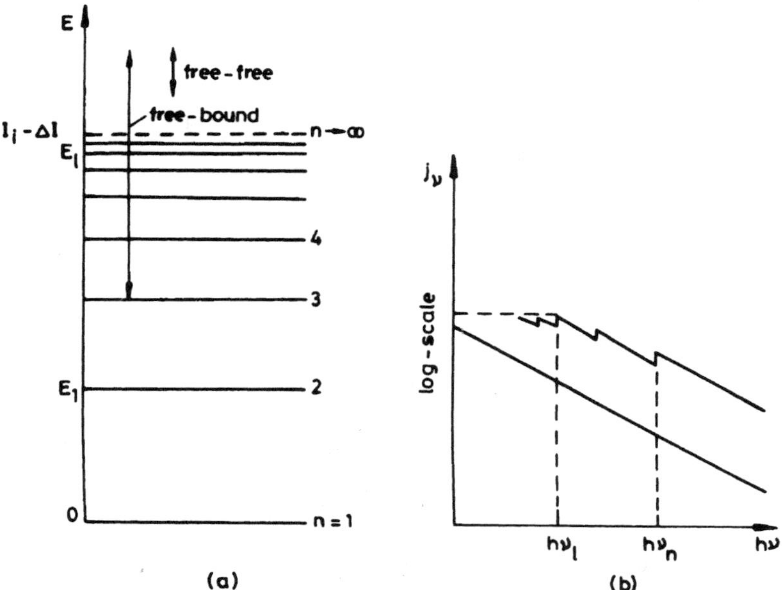

Fig. 4.6. (a) Energy levels and (b) Schematic distribution of spectral volumetric energy

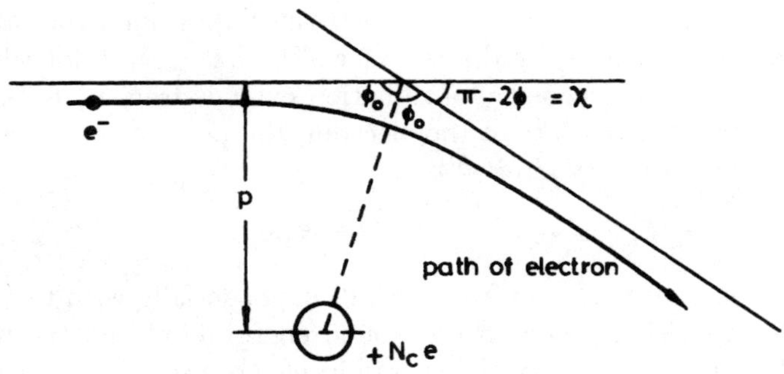

Fig. 4.7. Deflection of an electron in the field of an ion with charge N_c

For the free-free transition, which can be considered as transitions between hyperbolic orbits, *Kramer* (for detailed discussion, see *Finkelnburg* and
Peters[27]) derived an expression for the radiated energy following classical
electrodynamics and mechanics. According to the classical electrodynamics,
the radiated energy emitted per unit time due to movement of a free electron
around a nucleus with charge $+N_{ce}$ (Fig. 4.7) is given by the relation

$$\dot{e}_R = \frac{e^2 \dot{v}_e^2}{6\pi\epsilon_o r^2} \; , \; \text{W} \; .$$

(4.128)

Balancing the inertial and Coulomb forces,

$$M_e \dot{v}_e = \frac{N_c e^2}{4\pi\epsilon_o r^2}$$

(4.129)

where N_c is the effective (positive) multiple of nuclear charge, we get the
relation for the acceleration of the electron

$$\dot{v}_e = \frac{N_c e^2}{4\pi M_e \epsilon_o r^2} \; .$$

(4.130)

Substituting (4.130) into (4.128), we get

$$\dot{e}_R = \frac{N_c^2 e^6}{96\pi^2 M_e^2 \epsilon_o^3 c^3 r^4} \; , \; \text{W} \; .$$

(4.131)

Introducing polar coordinates and observing the *area rule*

$$r^2 \dot{\varphi} = p v_e = \text{constant}$$

(4.132)

where p is the smallest distance without deflection, we get from (4.131)

$$\dot{e}_R = \frac{N_c^2 e^6}{96\pi^2 M_e^2 \epsilon_o^3 c^3} \int \frac{1}{r^4} \mathrm{d}t = \frac{N_c^2 e^6}{96\pi^2 M_e^2 \epsilon_o^3 c^3 p v_e} \int \frac{\mathrm{d}\varphi}{r^2} \; .$$

(4.133)

In between the coordinates of hyperbolic orbits, there is further the relation

$$\frac{1}{r} = \frac{1 - \epsilon \cos \varphi}{p \tan \varphi_o} \; .$$
(4.134)

where $\epsilon = 1/\cos \varphi_o = $ *eccentricity*. In addition the *law of scattering* due to *Rutherford* gives the expression

$$\tan \varphi_o = M_e p v_e (4\pi \epsilon_o)/(N_c e^2)$$
(4.135)

where $(\pi - 2\varphi_o)$ is the *angle of deflection*. Substituting the above two expressions into (4.133), we get after some manipulation

$$\dot{e}_R = \frac{N_c^4 e^{10}}{1536\pi^3 c^3 M_e^4 \epsilon_o^5 p^5 v_e^5} \int_{\varphi_o}^{2\pi - \varphi_o} (1 - \epsilon \cos \varphi)^2 d\varphi$$

$$= \frac{N_c^4 e^{10}}{1536\pi^3 c^3 M_e^4 \epsilon_o^5 p^5 v_e^5} \left[2 \left(1 + \frac{1}{2} \frac{1}{\cos^2 \varphi_o} \right) (\pi - \varphi_o) + 3 \tan \varphi_o \right] \; .$$
(4.136)

Equation (4.136) can be evaluated easily for the limiting cases of $\varphi_o \to 0$ and $\varphi_o = 90^o$. In the first case ($\varphi_o \to 0$), the electron deviates approximately by 180° and the orbit can be considered as parabolic near the nucleus ($\epsilon = 1$).

Distribution of total radiation over frequencies is obtained from the time distribution of the acceleration as per the magnitude and direction of the velocity vector in the entire orbit. This orbital acceleration is split by *Kramer* in orbits parallel and normal to the entire orbit, and after considering various aspects of interaction between the electrons and the (positive) charge N_c with ion index designated with subscript i, *Kramer* derived the following expression for the volumetric energy emission in a free-free energy transfer:

$$\dot{j}_{ff\nu} = \frac{N_c^2 e^6}{6\sqrt{3}\pi c^3 (2\pi M_e \epsilon_o)^3} \frac{n_e n_i G_{ff}}{(k_B T)^{1/2}} \exp^{-h\nu/(k_B T)}$$

$$= 5.43 \times 10^{-52} \frac{N_c^2 n_e n_i G_{ff}}{\sqrt{T}} \exp^{-h\nu/(k_B T)} \; , \; \mathrm{Wsm^{-3} sterad.^{-1}}$$
(4.137)

where n_i and n_i are the number density for the i-th ions and electrons, respectively. Further in the above expression $G_{ff}(\nu)$ for free-free transition is the correction factor (*Gaunt factor*) that has to be brought in, since the initial calculation is done for the two limiting cases of *parabolic* and *hyperbolic* orbits. Normally $G_{ff} = 1$, except for $\nu \to 0$, which means, that for not too small radiative frequencies ($\nu > 0$), the main part of the free-free radiation is for strongly deflected (*parabolic*) orbits. For $\nu \to 0$ however, $G_{ff}(\nu)$ takes large values and the contribution of more and more hyperbolic orbits are added. Now for $\nu \to 0$, the value of G_{ff} can be obtained from the relation

$$G_{ff} = \frac{\sqrt{3}}{\pi} \ln \frac{\epsilon_o k_B T}{1.44 N_c e^2 n_i^{1/3}} = 0.55133 \ln \left(3306.6 \frac{T}{N_c n_i^{1/3}} \right)$$

$$= 4.4678 + 0.55133 \ln \left[\frac{T}{N_c n_i^{1/3}} \right] \; .$$

For the free-bound transitions when the bound energy levels are very near each other, the *volumetric energy emission* is given by the relation

$$j_{fb\nu} = \frac{N_c^2 e^6 n_i n e}{6\sqrt{3}\pi c^3 (2\pi M_e \epsilon_o)^{3/2}}$$

$$= 5.43 \times 10^{-52} n_e n_i \; , \; \mathrm{Wsm^{-3} sterad.^{-1}} \; (\text{for } \nu \leq \nu_l) \; . \quad (4.138)$$

In the range $\nu_l < \nu < \infty$ *volumetric energy emission* to each bound energy level is given by the relation

$$j_{fb\nu} = \frac{e^{10} M_e N_c^4}{24\sqrt{3}\pi c^3 h^5 \epsilon_o^5} \left(\frac{2Z_i}{Z_{i-1}} \right) \frac{n_{i-1}}{n^3} G_{fb} \exp^{-[h(\nu_n - \nu) + I_i - \Delta I]/(k_B T)} \; . \quad (4.139)$$

In equation (4.139), Z is the partition function of excitation, G_{fb} is the correction factor (*Gaunt factor*) for *free-bound transition* to energy E_n (corresponding to the *principal quantum number n* for a hydrogen-like atom) and ν_n is the frequency of radiation from the effective ionization energy $(I_i - \Delta I_i)$ to E_n. Now by using *Saha equation* for the single temperature case (discussed in Chap. 6),

$$S = \frac{n_e n_i}{n_a} = \frac{2Z_i}{Z_{i-1}} \left(\frac{2\pi M_e k_B T}{h^2} \right)^{3/2} \exp^{-(I_i - \Delta I)/(k_B T)} \; . \quad (4.140)$$

Equation (4.139) can now be written as

$$j_{fb\nu} = \frac{M_e N_c^4 e^{10} n_i n_e}{24\sqrt{3}\pi c^3 h^2 n^3 \epsilon_o^5 (2\pi M_e k_B T)^{3/2}} \exp^{-h(\nu - \nu_n)/(k_B T)} \; . \quad (4.141)$$

Now from (4.68)

$$\kappa_\nu = j_\nu / I_\nu^* = \frac{c^2 j_\nu}{2h\nu^3} \left[\exp^{h\nu/(k_B T)} - 1 \right] \approx \frac{c^2 j_\nu}{2h\nu^3} \exp^{h\nu/(k_B T)} \; . \quad (4.142)$$

By substituting the above equation into (4.139) or (4.141), we get

$$\kappa_{fb\nu} = \frac{e^{10} M_e N_c^4}{48\sqrt{3}\pi c h^6 \epsilon_o^5} \left(\frac{2Z_i}{Z_{i-1}} \right) \frac{n_{i-1}}{(n\nu)^3} G_{fb} \exp^{-[h\nu_n + I_i - \Delta I]/(k_B T)} \quad (4.143)$$

or

$$\kappa_{fb\nu} = \frac{e^{10} M_e N_c^4}{48\sqrt{3}\pi c h^3 \epsilon_o^5 (2\pi M_e k_B T)^{3/2}} \frac{n_i n_e}{(n\nu)^3} \exp^{-h\nu_n/(k_B T)} \; . \quad (4.144)$$

From more accurate quantum mechanical calculations, however, the frequency dependency of the absorption coefficient can deviate from the behavior of the absorption coefficient for hydrogen or similar molecules ($\kappa_\nu \propto \nu^{-3}$) quite differently. In some cases even this increases with ν, for example neon.

For sodium these decrease very steep and for potassium it decreases first to increase later. In any case the absorptions coefficient for neutral atoms and positive ions approach a finite value at $\nu \to \nu_n$, whereas for negative ions, they tend to zero.

By defining an absorption cross-section of the gas, in which the free-electrons are captured to reach (n,l) state, where n and l are the principal and the angular quantum number, respectively, we can write

$$j_{fb\nu} = \frac{2h\nu^3}{c^2} Q_{n,l} n_{n,l} \exp^{-h\nu/(k_B T)} = \frac{2hc}{\lambda^3} Q_{n,l} N_{n,l} \exp^{-h\nu/(k_B T)}$$

$$= 3.978 \times 10^{-25} \frac{1}{\lambda^3} Q_{n,l} n_{n,l} \exp^{-h\nu/(k_B T)} \ , \ \text{Wsm}^{-3}\text{sterad.}^{-1} \quad (4.145)$$

where λ is the *wave length* [m], $Q_{n,l}$ is the *absorption cross-section* [m^2] and $n_{n,l}$ is the number density of particles in the state (n,l). *Zhiguler*, et al. [110] have discussed the method of calculation of hydrogen like atoms. In (4.145) evaluation of the expression has to be done for $\nu > \nu_n$ and *overall volumetric energy* emission is given by the relation

$$j_\nu = j_{ff\nu} + \sum j_{fb\nu} \ . \quad (4.146)$$

Combining the free-free and free-bound emissions and integrating over all wavelengths, *Liu* et al. [79] give the relation for the *total volumetric energy*

$$j = \frac{64 N_c^2 \pi^{3/2} e^6 G}{3\sqrt{6} M_e^{3/2} c^3 h \epsilon_o^{3/2}} \frac{n_e n_i}{(k_B T_e)^{1/2}} (h\nu_c + k_B T_e)$$

$$= C N_c^2 \frac{n_e n_i}{T_e^{1/2}} (h\nu_c + k_B T_e) \ , \ \text{Wm}^{-3} \quad (4.147)$$

where G is the *average Gaunt factor* (may be taken equal to one) and ν_c is the cut-off frequency. Further N_c is the *"effective nuclear charge"* to modify the photo-ionization model for hydrogen to other gas plasmas. Now the rate of radiant energy loss per unit volume of a plasma consists of the rates of the energy loss through continuum radiation and line radiation. Therefore the above equation was multiplied by 2 and applied to argon plasma (*Liu*, et. al[79]), for which the values taken were $h\nu_c = 2.85$ eV (corresponding to 3p electrons) and $N_c^2 = 1.67$; a comparison with Fig. 4.8 and (4.147), specially applied for argon plasma, becomes now [Wm^{-3}]

$$j = 4.676 \times 10^{-14} \frac{n_e^2}{\sqrt{T_e}} \left(4.5657 \times 10^{-19} + 1.38 \times 10^{-23} T_e \right) \quad (4.148)$$

where n_e [m^{-3}] is the electron number density.

For a temperature range 10,000 to 20,000K and for plasma with maximum singly charged ion, *Zhigular* et al. [110] gave relations for hydrogen, oxygen and nitrogen in which the volumetric radiative energy and the number densities are in cgs-units. By multiplying with appropriate factors, the following relations [Wm^{-3}] are obtained.

For Hydrogen (at lower kinetic energies of free electrons):

$$j_H = n_i n_e \left[6.27 \times 10^{-35} T^{-0.76} + 9.73 \times 10^{-40} T^{-0.16} \right] \qquad (4.149)$$

for Oxygen:

$$j_O = n_i n_e \left[\frac{2.06 \times 10^{-36}}{\sqrt{T}} + 5.46 \times 10^{-41} \sqrt{T} + 5.31 \times 10^{-46} T^{3/2} \right] \qquad (4.150)$$

for Nitrogen:

$$j_M = n_i n_e \left[\frac{3.13 \times 10^{-36}}{\sqrt{T}} + 7.67 \times 10^{-41} \sqrt{T} + 7.74 \times 10^{-46} T^{3/2} \right] \qquad (4.151)$$

For the line and band spectra (*bound-bound radiation*), the distribution of I_ν for a spectral line is difficult to obtain since the line shape is determined by a combination of natural, collision and Doppler broadening. However, a combination of these factors can be taken care of in the *transition probability coefficient*, A_{mn}, and the *intensity of line radiation* is

$$I_L = \int_L I_\nu d\nu = \frac{1}{4\pi} A_{mn} \frac{n_r g_{r,m} \exp -E_{r,m}/(k_B T)}{\sum_m g_{r,m} \exp -E_{r,m}/(k_B T)} h\nu L \ , \ \mathrm{Wm}^{-2}\mathrm{sterad.}^{-1}$$

$$(4.152)$$

where L is the optical thickness of a radiating gas [m], $g_{r,m}$ is the statistical weight in the r-th ionization state (r = 0, neutral; r = 1, singly-charged ion; etc.) and the m-th energy level in that ionization state, E is the energy, and n_r is the total number density in the r-th ionization state. Taking various components of line and continuum radiation, radiation intensities and absorption coefficients, *Morris*[88, 87] and *Yos* and *Morris*, et al. [86] calculated the *spectral volumetric absorption coefficient* in a tabulated form for argon, nitrogen, oxygen and air for 1 to 30 bar and 9,000 to 30,000 K. These tabulated values of the absorption coefficient κ_ν [m^{-1}] have been plotted in Fig. 4.8 for argon as a function of wavelength (or frequency), for two pressures of 1 and 10 bars and temperature range between 10,000 to 20,000 K. From Fig. 4.3 it is shown, that one can consider gas radiation as *optically thin*, if the product of the absorption coefficient and the *non-dimensional optical path length* $(\kappa_\nu L) < 0.25$, and *optically thick* if the product $(\kappa_\nu L) > 5.0$. Thus it can be seen that, as a rule, the absorption at large wave lengths starts gaining more importance for smaller lengths of the optical path length L. It may, however, be recalled that as shown schematically in Fig. 4.6, the values of the absorptions coefficient may not be falling uniformly with frequency, but there is, in general, a saw-tooth like character corresponding to the transitions between the *free* and different *bound energy levels*. Further from Fig. 4.8 it can be seen for $p = 1$ bar and $T = 10,000$ K, that for argon plasma in the range

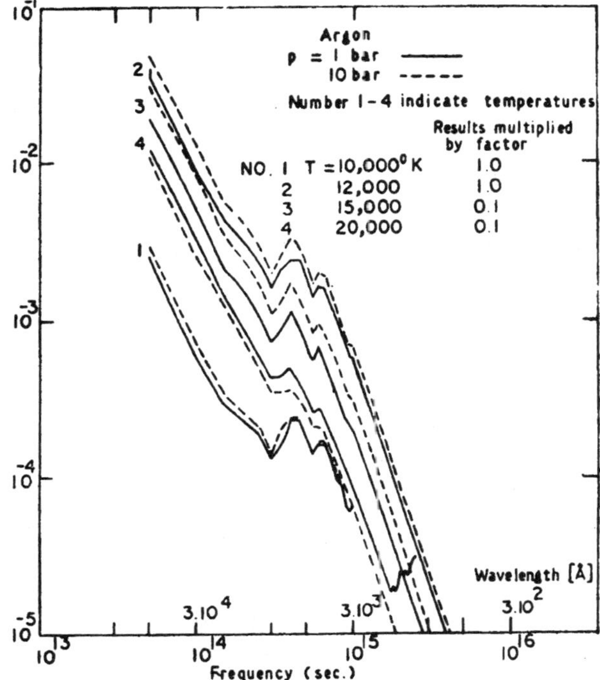

Fig. 4.8. Spectral volumetric radiation energy for argon [86]

considered, one can consider the radiation at 5×10^{13} Hz as *optically thin*, if $L = L_1 < 100$ m and *optically thick* if $L = L_2 > 2$ km, whereas at 2×10^{15} Hz *optically thin* if $L = L_1 < 250$ m and *optically thick* if $L = L_2 > 5$ km. For practical purposes, these results, however important, may be used only to examine the *limiting* cases, but it is more useful to consider values averaged over all frequencies or wavelengths.

Although we are going to write down the basic gas dynamic equations for high temperature gases at a later Chap. 11, it is now worthwhile to point out to the effect of radiation on the gas dynamic variables. It is pointed out by *Zhiguler* et al. [110] that the total momentum change due to spontaneous emission is zero, although absorption may cause directional change in the momentum. On the other hand, in setting up the energy equation we must consider, in addition to the volume force (corresponding to the momentum change) also the energy transfer of the particles by emission (and absorption).

From the values of the spectral absorption coefficient, the *spectral volumetric radiative energy*, is obtained from the formula

$$\dot{e}_\nu = \kappa'_\nu B^*_\nu \ . \tag{4.153}$$

It is found that in the temperature range around 10,000 to 20,000 K, the maximum $\dot{e}_\nu$ is in the visible range or near the infrared. Therefore, to determine the *total volumetric radiative energy* it is sufficient, for temperature below 20,000 K, to cut-off the upper frequency limit of integration at around 10^{16} Hz. Now the total *volumetric radiative energy* for *spontaneous radiation* is defined by the relation

$$\dot{e}_R = \int_0^\infty e_\nu \mathrm{d}\nu = \int_0^\infty \kappa'_\nu B_\nu^* \mathrm{d}\nu = \bar{\kappa}' B^*, \mathrm{Wm}^{-3} \tag{4.154}$$

and the *average absorptions coefficient* is given by the relation

$$\bar{\kappa}' = \frac{1}{B^*} \int_0^\infty \kappa'_\nu B_\nu^* \mathrm{d}\nu \ . \tag{4.155}$$

The difference between the average absorptions coefficient and the *Rosselend mean absorptions coefficient*, (4.115) may be noted. Values of $\dot{e}_R$ and $\bar{\kappa}'$ for argon, nitrogen and air are plotted in Figs. 4.9 and 4.10. If it is assumed

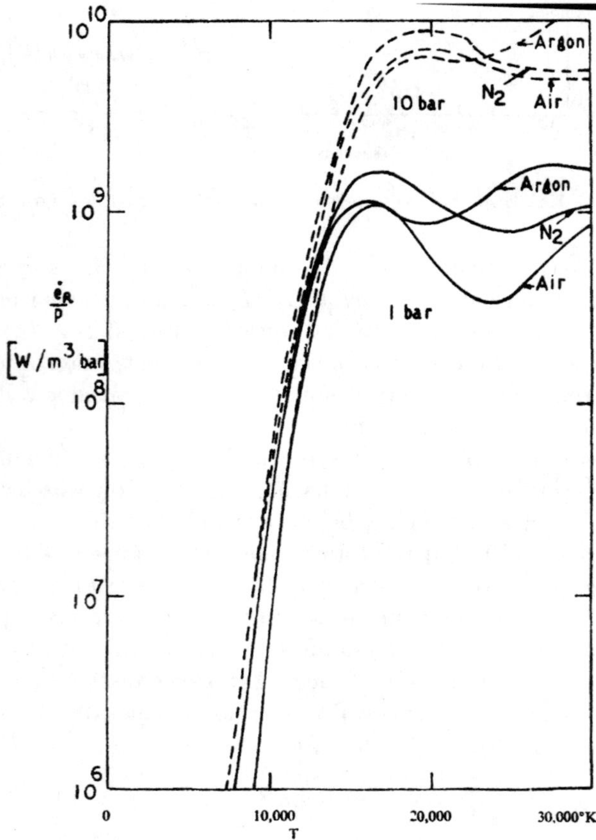

Fig. 4.9. Total volumetric radiation energy for argon, nitrogen and air

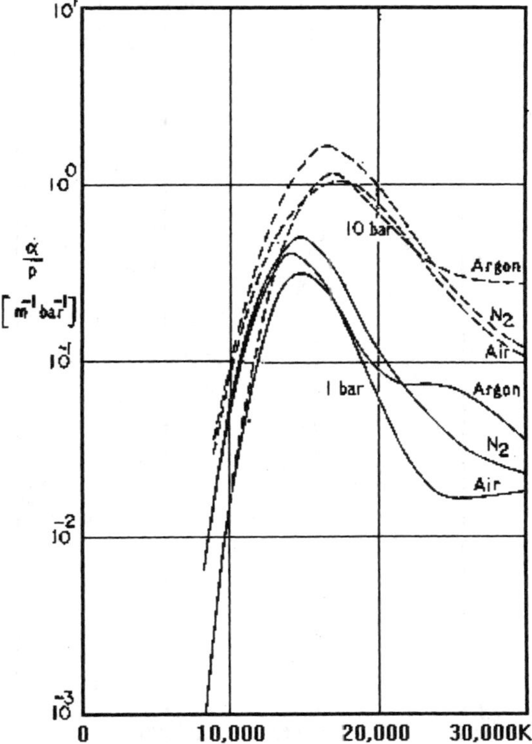

Fig. 4.10. Average absorption coefficient for argon, nitrogen and air

that the radiation may be considered as *optically thin*, that is, if $\bar{\kappa}' L_1 < 0.25$, the maximum length of path, L_1 for optically thin radiation are shown for these three gases in Fig. 4.10. It is evident, that the average L_1 is smaller than the range of L_1 obtained from Fig. 4.9. Thus, although for radiative heat transfer calculations the value of average L_1 may be somewhat adequate, for any spectral measurements the absorption at long wave lengths may be dominated. It can be seen further from Figs. 4.9 to 4.11, that the values of $\dot{e}_R$, $\bar{\kappa}'$ and L_1 are approximately dependent on the (3/2)-th power of pressure, in which direct proportionality relation should be observed for non-reacting gases, and the remaining (1/2) power of pressure is obtained from the shift in the equilibrium composition of the gases.

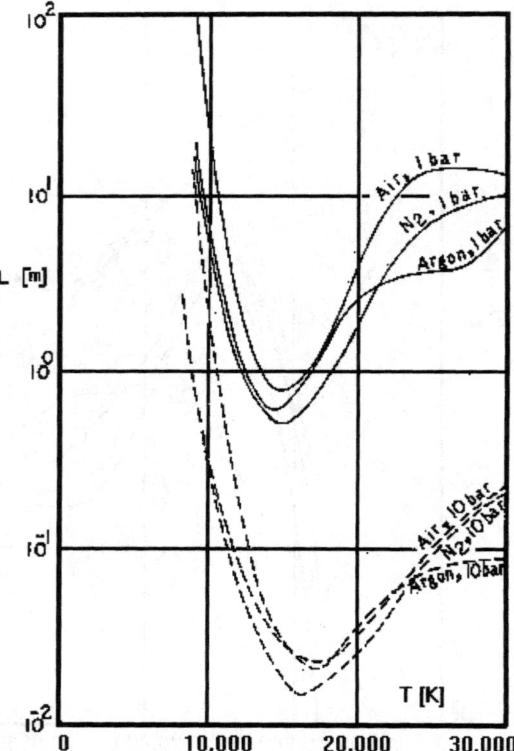

Fig. 4.11. Maximum length of optical path L_1 for validity of the concept of the optically thin radiation for argon, nitrogen and air

4.4 Evaluation of Radiation

In the previous two sections absorption coefficient data for some molecular gases and plasmas have been given. These are not for any specific wave length (or frequency) but are generally applicable for a *"gray"* gas, and hence these are averaged absorption coefficient data independent of the frequency. Taking such data, we would now discuss methods to evaluate volumetric emission and heat flux to the surrounding wall. For this we would first keep the *"scattering"* generally out of our consideration, and the starting point will be the equation of radiative transfer, (4.73) for a gray gas as

$$\frac{\partial I}{\partial s} = \kappa(I^* - I) \ . \tag{4.156}$$

Some of the methods we would now discuss requires integration in the path of the ray of radiation in straight line from wall to wall or free space, and hence it will be good to discuss first some geometrical relations. For

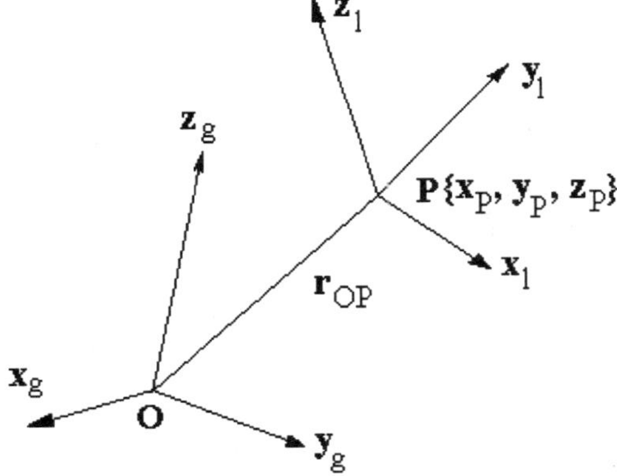

Fig. 4.12. Global and local coordinates

this purpose we may have computational grids in global coordinate system $(x_g,\ y_g,\ z_g)$, but we have to trace the rays originating from a surface with respect to *local coordinates* $(x_l,\ y_l,\ z_l)$ (Fig. 4.12), which may be given also in the local spherical $(r,\ \theta,\ \phi)$ system. We can differentiate between the two coordinate systems with an example as follows. Let us consider a motor car, in which we can have *global coordinates* of the car at the front surface at mid-width, where the x is in the direction of rear, y is going towards right and z direction goes to vertical above. This itself could be defined with respect to terrestrial directions like north, east and zenith. On the other hand, a local coordinate will be from individual surface element on the car surface (inside or outside), in center of which we put the origin of the local coordinates. The ray emanating from this surface can also be described in local spherical coordinates. Thus the cosines of the ray with respect to the local cartesian coordinate is given by

$$l_x = \sin\theta\cos\varphi,\ l_y = \sin\theta\sin\varphi,\ l_z = \cos\theta \tag{4.157}$$

which can be converted to cosines in global coordinates. For this we write first in terms of the local coordinates

$$\begin{pmatrix} x - x_P \\ y - y_P \\ z - z_P \end{pmatrix}_l = r \begin{pmatrix} l_x \\ l_y \\ l_z \end{pmatrix}_l \tag{4.158}$$

where r is the radial distance from the local origin. On the other hand the local cartesian coordinates with local unit vectors $\mathbf{i}$, $\mathbf{j}$, $\mathbf{k}$ may be written in terms of global cartesian coordinates as

$$\begin{pmatrix} x - x_P \\ y - y_P \\ z - z_P \end{pmatrix}_g = r \begin{pmatrix} l_x \\ l_y \\ l_z \end{pmatrix}_g = r \begin{bmatrix} i_x & i_y & iz \\ j_x & j_y & jz \\ k_x & k_y & kz \end{bmatrix} \begin{pmatrix} l_x \\ l_y \\ l_z \end{pmatrix}_l \tag{4.159}$$

and hence,

$$\begin{pmatrix} l_x \\ l_y \\ l_z \end{pmatrix}_g = \begin{bmatrix} i_x & i_y & iz \\ j_x & j_y & jz \\ k_x & k_y & kz \end{bmatrix} \begin{pmatrix} l_x \\ l_y \\ l_z \end{pmatrix}_l . \tag{4.160}$$

Example: Let us consider a rectangular box and the local coordinates on the six surfaces of the box may be defined by unit vectors in local coordinates as follows:

$$\begin{aligned}
\text{Bottom}: &\ \mathbf{i} = (1,0,0), & \mathbf{j} &= (0,1,0), & \mathbf{k} &= (0,0,1) \\
\text{Top}: &\ \mathbf{i} = (-1,0,0), & \mathbf{j} &= (0,1,0), & \mathbf{k} &= (0,0,-1) \\
\text{West}: &\ \mathbf{i} = (0,1,0), & \mathbf{j} &= (0,0,1), & \mathbf{k} &= (1,0,0) \\
\text{East}: &\ \mathbf{i} = (0,-1,0), & \mathbf{j} &= (0,0,1), & \mathbf{k} &= (-1,0,0) \\
\text{South}: &\ \mathbf{i} = (0,0,1), & \mathbf{j} &= (1,0,0), & \mathbf{k} &= (0,1,0) \\
\text{North}: &\ \mathbf{i} = (0,0,-1), & \mathbf{j} &= (1,0,0), & \mathbf{k} &= (0,-1,0)
\end{aligned}$$

Now let the ray (a straight line) originating at P pass through a point Q at distance one apart (for example, on the surface of another cell or wall surface). Therefore,

$$x_Q - x_P = l_x, y_Q - y_P = l_y, z_Q - z_P = l_z . \tag{4.161}$$

We compute the point Q. Let the straight line be at the intersection of any two planes out of the possible three planes which pass through the individual coordinates, and these are PQX, PQY and PQZ. Further let us consider a point on one plane, for example PQX, and coordinate value one, and we write the determinant as follows:

$$\begin{vmatrix} x - x_P & y - y_P & z - z_P \\ l_x & l_y & l_z \\ 1 & 0 & 0 \end{vmatrix} . \tag{4.162}$$

This is equivalent of the equation

$$l_z(y - y_P) - l_y(z - z_P) = 0 \tag{4.163}$$

Similarly for the PQY and PQZ planes we can write the two equations

$$l_x(z - z_P) - l_z(x - x_P) = 0$$
$$l_y(x - x_P) - l_x(y - y_P) = 0 .$$

Adding the three equations, we can write the *equation of a straight line* as

$$a(x - x_P) + b(y - y_P) + c(z - z_P) = 0 \tag{4.164}$$

where

$$a = l_y - l_z, b = l_z - l_x, c = l_x - l_y . \tag{4.165}$$

There will be a special case if one of the cosines is equal to zero, for example, $l_z = 0$, for which, $z - z_P = 0$ and we can put $c = 0$.

We determine $\{a_g, b_g, c_g\}$ from the coefficients for local coordinates $\{a_l, b_l, c_l\}$ with the help of the relation

$$\begin{pmatrix} a \\ b \\ c \end{pmatrix}_g = \begin{bmatrix} i_x & i_y & i_z \\ j_x & j_y & j_z \\ k_x & k_y & k_z \end{bmatrix} = \begin{pmatrix} a \\ b \\ c \end{pmatrix}_l . \tag{4.166}$$

Now the above ray may impinge on a plane, while the latter passes through a point $\{x_o, y_o, z_o\}$ and has a normal (unit) vector $\{k_x, k_y, k_z\}$. Thus the *equation of a plane* is defined by the equation

$$Ax + By + Cz + D = 0 \tag{4.167}$$

where $A = k_x$, $B = k_y$, $C = k_z$ and $D = -(k_x x_o + k_y y_o + k_z z_o)$.

Now a plane given by the above equation and a straight line originating at P be represented by

$$x = x_P + l_x r, y = y_P + l_y r, z = z_P + l_z r, . \tag{4.168}$$

Substituting the above three equations of (4.168) into (4.167) we get the *impingement distance* of the ray on to the impinging plane as

$$r = -\frac{Ax_P + By_P + Cz_P + D}{Al_x + Bl_y + Cl_z} \tag{4.169}$$

which, when after substituting back into the three equations of (4.168), gives the point of intersection of the ray into the impinging plane. However, the plane is subdivided into elements of cell or wall surface, identification of the particular element has, of course, to be done separately.

Having discussed the geometrical aspects of the ray of radiation, let us now discuss the computational aspects of the radiative transfer equation. Although a typical book on radiative transfer, for example *Siegal* and *Howell* [19] may discuss various methods of calculation of the radiative transfer including stochastic methods, one finds very little in the literature about calculation of radiative transfer for a highly emitting-absorbing gas. Herein we discuss now the following methods: (a) Optical thin model, (b) Rosseland model, (c) Multiple-flux model, (d) Ray model, and (e) Montecarlo model. A sixth model, the so-called P-N model, discussed by [19], has not been discussed here, since the required mathematical background is outside the scope of this book.

(a) Optical thin model: As discussed already, the model is applicable if the optical path length $\tau <0.1$. In such a case, the volumetric radiative energy released per unit volume is $4\pi\kappa I^*$, which will be taken into account in the overall gas energy equation (with opposite sign). The method, however, can not be used to determine the local heat flux distribution at the wall.

(b) Rosseland model: This model, discussed already in Sect. 4.2, is applicable if the optical path length $\tau > 5.0$, a requirement which will hardly be satisfied for terrestrial applications except at extreme pressures and temperatures. Under such conditions a radiative thermal conductivity coefficient can be defined by (4.117) which is added to the heat conductivity coefficient due to *"pure"* conduction and *"diffusion-reaction"*.

(c) Multiple-flux model: For this model let the arbitrary path s be at an angle θ from the positive direction. We also introduce superscripts $+$ or $-$, corresponding to directions with positive or negative $\cos\theta$; I^+ corresponds to $0 \leq \theta \leq \pi/2$ and I^- corresponds to $\pi/2 \leq \theta \leq \pi$. Further let the optical path τ_x be defined along the x-coordinate as

$$\tau_x = \int \kappa(x)\mathrm{d}x \tag{4.170}$$

Therefore,

$$\tau_s = \int \kappa(s)\mathrm{d}s = \frac{1}{\cos\theta} \int \kappa(s)\mathrm{d}x = \frac{\tau_x}{\cos\theta} . \tag{4.171}$$

Similarly in the negative direction, $\mathrm{d}s = -\mathrm{d}x/\cos(\pi - \theta) = \mathrm{d}x/\cos\theta$ and again $\tau_s = \tau_x/\cos\theta$. Thus (4.156) can now be split into two equations

$$\cos\theta\frac{\mathrm{d}I^+}{\mathrm{d}x} + \kappa I^+ = \kappa I^*, (0 \leq \theta \leq \pi/2) \tag{4.172}$$

and

$$\cos\theta\frac{\mathrm{d}I^-}{\mathrm{d}x} + \kappa I^- = \kappa I^*, (\pi/2 \leq \theta \leq \pi) . \tag{4.173}$$

The above two first order differential equations can have only one boundary condition each starting from opposite walls. The pair of equations are multiplied with $\mathrm{d}\Omega = 2\pi\sin\theta\,\mathrm{d}\theta$ and integrated over the respective solid angle. Further we multiply the resultant equation with π and noting that $B = \pi I$, we get the following pair of equations:

$$\frac{\mathrm{d}B^+}{\mathrm{d}x} = 2\kappa(B^* - B^+) \tag{4.174}$$

and

$$\frac{\mathrm{d}B^-}{\mathrm{d}x} = 2\kappa(B^* - B^-) . \tag{4.175}$$

where B is the radiative energy flux $[\mathrm{Wm}^{-2}]$ and B^* is the radiative flux of the *"black body radiation"*.

Since the above pair of equations are of first order, integration has to be done from a wall in the direction away from the wall. Thus we may have B^+ in one direction and B^- in the opposite direction and falling either on a wall or escaping all together from the computational domain if there is no wall. In the former case and if the wall, on which the radiative energy is falling, is perpendicular to the coordinate direction (wall normal aligned to the coordinate direction), then the entire incoming energy will be incident on it. If, however, the wall surface normal makes an angle θ_w with respect to the coordinate direction, then effectively only a smaller surface area is available and hence a smaller energy flux, $B_{in} = B_i \cos\theta_w$ is to fall with respect to the original surface area. This basically an one-dimensional analysis is modified in the multi-dimensional case as $B_{in} = \sum B_i \, l_i$, where l_i is the cosine of the angle between the surface normal and the respective coordinate direction. At any wall, the required relation between the incident and outgoing radiative flux is now

$$|B_{\text{out}}| = \epsilon_w \sigma T_w^4 + (1 - \epsilon_w)|B_{in}| \tag{4.176}$$

which is in the direction of the surface normal and the energy flux into the wall is

$$q_w = \epsilon_w \left[|B_{in}| - \sigma T_w^4\right] . \tag{4.177}$$

Herein ϵ_w is the emissivity of the wall. For an adiabatic wall, of course, the wall temperature is determined by the incoming energy flux. The component of the outgoing energy flux in the i-th coordinate direction is obtained by multiplying the right hand side of (4.176) with l_i.

The volumetric radiative source term (negative of energy sink in fluid energy equation) is obtained by integrating the right hand side of (4.174, 4.175) for each direction in a volume element by considering the absolute values of B and B^*. The total volumetric radiative source is now the average of all these *"directional"* volume source term where the sum is divided by 4 for a two-dimensional case, or by 6 for a three-dimensional case, to take care of the overlapping of diffusive radiation integral in hemisphere for each coordinate direction.

(d) *Ray model*: The method is also called the *"Discrete Transfer Radiation Model"* or *DTRM Model*, and is based on tracing diffuse radiative energy packages originating from a surface and moving in directions defined by the two *random numbers* R_θ and R_φ, through an emitting-absorbing gas. The convenient functions relating the random numbers (distributed uniformly between 0 and 1) with the respective angles within the small angles of interval are:

$$\text{Cone angle}: \sin\theta = \sqrt{R_\theta}$$

$$\text{Azimuthal angle}: \varphi = 2\pi R_\varphi .$$

Regarding determination of θ it may be observed that the operations involving arc of a sine function may be very time-consuming to be practical when a large number of photon bundles are to be processed. Instead one may pre-calculate a table of θ versus R_θ to get a faster solution. On the other hand determination of φ direction is straight forward requiring only a simple multiplication with the corresponding random number.

Now let there be N rays of the above radiative energy transmitted per ray (at the wall i) in a straight line during transmittal of the energy from the one surface to the other. For each ray, during transmittal through the hot gas, the above intensity is no doubt be modified till it reaches another wall or is allowed to escape in an open boundary. Integrating (4.156) over the path length inside a cell the change in the intensity is

$$\Delta I_{\text{cell}} = \int_{\text{cell}} \kappa(I^* - I)ds \ , \ \text{Wm}^{-2}\text{ster.}^{-1} \tag{4.178}$$

which is added inside a cell. On reaching a wall the impinging intensity of radiation is modified by the ratio of the surface area of the ray emanating wall to that of the absorbing wall, is tallied (added and number of radiative flux is counted), which is then used to obtain the incoming energy flux to the wall, $q_{in} = \pi \left|\bar{I}_w\right|_{in}$ [Wm^{-2}]. The heat flux to the wall is then obtained from (4.177) and for the next set of radiative calculations the outgoing radiative intensity $|B_{\text{out}}|/\pi$ [in Wm^{-2}.ster.$^{-1}$] is calculated from (4.176). Also tallied in a cell is the value of the above ΔI_{cell} to obtain the average of the value of the volumetric energy source per unit solid angle. This is multiplied with (4π) and we get, with opposite sign, the volumetric radiation energy source term [Wm^{-3}] in the fluid energy equation.

(e) *Montecarlo model*: We have already discussed how two random numbers in the range 0 to 1 can be used to determine the ray direction, but otherwise the ray is traced in a straight line and gets modified during the transmittal. In a pure "*Monte carlo method*", one can model further the absorption and scattering generated by a third random number.

The emission per unit time and area of a surface element, B_{out}, is given by (4.176), which in a first instance may be put equal to $\epsilon_w \sigma T_w^4$. If the surface area of the element is A and N packets are emitted per unit time, then each packet must carry a quantity of energy $w = AB_{\text{out}}/N$ [W]. The packets are emitted at a cone angle $\theta = \sin^{-1}\sqrt{R_\theta}$ and azimuthal angle $\varphi = 2\pi R_\varphi$, where the R's are the two random numbers. A typical packet will now travel a distance l or optical path length

$$\tau = \int_0^l \kappa ds \tag{4.179}$$

before getting absorbed. The probability of travelling this distance, R_l, is obtained from the relation

$$R_l = 1 - \exp^{-\tau} \ \text{ or } \ \tau = -\ln(1 - R_l) \ . \tag{4.180}$$

Herein again R_l is another random number with value between 0 and 1. If it is found that the distance l is so large, that the packet is going to heat a wall, then the counter for the particular wall surface, N_w, is increased by 1 and the energy of the packet, w, is added to that wall surface cell element. However if the energy packet is absorbed in a cell before reaching a wall surface, then the counter for the cell N_{cell}, where the absorption takes place, is also increased by one and the energy absorbed in the cell is tallied (added to the absorbed energy at the cell). However it is assumed that at each absorption in a volume element it is emitted immediately from the same element to conserve a radiative equilibrium in the first instance. The new angles of emission are obtained again by two new random numbers, R_θ and R_φ, except that now $\theta = \cos^{-1}(1 - 2R_\theta)$.

It is now time to make an estimate of the volumetric energy release or absorption in each cell. While the total energy absorbed in the cell, $\sum w$, is the sum of all energy that has been absorbed [W], the energy emitted from the cell is $4\kappa_{\text{cell}}\sigma T_{\text{cell}}^4 V_{\text{cell}}$ [W], where V_{cell} is the volume of the cell. The volumetric energy source term (with an opposite sign) for the energy equation of the fluid is given by

$$\frac{1}{V_{\text{cell}}}\left(4\kappa_{\text{cell}}\sigma T_{\text{cell}}^4 V_{\text{cell}} - \sum w\right) . \tag{4.181}$$

Obviously the method will be difficult to implement for a gas volume with very strong difference in absorption coefficient, since calculation of the optical path length will depend very much on the path direction.

4.5 Exercise

4.5.1 The following information are available in connection with the solar radiation on the earth: a) sun diameter $= 1.39 \times 10^6$ km; b) earth diameter $= 1.27 \times 10^4$ km; c) distance sun-earth $= 1.495 \times 10^8$ km; d) the sun subtends an angle $32'$ $(= 0.018643$ radian) on earth; e) the sun's disk temperature may be approximated as 5,780K; and f) the solar constant, that is the energy from the sun per unit time and at mean sun-earth distance outside the earth atmosphere $= G = 1367$ Wm^{-2}.
Verify the above subtending angle by the sun on the earth, which is different from the solid angle of the sun disk on the earth. What is the latter value? [Ans: 6.7895e-5 steradian]
Also compute the energy emanating from the sun in all directions from the sun disk, and the energy that is falling on the earth (solar constant!). Verify the solar disk temperature.

4.5.2 The mid wavelengths of vibration-rotation bands of CO_2 are in 2.99μ, 4.3μ and 15μ. Calculate the characteristic vibration temperatures. Similarly the mid wave lengths of the vibration-rotation bands of OH, NO, HF, CN, HCl and CO are 4.82μ, 8.72μ, 4.69μ, 5.81μ, 6.55μ and 6.67μ, respectively. Compute the characteristic vibration temperatures.

4.5.3 A tubular chamber of length 1m contains CO_2 at $p = 1$ bar and uniform temperature $T = 1,500$K (overall absorption coefficient $= 0.2$ m^{-1}). a) Compute the emissivity of gas at one end of the chamber, and b) if the chamber length is doubled, then compute the emissivity.

4.5.4 Estimate the volumetric energy release for gases given in Table 4.1 at given pressure and temperature.

4.5.5 Compare the *Rayleigh-Jean law* and *Wien's law* with the *Prandtl radiation law* for the range of λT larger than 2898 mK.

4.5.6 In a solar collector, heat coming from the sun, is forwarded by a reflector to a small area, which is used to heat a fluid inside that area to run a turbine. If the sun's disk temperature is given as 5,780K, can the temperature at the focussing point be larger than this temperature, if the diameter of the focussing mirror is made very large?

5 Collision Processes for High Temperature Gases

In Chap. 3 it has been shown that the individual molecules in a gas are continuously at random motion, whose velocity distribution at a given instant defines the translational temperature of the particles. It is evident that the individual particles do not retain their original speed and direction of motion at all times, and only for rigid particles and in between collisions these may be considered constant with time. Although the particles may not be considered to be rigid sphere, such an assumption simplifies the mathematical procedure considerably, and any deviation from the rigid sphere model can be taken care of by using correction factors. While the temperature is considered to be linked with the energy distribution and hence requires the concept of equilibrium, one can assume local equilibrium to define the local temperature. Any deviation of the temperature from place to place is then considered as a deviation of the energy distribution, which causes transport of mass, momentum and energy from place to place. In this chapter, therefore, we consider the collision between the particles.

5.1 Dynamics of Binary Collision

Let us now consider two rigid spherical molecules 1 and 2, with mass M_1 and M_2, and let them collide in line with direction of motion of each of them (Fig. 5.1). Let v_1' and v_2' be the respective speeds before collision, and v_1'' and v_2'' after the collision. Therefore the principle of conservation of momentum and energy give the following two equations:

$$M_1 v_1' + M_2 v_2' = M_1 v_1'' + M_2 v_2'' \tag{5.1}$$

$$M_1 v_1'^2 + M_2 v_2'^2 = M_1 v_1''^2 + M_2 v_2''^2 \ . \tag{5.2}$$

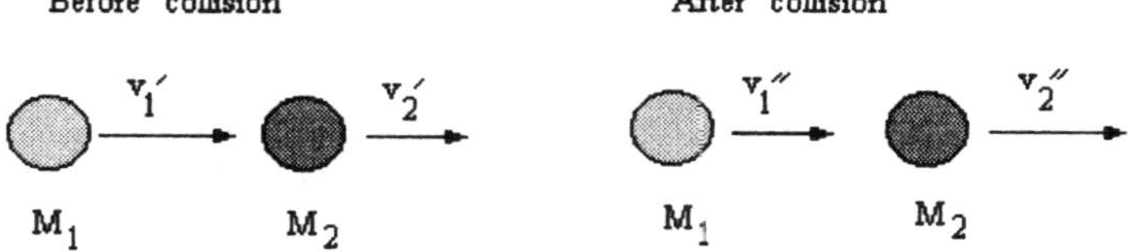

Fig. 5.1. Binary collision of non-attracting rigid sphere

From the above two equations, we can write further

$$M_1(v_1' - v_1'') = -M_2(v_2' - v_2'') \quad \text{and}$$
$$M_1(v_1'^2 - v_1''^2) = -M_2(v_2'^2 - v_2''^2) \ . \tag{5.3}$$

Dividing one by the other, we get

$$v_1' + v_1'' = v_2' + v_2'' \ . \tag{5.4}$$

Hence,

$$v_1' - v_2' = v_2'' - v_1'' \ . \tag{5.5}$$

Introducing the concept of the *relative speed*

$$g_{21}' = v_2' - v_1' = -g_{12}'$$
$$g_{21}'' = v_2'' - v_1'' = -g_{12}'' \tag{5.6}$$

we can write, therefore,

$$g_{21}'' = g_{12}', g_{12}'' = g_{21}' \tag{5.7}$$

and can state further, that the relative motion of the particles change in direction only, but not in magnitude after collision. By simple manipulation of (5.1, 5.2), we get further

$$v_1'' = \frac{2M_2 v_2'}{M_1 + M_2} + \frac{(M_1 - M_2)v_1'}{M_1 + M_2} \tag{5.8}$$

and

$$v_2'' = \frac{2M_1 v_1'}{M_1 + M_2} + \frac{(M_2 - M_1)v_2'}{M_1 + M_2} \ . \tag{5.9}$$

From (5.8, 5.9) two special situations may be noted. For collision between two particles of equal mass ($M_1 = M_2$), we get the relation $v_1'' = v_2'$ and $v_2'' = v_1'$. Therefore, in this particular case, the particles exchange their velocities on collision. In the second case, if one of the particles is very light with respect to the other (for example, subscript 1 may denote an electron and 2 a heavy particle), then $v_1'' = 2v_2' - v_1' \approx -v_1'$ and $v_2'' = (2M_1/M_2)v_1' + v_2' \approx v_2'$ (unless $v_1' \gg v_2'$). Therefore, the heavy particle continues to move after collision in the same direction and with the same speed as before the collision, whereas the light electron just bounces back like an elastic ball.

At this stage we introduce the concept of the *mass-averaged speeds* from the definition

$$G' = (M_1 v_1' + M_2 v_2')/(M_1 + M_2) \tag{5.10}$$

and

$$G'' = (M_1 v_1'' + M_2 v_2'')/(M_1 + M_2) \ . \tag{5.11}$$

Substituting (5.10, 5.11) into (5.9) and after some manipulation, one can show easily that

$$G' = G'' = G \tag{5.12}$$

where

$$G = Y_1 v_1' + Y_2 v_2' = Y_1 v_1'' + Y_2 v_2'' \tag{5.13}$$

and the respective mass fractions are given by relations

$$Y_1 = \frac{M_1}{M_1 + M_2} \text{ and } Y_2 = \frac{M_2}{M_1 + M_2} \; . \tag{5.14}$$

From (5.8, 5.9) and the definition of the relative speed and the mass-averaged speed, (5.6, 5.7) and (5.10, 5.11), we can, write, therefore, the following four expressions

$$v_1' = G - Y_2 g_{21}'; v_1'' = G - Y_2 g_{21}''$$
$$v_2' = G - Y_1 g_{12}'; v_2'' = G - Y_1 g_{12}'' \; .$$

From the above equations we can now find the expressions for the change of momentum

$$\Delta p_1 = M_1(v_1'' - v_1') = \mu(v_2' - v_1') = \mu g_{21}' \tag{5.15}$$

and

$$\Delta p_2 = M_2(v_2'' - v_2') = \mu(v_1' - v_2') = \mu g_{12}' \tag{5.16}$$

and the change of energy of each particle is

$$\Delta E_1 = \frac{M_1}{2}(v_1''^2 - v_1'^2) = \mu G g_{21}' \tag{5.17}$$

and

$$\Delta E_2 = \frac{M_2}{2}(v_2''^2 - v_2'^2) = \mu G g_{12}' \tag{5.18}$$

where the *reduced mass* is

$$\mu = \frac{2M_1 M_2}{M_1 + M_2} \; . \tag{5.19}$$

On the other hand, the total energy before collision is

$$E' = \frac{1}{2} M_1 v_1'^2 + \frac{1}{2} M_2 v_2'^2 = \frac{M_1}{2}[G - Y_2 g_{21}']^2 + \frac{M_2}{2}[G - Y_1 g_{12}']^2$$
$$= \frac{1}{2} G^2 (M_1 + M_2) + \frac{1}{4} \mu g_{12}'^2 \tag{5.20}$$

which remains conserved after collision.

Substituting (5.8) into (5.17) we can write, after some manipulation as

$$\Delta E_1 = \frac{2M_1 M_2}{(M_1 + M_2)^2}(M_1 v_1' + M_2 v_2')(v_2' - v_1') \; . \tag{5.21}$$

If we relate average kinetic energy of each particle with temperature of the particle by the relation

$$\frac{1}{2}Mv'^2 = \frac{3}{2}k_B T \tag{5.22}$$

we can write (5.21) approximately as

$$\Delta E_1 = \frac{3M_1 M_2}{(M_1 + M_2)^2} k_B (T_2 - T_1) \ . \tag{5.23}$$

Thus, the energy exchange between the particles depends on the value of the mass ratio (M_1/M_2), and is considerable reduced if $M_1 \ll M_2$. Based on a rigid analysis, *Sutton and Sherman* [21] found that that the average change in the kinetic energy per collision between the electron (subscript "e") and the heavy particle (subscript "h") is given, as also from (5.23), by the relation

$$\Delta E = 3\frac{M_e}{M_h} k_B (T_e - T_h) \ . \tag{5.24}$$

Consideration of the relative speed instead of the absolute speeds simplifies the analysis considerably and is often used in practice. Thus it is assumed that a particle at rest is approached by another particle of mass μ and relative speed before collision g'_{12} in such a manner that the centers of two particles are separated by a distance b_o (Fig. 5.2a). After collision the relatively moving particle is deflected by an angle χ. Please note that the angle of deflection χ and the collision angle ψ for a collision between rigid spheres are related by the relation $\chi = \pi - 2\psi$. It is evident that for rigid spherical molecules the angle of deflection is dependent on the value of b_o, and will be equal to zero, if $b_o \geq (d_1 + d_2)/2$. It is not dependent on the value of the relative speed before collision g_{12}.

For collisions between molecules in which the long-distance forces act before and after collision, the flight path of the approaching molecule (when

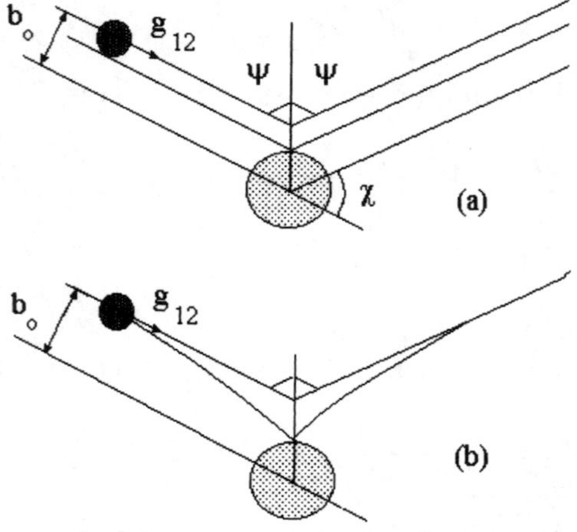

Fig. 5.2. Binary collision between two spheres: **(a)** rigid, and **(b)** attracting

the other molecule is assumed to be relatively at rest) changes its direction of motion even before and after collision, as shown for interaction between attracting spheres in Fig. 5.2b. As a result of the force of attraction the path of the colliding molecule is curved, whereas the point of collision for an apparent straight path before collision lies at a larger distance than the actual point of collision. Thus it is concluded that for two colliding particles for which the nature of forces is known (for example, a Coulomb force acts at the time of binary collision between electrons and other charged particles. For collisions between non-similar charges the particles attract each other, and for collisions between similar charges they repel each other; the one with an attracting force between the molecules leads to a larger value of the collision cross-section than the one with a repulsive force. This can be noted from the calculated values of the collision cross-sections (for diffusion) for argon between 5,000 and 20,000 K, as given in Table 5.1. It is worth noting, that the collision cross-sections between electrons and neutral atoms are of the same order of magnitude as that given for hydrogen atom given in chapter 2 by estimating it from the *electron orbit radius*, whereas for neutral-neutral collision these are just about an order of magnitude larger than those for electron-atom collision cross-section. The collision cross-section between charged particles are, however, 3 to 4 order of magnitude larger than those for the electron-atom, which are due to long-distance forces acting between them. As a result of these long distance forces, the collision cross-section has effectively a value Q_{eff}, different from the one that is calculated from the rigid spherical model Q_{rigid} and they are related by the expression

Table 5.1. Computed collision cross-section $[\text{Å}]^2$ for argon

T[K]	Ar-Ar	Ar$^+$	e-Ar	e-Ar$^+$	e-e
5000	17.4	100.0	1.48	1.74e4	1.67e4
6000	16.7	96.3	1.87	6.87e3	6.82e3
7000	16.1	93.1	2.26	4.28e3	4.19e3
8000	15.7	90.4	2.66	2.87e3	2.77e3
9000	15.2	88.1	3.05	2.03e3	1.92e3
10000	14.9	86.0	3.44	1.50e3	1.39e3
11000	14.5	84.2	3.84	1.15e3	1.05e3
12000	14.2	82.5	4.23	918.1	825.4
13000	14.0	81.0	4.62	753.7	671.0
14000	13.7	79.6	5.00	640.1	567.5
15000	13.5	78.3	5.37	560.3	497.4
16000	13.3	77.1	5.74	502.1	448.0
17000	13.1	76.0	6.09	456.7	410.4
18000	12.9	74.9	6.44	418.5	378.8
19000	12.7	74.0	6.77	386.0	351.7
20000	12.5	73.0	7.08	357.8	328.2

$$Q_{\text{eff}} = Q_{\text{rigid}} \left[1 + \left\{ \frac{4}{\mu} \frac{1}{\bar{g}^2} \int_0^\infty -F(r)\mathrm{d}r \right\} \right] \qquad (5.25)$$

where $Q_{\text{rigid}} = \pi d^2$, $d = (d_1 + d_2)/2$ is the average of diameters of the two colliding particles, F is the force acting between the two particles, μ is the reduced mass and $\bar{g}^2$ is the mean of (g^2), that is the square of the relative speed, as the particles approach each other. For large values of g, it is possible to assume $g^2 = \bar{g}^2$. Since the force and *potential distribution* are related, $F = -\nabla\phi$, where ϕ is the potential, then

$$\int_d^{-\infty} -F(r)\mathrm{d}r = \phi_\infty - \phi_d \ . \qquad (5.26)$$

Usually, the potential at infinity is assumed to be zero, and it is enough to know the potential ϕ_d at the time of collision. This is further explained in the next section and the methods to compute the collision cross-section are given.

5.2 Collision Cross-Section

5.2.1 Collision Between Neutrals

For a collision between two neutral particles several models are available, some of which are given in Fig. 5.3, and from which the collision cross-sections $Q^{(l,s)}$ are obtained. The superscript (l,s) refer to certain values that are given in terms of the so-called *sonine polynomials*. Among the various potential models, Lennard-Jones 6-12 model and exponential repulsive model are both used for neutral-neutral collision. The Lennard-Jones 6-12 model, so-called, because of the exponents 6 and 12 in the equation for potential, has the advantage of being a two-parameter model, for which the mole mass and both the parameters $(d_j, \epsilon/k_B)$ are tabulated for a large number of gases (Table 5.2). The collision cross-section is obtained by defining a reduced temperature T^* $= k_B T/\epsilon$, which is further used to obtain the reduced collision cross-section $Q^{*(l,s)}$ by interpolating the tabulated reduced cross-section values (for example, by Hirschfelder, et al.), part of which is reproduced in Table 5.4. For collisions between particles of different specie, the potential parameter and the average particle diameter are obtained from the relations

$$\epsilon_{jk} = \sqrt{\epsilon_j \epsilon_k} \quad \text{and} \quad d_{jk} = (d_j + d_k)/2 \ . \qquad (5.27)$$

The Lennard-Jones potential model, because of the availability of the collision cross-section data for a large number of gases, is easy to use. Even when they are not tabulated, they can be obtained easily from the critical states or state values at one atmospheric pressure. For this purpose *Guldberg's rule* states that at one atmospheric pressure the boiling point temperature T_b is

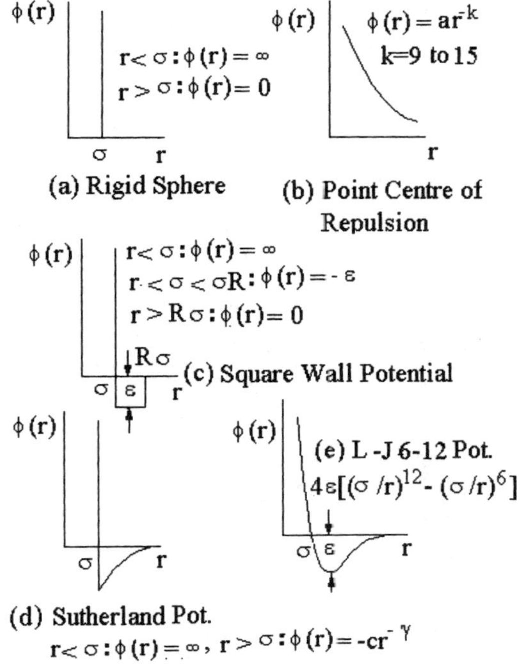

Fig. 5.3. Some commonly used potentials for collision between two particles

approximately two-thirds of the critical temperature T_c. Further from the principle of corresponding states, the *critical specific volume* v_c is about the three times the liquid specific volume v_b', which is the inverse of the liquid density ρ_b' at the boiling point temperature, whilst the compressibility factor at the critical state, $p_c v_c/(RT_c)$, is constant and is equal to 0.293 for non-polar molecules. Incidentally, some of these properties for metals given in Table 5.3 were obtained by using these rules if the data were not otherwise available. In addition, from the correspondence principle of transport properties, the two parameters of Lennard-Jones model, potential parameter and molecular diameter, are related to the critical temperature and critical specific volume. The potential parameter $(\epsilon/k_B) = 0.77\,T_c$. and the molecular diameter can be determined, from a dimensional analysis, as

$$d = (M\{\text{kg}\}v_b'\{\text{m}^3\text{kg}^{-1}\})^{1/3} = \left(\frac{mv_b'}{N_A}\right)^{1/3} \tag{5.28}$$

where the result is in [m]. If it is divided with 10^{-10} then we get the relation for molecular diameter $d = 11.843(mv_b')^{1/3}$ [Å]. In literatures the constant is put slightly different and the molecular diameter is given by the relation

$$d = 11.659(mv_b')^{1/3}, \text{ Å}. \tag{5.29}$$

Table 5.2. Lennard-Jones potential constants for selected molecules

Gas	Molemass	$d_j(\text{Å})$	ϵ/k_B
He	4.003	2.576	10.22
Ne	20.18	2.858	27.5
Ar	39.94	3.421	119.5
Kr	83.80	3.610	190.0
Xe	131.30	4.055	229.0
Li	6.94	2.970	1848.0
Na	22.99	3.665	1333.0
K	39.10	4.545	1190.0
Cs	132.90	5.005	1097.0
Mg	24.32	2.897	1612.0
Al	26.98	2.615	3003.0
Cu	63.54	2.332	3312.0
Hg	200.6	2.923	728.0
H	1.008	2.680	38.0
N	14.01	3.100	91.5
O	16.0	2.900	100.0
Cl	35.5	3.600	316.0
H_2	2.016	2.915	38.0
N_2	28.02	3.681	91.5
O_2	32.0	3.499	100.0
Cl_2	70.9	4.217	316.0
NO_2	30.01	3.481	121.0
N_2O	44.02	3.816	237.0
NH_3	17.03	2.902	692.0
H_2O	18.016	2.520	775.0
CO	28.01	3.678	94.5
HCl	36.47	3.305	360.0
CH_4	16.04	3.796	144.0
CO_2	44.01	3.952	200.0

Table 5.3. Critical states for some metals

Element	$T_b[\text{K}]$	$\rho_b[\text{kgm}^{-3}]$	$T_c[\text{K}]$	$v_c[\text{m}^2.\text{kg}^{-1}]$	$p_c[\text{bar}]$
Li	1600	420	2400	7.14e-3	1180
Na	1155	740	1732	4.05e-3	453
K	1030	660	1545	4.54e-3	212
Cs	950	1680	1425	1.78e-3	147
Mg	1363	1585	2093	1.89e-3	1108
Al	2600	2390	3900	1.25e-3	2805
Cu	2868	7940	4302	3.78e-4	4365
Hg	630	12740	945	2.35e-4	728

Herein, m is the mole mass and v_b' is the liquid molar specific volume $[\mathrm{m}^3\mathrm{kg}^{-1}]$ at normal boiling temperature (equivalent of one atmosphere pressure).

We would examine the above relationship on the basis of existing data for water, although it is a polar gas. The existing data in literature for water are as follows:

$$v_b' = 0.0010435 \ [\mathrm{m}^3\mathrm{kg}^{-1}], v_c = 0.00326 \ [\mathrm{m}^3\mathrm{kg}^{-1}], T_b = 373.15 \ [\mathrm{K}]$$
$$T_c = 647.30 \ [\mathrm{K}] \text{ and } p_c = 220.64 \ [\mathrm{bar}].$$

Accordingly, $p_c v_c/(RT_c) = 0.241$ (from van der Waals equation, it should be $3/8$ or 0.375). Further from the above sketched estimates for water, $(\epsilon/k_B) = 498$ K and $d = 3.100$ Å. However the data by Brokaw [51] for these, from viscosity data, are: $\epsilon/k_B = 775$ K and $d = 2.520$ Å.

For some of the technically important metals the values of the critical state are given in Table 5.3 and the corresponding Lennard-Jones potential parameter are given in Table 5.2. Table 5.4 contains the non-dimensional cross-sectional data $Q^{*(l,s)}$ as a function of T^* to Lennard-Jones 6-12 potential reproduced from *Hirschfelder*, et al. [10]. The appropriate $Q^{(l,s)}$ data is obtained by multiplying $Q^{*(l,s)}$ with the cross-sectional area for rigid-body collision, $Q_{\mathrm{rigid}} = \pi d_{jk}^2$. However, *Amdur* and *Mason* found from a series of experiments, that, instead of the Lennard-Jones 6-12 model a two parameter exponential repulsive model

$$\phi = \phi_o \exp^{-r/r_o} \tag{5.30}$$

requiring the two values of ϕ_o and r_o gives a fair fit to both the potential obtained from the scattering data and from the measurements of viscosity at lower temperatures. These values are given in Table 5.5. For argon, the values were subsequently modified to $\phi_o = 7100$ eV and $r_o = 0.258$ Å. For collision between two different particles, the parameters ϕ_o and r_o for each particle are replaced by the following expressions:

$$\phi_o = \sqrt{\phi_{oj}\phi_{ok}} \tag{5.31}$$

and

$$\frac{1}{r_o} = \frac{1}{2}\left(\frac{1}{r_{oj}} + \frac{1}{r_{ok}}\right). \tag{5.32}$$

Actual determination of the collision cross-section for the exponential repulsive potential is very simple. A collision parameter

$$\alpha = \ln[\phi_o/(k_B T_h)] \tag{5.33}$$

is used to determine the tabulated values of a function $I^{(l,s)}$ in Monchik Tables [85] reproduced in Table 5.6 and T_h is the temperature of the heavy particles. Collision cross-section is then obtained from the relation

$$Q^{(l,s)} = \frac{8\pi\alpha^2 r_o^2 I^{(l,s)}}{(s+1)!\left[1 - \dfrac{1+(-1)^l}{2(l+1)}\right]}. \tag{5.34}$$

Table 5.4. $Q^{*(l,s)}$ for Lennard-Jones potential

$k_B T/\epsilon$	$Q^{*(1,1)}$	$Q^{*(1,2)}$	$Q^{*(1,3)}$	$Q^{*(1,4)}$	$Q^{*(1,5)}$	$Q^{*(2,2)}$	$Q^{*(2,3)}$	$Q^{*(2,4)}$	$Q^{*(3,1)}$
0.3	2.662	2.256	1.962	2.785	2.535	2.333	2.152	1.990	2.557
0.4	2.318	1.931	1.663	2.492	2.232	2.016	1.883	1.682	2.223
0.5	2.066	1.705	1.468	2.257	1.992	1.781	1.614	1.486	1.975
0.6	1.877	1.543	1.336	2.065	1.806	1.610	1.463	1.356	1.788
0.7	1.729	1.423	1.242	1.908	1.661	1.484	1.357	1.267	1.645
0.8	1.612	1.332	1.172	1.780	1.549	1.389	1.278	1.201	1.535
0.9	1.517	1.261	1.119	1.675	1.460	1.316	1.219	1.152	1.447
1.0	1.439	1.204	1.076	1.587	1.388	1.258	1.172	1.113	1.377
1.1	1.375	1.157	1.041	1.514	1.329	1.212	1.135	1.082	1.319
1.2	1.320	1.119	1.013	1.452	1.280	1.174	1.104	1.056	1.272
1.3	1.273	1.086	0.989	1.399	1.239	1.142	1.078	1.035	1.232
1.4	1.233	1.059	0.968	1.353	1.205	1.115	1.057	1.016	1.198
1.5	1.193	1.034	0.950	1.314	1.175	1.092	1.037	1.000	1.169
1.6	1.167	1.013	0.934	1.279	1.149	1.072	1.022	0.986	1.144
1.7	1.140	0.995	0.920	1.248	1.126	1.054	1.007	0.973	1.122
1.8	1.116	0.978	0.908	1.221	1.106	1.038	0.994	0.962	1.103
1.9	1.094	0.963	0.897	1.197	1.088	1.024	0.982	0.952	1.085
2.0	1.075	0.950	0.887	1.175	1.073	1.012	0.972	0.943	1.070
2.2	1.041	0.927	0.869	1.138	1.045	0.989	0.952	0.926	1.043
2.4	1.012	0.907	0.854	1.107	1.022	0.971	0.937	0.912	1.021
2.6	0.988	0.891	0.841	1.081	1.002	0.955	0.923	0.900	1.003
2.8	0.967	0.877	0.829	1.058	0.985	0.941	0.912	0.889	0.986
3.0	0.949	0.864	0.819	1.039	0.971	0.929	0.901	0.879	0.972
3.2	0.933	0.852	0.809	1.022	0.958	0.918	0.891	0.870	0.960
3.4	0.919	0.842	0.801	1.007	0.946	0.908	0.882	0.862	0.948
3.6	0.906	0.833	0.793	0.993	0.936	0.899	0.874	0.854	0.938
3.8	0.894	0.824	0.786	0.981	0.926	0.891	0.867	0.847	0.929
4.0	0.884	0.817	0.779	0.970	0.917	0.884	0.859	0.840	0.920
4.2	0.874	0.809	0.773	0.960	0.909	0.877	0.853	0.834	0.912
4.4	0.865	0.803	0.767	0.951	0.902	0.870	0.847	0.828	0.905
4.6	0.857	0.796	0.761	0.942	0.895	0.864	0.841	0.823	0.899
4.8	0.849	0.791	0.756	0.934	0.888	0.858	0.836	0.818	0.892
5.0	0.842	0.785	0.751	0.927	0.882	0.853	0.831	0.813	0.886
6.0	0.812	0.761	0.729	0.896	0.856	0.829	0.808	0.791	0.861
7.0	0.790	0.742	0.712	0.873	0.836	0.810	0.790	0.774	0.841
8.0	0.771	0.726	0.697	0.854	0.819	0.794	0.775	0.759	0.825
9.0	0.756	0.713	0.685	0.838	0.805	0.781	0.762	0.746	0.811
10.	0.742	0.701	0.673	0.824	0.792	0.769	0.750	0.734	0.799
20.	0.664	0.629	0.605	0.743	0.716	0.695	0.678	0.664	0.724
30.	0.623	0.591	0.568	0.700	0.675	0.655	0.640	0.626	0.684
40.	0.596	0.565	0.543	0.672	0.647	0.628	0.613	0.601	0.657
50.	0.576	0.546	0.525	0.650	0.627	0.608	0.594	0.582	0.637
60.	0.560	0.531	0.510	0.633	0.610	0.593	0.578	0.566	0.621
70.	0.546	0.518	0.498	0.619	0.597	0.579	0.566	0.554	0.607
80.	0.535	0.507	0.488	0.608	0.585	0.568	0.555	0.543	0.596
90.	0.526	0.498	0.479	0.597	0.575	0.559	0.545	0.534	0.587
100.	0.517	0.490	0.471	0.588	0.567	0.550	0.537	0.526	0.578
200.	0.464	0.440	0.423	0.532	0.513	0.498	0.486	0.476	0.525
300.	0.436	0.413	0.397	0.502	0.483	0.469	0.458	0.449	0.495
400.	0.417	0.395	0.380	0.481	0.464	0.450	0.439	0.430	0.476

Table 5.5. Exponential repulsive potential constants for selected gases

Gas	ϕ_o(eV)	r_o(Å)
He	386	0.220
Ne	8680	0.196
Ar	3.23e4	0.224
Kr	3.87e5	0.202
Xe	3.11e6	0.208
N2	1.35e4	0.263

Table 5.6. Monchik Table for Exponential repulsive potential for neutral-neutral collision

α	$I^{(1,1)}$	$I^{(1,2)}$	$I^{(1,3)}$	$I^{(2,2)}$	$I^{(2,3)}$	$I^{(3,3)}$	$I^{(1,4)}$	$I^{(1,5)}$	$I^{(2,4)}$
3.5	0.188	0.445	1.465	0.503	1.713	2.242	6.189	31.93	7.447
4.0	0.193	0.472	1.602	0.509	1.772	2.327	6.958	36.85	7.872
4.5	0.197	0.495	1.718	0.512	1.816	2.396	7.614	41.09	8.195
5.0	0.201	0.515	1.817	0.515	1.849	2.453	8.176	44.74	8.447
5.5	0.204	0.532	1.902	0.516	1.875	2.501	8.660	47.91	8.646
6.0	0.207	0.546	1.976	0.517	1.895	2.541	9.082	50.68	8.807
6.5	0.209	0.559	2.040	0.518	1.911	2.575	9.451	53.11	8.939
7.0	0.212	0.570	2.097	0.518	1.925	2.605	9.777	55.25	9.049
7.5	0.214	0.580	2.147	0.518	1.935	2.630	10.067	57.17	9.141
8.0	0.215	0.589	2.192	0.518	1.944	2.653	10.326	58.89	9.219
8.5	0.217	0.597	2.232	0.518	1.952	2.673	10.560	60.43	9.286
9.0	0.218	0.604	2.269	0.518	1.958	2.691	10.770	61.83	9.344
9.5	0.220	0.611	2.302	0.518	1.963	2.707	10.962	63.10	9.394
10.0	0.221	0.617	2.332	0.518	1.968	2.722	11.137	64.26	9.438
10.5	0.222	0.622	2.360	0.517	1.972	2.735	11.297	65.32	9.477
11.0	0.223	0.627	2.385	0.517	1.975	2.747	11.444	66.30	9.511
11.5	0.224	0.632	2.409	0.517	1.978	2.758	11.580	67.20	9.542
12.0	0.225	0.636	2.430	0.517	1.981	2.768	11.705	68.04	9.569
12.5	0.226	0.640	2.451	0.516	1.983	2.777	11.822	68.82	9.594
13.0	0.227	0.644	2.469	0.516	1.985	2.786	11.931	69.54	9.616
13.5	0.227	0.647	2.487	0.516	1.987	2.794	12.032	70.21	9.636
14.0	0.228	0.651	2.503	0.515	1.988	2.801	12.127	70.84	9.654
14.5	0.229	0.654	2.518	0.515	1.990	2.808	12.215	71.44	9.671
15.0	0.229	0.656	2.533	0.515	1.991	2.814	12.299	71.99	9.686
15.5	0.230	0.659	2.546	0.514	1.992	2.820	12.377	72.51	9.700
16.0	0.230	0.662	2.559	0.514	1.993	2.826	12.451	73.01	9.713
17.0	0.231	0.666	2.583	0.514	1.994	2.836	12.587	73.91	9.736
18.0	0.232	0.670	2.604	0.513	1.996	2.845	12.709	74.73	9.756
19.0	0.233	0.674	2.623	0.513	1.997	2.853	12.819	75.46	9.773
20.0	0.234	0.678	2.640	0.512	1.998	2.860	12.919	76.13	9.788
21.0	0.235	0.681	2.655	0.512	1.998	2.867	13.010	76.73	9.802
22.0	0.235	0.684	2.670	0.511	1.999	2.873	13.093	77.29	9.813
23.0	0.236	0.686	2.683	0.511	2.000	2.879	13.169	77.80	9.824
24.0	0.236	0.689	2.695	0.511	2.000	2.884	13.240	78.27	9.833
25.0	0.237	0.691	2.706	0.510	2.000	2.888	13.305	78.70	9.841
26.0	0.237	0.693	2.717	0.510	2.001	2.892	13.365	79.11	9.849
27.0	0.238	0.695	2.727	0.510	2.001	2.896	13.422	79.48	9.856
28.0	0.238	0.697	2.736	0.509	2.001	2.900	13.474	79.83	9.862
28.5	0.238	0.698	2.740	0.509	2.001	2.902	13.499	80.00	9.865

5.2.2 Collision Between Electrons and Neutrals

The collision cross-section for collisions between the electrons and neutral particles are determined from the experimental values of the of the gas-kinetic cross-sections, $Q^{(1)}$, values of which are given by *Kollath* [29] and *Massey* and *Burhop* [14] as a function of the kinetic energy E of the colliding electrons. These data for different gases are given in Table 5.7 and 5.8. $Q^{(1,s)}$ is now determined with the help of the equation

$$Q^{(l,s)} = \frac{(k_B T_e)^{-(s+2)}}{(s+1)!} \int_0^\infty \exp^{-E/(k_B T_e)} Q^{(1)} E^{(s+1)} \mathrm{d}E \ . \tag{5.35}$$

However, $Q^{(1)}$ values, which are needed for integration in the kinetic energy range from zero to infinity, are available experimentally only in a limited range of the kinetic energy, and the classical methods to determine the collision cross-section by assuming the potential with distance can not be used because of the so-called *Ramsauer effect*. According to this effect, at low values of the kinetic energy the participating neutrals in the collision become almost transparent to the electrons. In order to estimate the effect of truncation in the experimental gas-kinetic cross-section data, we introduce a non-dimensional kinetic energy $E^* = E/(k_B T_e)$. Thus (5.35) becomes

$$Q^{(1,s)} = \frac{1}{(s+1)!} \int_0^\infty \exp^{-E^*} Q^{(1)} E^{*(s+1)} \mathrm{d}E^* \ . \tag{5.36}$$

Assuming $Q^{(1)}$ to be constant in the entire range of the experimentally determined tabulated data, (5.36) can be integrated within two limits E_1^* and E_2^*, and it becomes

$$\frac{Q^{(1,s)}}{Q^{(1)}} = -\exp^{-E^*} \left(1 + E^* + \frac{1}{2} E^{*2} + \ldots + \frac{1}{(s+1)!} E^{*(s+1)} \right) \Bigg|_{E_1^*}^{E_1^*} \ . \tag{5.37}$$

The maximum contribution in the integral comes from around the non-dimensional energy $E^* = E_m^* = (s+1)$. Assuming T_e between 5,000 and 20,000K and the values of s needed to be considered between 1 and 7, the maximum contribution to the integral should be around $E_m = 1\mathrm{eV}$ and $16\mathrm{eV}$. Therefore, during evaluation of (5.35), the experimental $Q^{(1)}$ data were extrapolated beyond the experimentally given range of E in order to compute $Q^{(1,s)}$. Integration was done by Simpson's rule and the error due to extrapolation of the experimentally determined gas-kinetic cross-section is estimated to be less than five percent.

Table 5.7. Tabulated value of $Q_{(1)}$ [Å^2] for selected electron-atom collision

E(eV)	He	Ne	Ar	Kr	Xe	Li	Na	K	Cs	H	N	O
0.00	5.0	0.2	8.1	30.7	176.0	299.	338.	494.	429.	17.6	0.00	3.7
0.01	5.2	0.3	6.1	26.0	116.0	272.	317.	481.	397.	17.6	0.01	3.7
0.02	5.3	0.3	3.7	19.7	80.0	259.	301.	476.	373.	17.6	0.03	3.7
0.03	5.5	0.4	2.8	16.0	61.3	252.	292.	473.	359.	17.6	0.04	3.7
0.04	5.5	0.5	2.3	13.5	48.0	247.	286.	471.	350.	17.6	0.05	3.7
0.05	5.6	0.5	1.8	11.4	39.5	243.	281.	469.	343.	17.5	0.07	3.7
0.06	5.7	0.6	1.5	10.0	33.4	240.	277.	468.	337.	17.5	0.08	3.8
0.07	5.7	0.6	1.1	9.1	29.0	238.	274.	467.	332.	17.5	0.10	3.8
0.08	5.8	0.6	0.9	8.2	25.6	236.	271.	466.	328.	17.5	0.11	3.8
0.09	5.8	0.7	0.6	7.5	23.4	234.	269.	465.	325.	17.5	0.12	3.8
0.10	5.9	0.7	0.5	6.8	20.4	232.	267.	464.	321.	17.5	0.14	3.8
0.11	5.9	0.7	0.3	6.3	18.7	231.	265.	464.	319.	17.4	0.15	3.8
0.12	5.9	0.8	0.3	5.8	16.9	229.	263.	463.	316.	17.4	0.16	3.9
0.13	6.0	0.8	0.3	5.3	15.1	228.	262.	463.	314.	17.4	0.18	3.9
0.14	6.0	0.8	0.2	4.9	14.1	227.	260.	462.	312.	17.4	0.19	3.9
0.15	6.0	0.8	0.2	4.4	13.0	226.	259.	462.	310.	17.4	0.20	3.9
0.16	6.1	0.9	0.2	4.0	12.0	225.	257.	461.	308.	17.3	0.22	3.9
0.17	6.1	0.9	0.2	3.6	11.1	224.	256.	461.	306.	17.3	0.23	4.0
0.18	6.1	0.9	0.2	3.2	10.2	223.	255.	460.	304.	17.3	0.25	4.0
0.19	6.1	0.9	0.2	3.0	9.3	222.	254.	460.	303.	17.3	0.26	4.0
0.20	6.2	0.9	0.2	2.5	8.4	221.	253.	460.	302.	17.2	0.27	4.0
0.25	6.3	1.0	0.2	1.5	5.3	218.	249.	458.	295.	17.1	0.34	4.1
0.30	6.3	1.1	0.2	1.0	3.2	215.	246.	457.	290.	17.0	0.41	4.2
0.35	6.4	1.1	0.2	0.8	2.5	218.	243.	456.	286.	17.0	0.48	4.3
0.40	6.5	1.2	0.2	0.6	1.7	222.	240.	455.	283.	16.9	0.55	4.3
0.45	6.5	1.2	0.2	0.6	1.6	226.	238.	454.	280.	16.8	0.62	4.4
0.50	6.6	1.3	0.3	0.5	1.4	229.	236.	453.	277.	16.7	0.69	4.5
0.55	6.6	1.3	0.3	0.5	1.3	232.	235.	453.	275.	16.6	0.75	4.6
0.60	6.7	1.4	0.4	0.5	1.3	235.	233.	452.	272.	16.5	0.82	4.7
0.65	6.7	1.4	0.5	0.5	1.4	238.	232.	451.	270.	16.4	0.89	4.8
0.70	6.7	1.4	0.5	0.5	1.5	241.	230.	451.	269.	16.3	0.96	4.9
0.75	6.7	1.5	0.6	0.5	1.5	243.	229.	451.	267.	16.2	1.03	5.0
0.80	6.8	1.5	0.7	0.5	1.6	245.	228.	450.	265.	16.1	1.10	5.0
0.85	6.8	1.5	0.8	0.5	1.8	248.	227.	450.	264.	16.0	1.17	5.1
0.90	6.8	1.5	0.9	0.6	2.0	250.	226.	449.	262.	15.9	1.23	5.2
0.95	6.8	1.6	1.0	0.6	2.3	252.	225.	449.	261.	15.8	1.30	5.3
1.00	6.6	1.6	1.0	0.6	2.5	254.	224.	449.	260.	15.7	1.37	5.4
1.50	6.4	1.8	1.7	1.3	5.4	273.	228.	403.	348.	14.9	2.03	5.9
2.00	6.3	2.0	2.5	2.1	8.2	257.	374.	358.	520.	14.0	2.60	6.3
2.50	6.2	2.2	3.3	3.5	12.6	224.	345.	345.	572.	13.2	3.14	6.7
3.00	5.9	2.2	4.1	4.8	17.0	211.	192.	335.	553.	12.5	3.68	7.0
3.50	5.7	2.3	4.9	7.4	20.9	192.	182.	309.	455.	11.8	4.20	7.2
4.00	5.6	2.5	5.8	10.0	24.8	179.	179.	280.	358.	11.1	4.71	7.5
6.00	5.0	2.7	8.7	12.6	29.2	153.	164.	226.	226.	8.9	5.34	7.7
8.00	4.7	3.0	11.7	15.9	33.7	130.	150.	202.	202.	7.0	5.85	7.9
10.00	4.4	3.1	13.8	19.3	32.0	105.	128.	176.	176.	5.6	6.27	8.1
12.00	3.7	3.2	14.5	22.0	29.7	91.	115	163.	163.	4.4	6.64	8.2
14.00	3.4	3.2	13.5	21.0	27.5	80.	105.	152.	152.	3.5	6.97	8.3
16.00	3.1	3.2	12.4	20.0	25.2	72.	97.	143.	143.	2.8	7.27	8.4
18.00	2.9	3.3	11.4	19.0	22.9	65.	89.	131.	131.	2.2	7.55	8.5

Table 5.8. Tabulated value of $Q^{(1)}$ [Å^2] for selected electron-molecule collision

E(eV)	H	O_2	N_2	Cl_2	NO	N_2O	NH_3	H_2O	CO	HCl	CH_4	CO_2
0.00	30.4	12.9	27.5	0.	27	80.0	4730	6683	25.8	0.0	5.4	266
0.01	29.3	11.7	26.9	200	26	41.4	1830	3020	24.6	1.7	4.4	150
0.02	28.1	10.0	25.7	514	25	29.9	896	1496	22.9	2.4	3.7	112
0.03	27.4	9.2	25.0	583	25	25.3	596	1059	21.9	3.0	3.3	95
0.04	26.9	8.6	24.5	583	25	21.3	427	794	21.3	3.4	3.1	85
0.05	26.5	8.2	24.2	583	25	19.8	331	635	20.8	3.8	3.0	75
0.06	26.2	7.9	23.9	583	24	18.0	280	536	20.4	4.2	2.8	66
0.07	26.0	7.6	23.6	583	24	16.6	244	464	20.1	4.5	2.7	59
0.08	25.8	7.4	23.4	583	24	15.5	216	410	19.8	4.9	2.6	53
0.09	25.6	7.2	23.2	583	24	14.5	194	367	19.6	5.1	2.5	49
0.10	25.4	7.1	23.1	583	24	13.8	176	333	19.4	5.4	2.5	45
0.11	25.3	6.9	22.9	583	23	13.1	161	305	19.2	5.7	2.4	42
0.12	25.1	6.8	22.8	583	23	12.5	149	281	19.0	5.9	2.4	40
0.13	25.0	6.7	22.7	581	23	12.0	139	261	18.8	6.2	2.3	37
0.14	24.9	6.6	22.6	580	23	11.5	130	243	18.7	6.4	2.3	36
0.15	24.8	6.5	22.5	579	23	11.1	122	228	18.6	6.6	2.2	34
0.16	24.7	6.4	22.4	578	23	10.7	115	215	18.4	6.9	2.2	32
0.17	24.6	6.3	22.3	578	23	10.4	109	203	18.3	7.1	2.2	31
0.18	24.5	6.2	22.2	576	22	10.0	103	192	18.2	7.3	2.1	30
0.19	24.4	6.1	22.1	575	22	9.8	98	183	18.1	7.5	2.1	29
0.20	24.3	6.0	22.0	571	22	9.5	94	174	18.0	7.7	2.1	27
0.25	24.0	5.8	21.7	535	22	8.5	76	142	17.6	8.6	2.0	23
0.30	23.3	6.1	20.0	507	21	7.7	65	120	16.7	9.4	1.9	20
0.35	22.8	6.4	18.6	484	21	7.1	56	104	15.9	10.2	1.8	18
0.40	22.3	6.7	17.5	466	20	6.6	50	91	15.3	10.9	1.7	17
0.45	21.9	6.9	16.6	450	20	6.2	45	82	14.7	11.5	1.7	15
0.50	21.6	7.2	15.8	436	19	5.9	41	74	14.3	12.1	1.9	14
0.55	21.3	7.4	15.2	424	19	5.6	37	68	13.9	12.7	2.0	13
0.60	21.0	7.6	14.6	413	19	5.5	34	63	13.5	13.3	2.2	12
0.65	20.7	7.9	14.3	404	18	5.4	32	58	12.6	13.8	2.4	12
0.70	20.5	8.2	14.0	395	18	5.6	30	54	11.9	14.4	2.6	11
0.75	20.1	8.2	13.6	387	18	5.8	28	51	12.5	14.9	2.7	11
0.80	19.7	8.2	13.2	380	18	6.3	26	48	13.0	15.4	2.9	10
0.90	19.0	8.3	12.6	367	17	6.9	24	43	14.1	16.3	3.2	9
0.95	18.7	8.3	12.3	307	17	7.1	23	41	14.6	16.7	3.4	9
1.00	18.5	8.4	12.0	355	16	7.2	22	39	19.9	17.2	3.6	9
1.50	17.9	9.4	17.6	315	14	11.2	15	27	31.0	21.0	5.2	6
2.00	19.0	9.4	29.6	290	12	17.4	11	20	44.5	28.5	6.8	7
2.50	19.9	9.6	35.1	271	10	22.0	12	18	46.8	24.3	8.3	9
3.00	21.1	9.9	29.3	257	10	16.5	12	17	36.3	29.7	9.9	11
3.50	22.0	10.2	18.1	266	10	11.8	12	16	26.9	32.1	11.4	14
4.00	19.9	10.3	17.0	300	10	9.3	13	16	23.4	34.3	12.9	15
6.00	16.4	10.5	16.4	531	10	9.9	17	16	17.6	41.0	18.7	9
8.00	13.5	11.3	11.1	651	10	11.2	20	16	17.0	45.4	23.5	10
10.00	11.7	15.2	14.0	609	10	12.8	20	17	16.5	44.5	23.8	12
12.00	9.9	15.0	14.2	549	12	14.1	20	17	16.4	39.8	21.8	13
14.00	9.4	14.0	15.0	497	12	15.1	19	17	17.3	37.5	20.2	14
16.00	8.8	13.7	15.2	454	13	16.1	18	16	18.7	34.2	19.0	15
18.00	8.4	13.7	15.5	429	13	17.0	17	15	18.7	32.2	17.9	16
20.00	8.0	13.7	15.8	411	13	17.2	16	15	18.7	30.4	17.0	16
22.00	7.5	13.7	15.1	403	14	17.8	16	14	18.7	29.3	16.3	17
24.00	7.0	13.7	16.4	394	14	17.8	15	13	18.7	28.1	15.6	17
26.00	6.7	13.7	16.3	386	14	17.8	14	12	18.7	26.8	15.0	17
28.00	6.5	13.7	16.1	387	14	17.9	14	12	18.7	25.6	14.5	17
30.00	6.3	13.7	15.9	389	14	17.8	14	12	18.7	23.6	14.0	17
40.00	5.8	13.7	15.0	360	13	16.7	13	10	18.7	17.8	12.2	16
50.00	5.8	13.7	12.9	291	12	15.1	12	10	18.7	14.0	11.2	14

5.2.3 Ion-Neutral Collision

In principle, experimentally determined gas-kinetic data for ion-atom collision similar to the collision between the electrons and neutrals should be used. It has, however, been noted by Kollath [29], that the light ions moving in a light or heavy gas show only scattering (for example, Li^+ in He and in Hg-vapor) without any loss of energy. On the other hand, the heavier ions in heavy neutrals exchange the charge preferentially and in lighter gases there can be losses in energy. Thus various different collisional models are applicable, and although there is a large quantity of data for the ion-atom collision available to-day, it is still not sufficient to explain the entire phenomenon. For some noble and other gases the charge transfer cross-section for the ion-atom collision, Q_{tr}, is available (please see, Hasted [9]), which is related to the relative speed g by the relation

$$Q_{tr} = \frac{1}{2}(A - B \ln g)^2 \tag{5.38}$$

where A and B are constants. These constants are obtained either from the direct measurement of the charge transfer cross-section or the measurement of the mobility coefficient. The value of these constants are given in Table 5.9 for some of the gases. The collision cross-section $Q^{(1,s)}$, which is needed mainly for calculation of the diffusion coefficient, is then obtained from the relation

$$Q^{(1,s)} = (96B^2 - 19.62AB + A^2) + (9.8B^2 + B^2\beta/2 - AB)\ln(T_h/m_h)$$
$$+ B^2[\ln(T_h/m_h)]^2/4 + B^2(\beta^2 + 37.2\beta + 1.644 - \eta) - AB\beta \tag{5.39}$$

where

Table 5.9. Constants A and B for charge transfer cross-section determination

Gas	A	B
Cs	46.25	1.99
K	43.79	1.90
Hg	28.43	1.29
Xe	27.24	1.26
H	24.26	1.10
Kr	23.78	1.07
	26.10	1.13
Ar	22.24	1.01
	36.74	2.11
Ne	17.39	0.71
He	18.04	0.83
	18.68	2.11

$$\beta = \sum_{1}^{s+1} \left(\frac{1}{n}\right) - \gamma; \eta = \sum_{1}^{s+1} \frac{1}{n^2} \qquad (5.40)$$

$$(\gamma = \text{Euler's constant} = 0.577215665) .$$

Further, $Q^{(2,s)}$ is sometimes determined by taking an average of the two collision cross-sections $Q_1^{(2,s)}$ and $Q_2^{(2,s)}$, where Q_1 and Q_1 are determined separately for the *exponential repulsive potential* with potential constants given in Table 5.10 using the Monchik or Lennard-Jones tables described earlier for the atom-atom collision. Thus,

$$Q^{(2,s)} = \frac{1}{3}Q_1^{(2,s)} + \frac{2}{3}Q_2^{(2,s)} . \qquad (5.41)$$

Table 5.10. Exponential repulsive potential data for atom-ion collision for noble gases

Gas	ϕ_{o1}	ρ_{o1}	ϕ_{o2}	ρ_{o1}
Ar	900.0	0.431	4640.0	0.306
He	179.7	0.344	179.7	0.344
Ne	734.0	0.322	3460.0	0.277
Xe	3.87e5	0.202	3.87e5	0.202
Kr	3.11e6	0.208	3.11e6	0.208

However since the potential data for the exponential repulsive potential are known for a small number of pure gases and not for the collision between different gases, it is convenient to apply the Lennard-Jones potential for both the atom-atom and the atom-ion collisions. Results of computation for argon, as calculated for the different potentials (the Lennard-Jones potential, exponential repulsive potential, the weighted average exponential repulsive potential) and (5.39) have been plotted for argon in Fig. 5.4, in which the lines denoted by 1 to 3 are $Q^{(1,1)}$ and by 4 to 6 are $Q^{(2,2)}$. The line 1 is the charge transfer cross-section, calculated by (5.39) by taking $A = 36.74$ and $B = 2.109$, line 2 is the cross-section calculated using the exponential repulsive potential, (5.34), by taking $\phi_o = 7100$ and $r_o = 0.258$ Å, and line 3 is from the Lennard-Jones potential with $\epsilon/k_B = 119.5$K and $d = 3.421$Å. Similarly, line 4 is for the cross-section by the weighted exponential repulsive potential, (5.41), taking $\phi_{o1} = 900$, $\rho_{o1} = 0.258$, $\phi_{o2} = 4640$ and $\rho_{o2} = 0.306$, line 5 is from the exponential repulsive potential and line 6 from the Lennard-Jones potential. It can be seen that for both $Q^{(1,1)}$ and $Q^{(2,2)}$, the cross-section calculated by the exponential repulsive potential and the Lennard-Jones 6-12 potential are in fair agreement with each other, and even $Q^{(2,2)}$ calculated by the weighted exponential repulsive potential is little different. The only large discrepancy is in the $Q^{(1,1)}$ values calculated for the

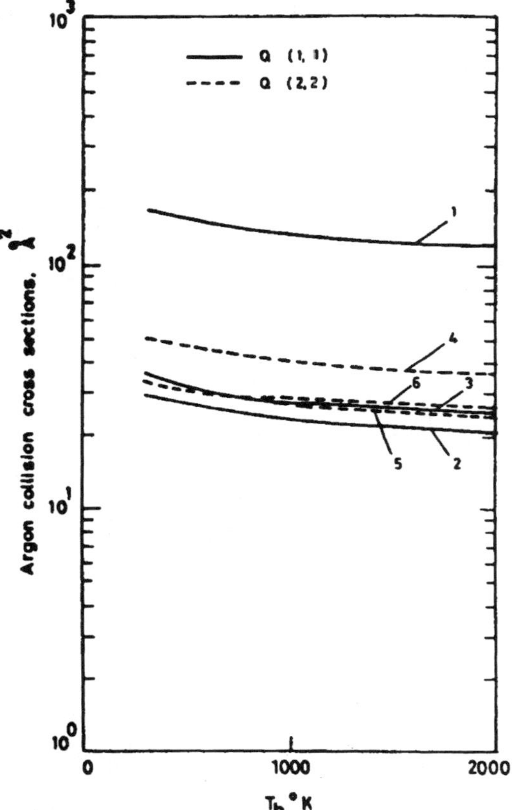

Fig. 5.4. Comparison of collision cross-section calculations by different methods for argon (for explanation of numbers, see text)

charge transfer cross-sections, which are needed for the calculation of the diffusion coefficients and related properties. For further calculations, therefore, the atom-atom collision cross-section data determined with the help of the Lennard-Jones potential can be used for atom-ion collisions also.

5.2.4 Charged Particle Collision

The collision cross-section between the charged particles (electrons, ions) is determined from the shielded Coulomb potential. For this purpose a dimensionless temperature

$$T^* = \left(\frac{4\pi\epsilon_o k_B \lambda_{De} T_{\mathrm{ref}}}{e^2 \mid N_{cj} N_{ck} \mid} \right) \tag{5.42}$$

is defined in which N_{cj} and N_{ck} are the charge of the j-th and k-th specie, respectively (for example the charge number for electrons is -1, for singly-charged ion, it is $+1$, etc.), T_{ref} is the reference temperature ($= T_e$, the *electron temperature*, if the electrons are one of the colliding partners, and $=$ T_h, *heavy particles translational temperature*, for the ion-ion collision) and λ_{De} is the *Debye shielding distance* given by the equation ($\theta = T_e/T_h$)

$$\lambda_{De} = \left(\frac{2\epsilon_o k_B T_e}{e^2 \left[n_e + \theta \sum_h N_{cj} n_j \right]} \right)^{1/2} \tag{5.43}$$

which has been described in detail in a later chapter. Separate tables for the attracting and the repulsive *Coulomb potential* can now be used for non-dimensional collisional integrals $Q^{*(l,s)}$, from which the collision cross-section is obtained from the relation

$$Q^{(l,s)} = \pi \lambda_{De}^2 Q^{*(l,s)} . \tag{5.44}$$

For this purpose the following values are obtained as a function of T^* (Table 5.11 and 5.12):

$$A = T^{*2} Q^{*(1,1)}, B = T^{*2} Q^{*(2,2)}, C = Q^{*(2,2)}/Q^{*(1,1)}$$

Table 5.11. Collision cross-section data for Coulomb potential (attracting)

T^*	A	B	C	D	E	F	G	H	I	J	K
1.E-01	0.06	0.04	0.61	1.47	0.76	0.84	0.68	0.86	0.03	0.02	0.03
2.E-01	0.14	0.10	0.71	1.46	0.66	0.81	0.60	0.80	0.05	0.04	0.07
6.E-01	0.33	0.30	0.92	1.37	0.55	0.71	0.59	0.68	0.08	0.06	0.16
8.E-01	0.40	0.38	0.96	1.34	0.54	0.69	0.60	0.66	0.09	0.07	0.19
1.E+00	0.45	0.45	0.99	1.33	0.52	0.67	0.60	0.64	0.10	0.07	0.22
2.E+00	0.65	0.68	1.06	1.28	0.49	0.63	0.61	0.62	0.13	0.09	0.30
3.E+00	0.77	0.84	1.08	1.26	0.48	0.62	0.62	0.60	0.14	0.10	0.35
4.E+00	0.87	0.95	1.09	1.24	0.47	0.61	0.62	0.60	0.16	0.11	0.39
6.E+00	1.02	1.12	1.10	1.23	0.45	0.60	0.62	0.59	0.17	0.12	0.45
8.E+00	1.13	1.25	1.11	1.21	0.45	0.59	0.61	0.58	0.19	0.13	0.49
1.E+01	1.21	1.35	1.11	1.20	0.44	0.58	0.61	0.58	0.20	0.14	0.52
2.E+01	1.50	1.66	1.11	1.18	0.43	0.57	0.60	0.57	0.23	0.16	0.62
3.E+01	1.67	1.85	1.11	1.17	0.42	0.56	0.60	0.56	0.25	0.17	0.67
4.E+01	1.80	1.99	1.11	1.16	0.42	0.56	0.59	0.56	0.26	0.18	0.71
6.E+01	1.98	2.18	1.10	1.15	0.41	0.55	0.59	0.55	0.28	0.20	0.77
8.E+01	2.11	2.32	1.10	1.14	0.41	0.55	0.59	0.55	0.30	0.21	0.82
1.E+02	2.22	2.43	1.10	1.14	0.40	0.55	0.58	0.55	0.31	0.21	0.85
2.E+02	2.54	2.77	1.09	1.12	0.40	0.54	0.58	0.54	0.34	0.24	0.96
3.E+02	2.74	2.97	1.08	1.12	0.39	0.54	0.57	0.54	0.36	0.25	1.02
4.E+02	2.88	3.11	1.08	1.11	0.39	0.54	0.57	0.54	0.38	0.26	1.06
6.E+02	3.08	3.31	1.08	1.10	0.39	0.54	0.57	0.54	0.40	0.27	1.12
8.E+02	3.22	3.46	1.07	1.10	0.38	0.54	0.56	0.54	0.41	0.28	1.16
1.E+03	3.33	3.57	1.07	1.10	0.38	0.54	0.56	0.54	0.42	0.29	1.20
1.E+04	4.48	4.72	1.05	1.07	0.37	0.53	0.55	0.53	0.59	0.37	1.54

Table 5.12. Collision cross-section data for Coulomb potential (repulsive)

T^*	A	B	C	D	E	F	G	H	I	J	K
1.E-01	0.02	0.03	1.36	1.36	0.75	0.81	0.97	0.84	0.01	0.01	0.02
2.E-01	0.05	0.07	1.36	1.39	0.71	0.78	0.95	0.81	0.02	0.02	0.04
3.E-01	0.08	0.11	1.36	1.39	0.69	0.76	0.93	0.78	0.03	0.03	0.07
4.E-01	0.11	0.15	1.36	1.40	0.67	0.75	0.91	0.77	0.04	0.03	0.09
6.E-01	0.16	0.21	1.35	1.40	0.64	0.73	0.88	0.74	0.05	0.04	0.12
8.E-01	0.20	0.28	1.35	1.39	0.62	0.71	0.87	0.73	0.06	0.05	0.15
1.E+00	0.25	0.33	1.34	1.39	0.61	0.70	0.85	0.71	0.07	0.06	0.17
2.E+00	0.42	0.55	1.31	1.36	0.56	0.66	0.80	0.67	0.11	0.08	0.26
3.E+00	0.54	0.70	1.29	1.34	0.54	0.64	0.77	0.65	0.13	0.09	0.32
4.E+00	0.65	0.82	1.27	1.33	0.52	0.63	0.75	0.64	0.14	0.10	0.36
6.E+00	0.80	1.00	1.25	1.30	0.50	0.62	0.72	0.62	0.16	0.12	0.42
8.E+00	0.92	1.14	1.23	1.28	0.49	0.61	0.70	0.61	0.18	0.13	0.46
1.E+01	1.02	1.24	1.22	1.27	0.48	0.60	0.69	0.60	0.19	0.13	0.50
2.E+01	1.34	1.59	1.18	1.23	0.45	0.58	0.66	0.58	0.23	0.16	0.60
3.E+01	1.54	1.80	1.17	1.21	0.44	0.57	0.64	0.57	0.25	0.17	0.66
4.E+01	1.68	1.94	1.15	1.19	0.43	0.57	0.63	0.57	0.26	0.18	0.71
6.E+01	1.89	2.15	1.14	1.18	0.42	0.56	0.62	0.56	0.28	0.20	0.77
8.E+01	2.04	2.30	1.13	1.17	0.42	0.56	0.61	0.56	0.30	0.21	0.81
1.E+02	2.15	2.42	1.12	1.16	0.41	0.55	0.60	0.55	0.31	0.21	0.85
2.E+02	2.51	2.77	1.10	1.14	0.40	0.55	0.59	0.55	0.34	0.24	0.95
3.E+02	2.72	2.97	1.09	1.13	0.40	0.54	0.58	0.54	0.36	0.25	1.01
4.E+02	2.86	3.12	1.09	1.12	0.39	0.54	0.57	0.54	0.38	0.26	1.06
6.E+02	3.07	3.32	1.08	1.11	0.39	0.54	0.57	0.54	0.40	0.27	1.12
8.E+02	3.21	3.46	1.08	1.10	0.38	0.54	0.56	0.54	0.41	0.28	1.16
1.E+03	3.33	3.57	1.07	1.10	0.38	0.53	0.56	0.53	0.42	0.29	1.20
1.E+04	4.48	4.72	1.05	1.07	0.37	0.53	0.55	0.53	0.59	0.37	1.54

$$D = [5Q^{*(1,2)} - 4Q^{*(1,3)}]/Q^{*(1,1)}$$

$$E = Q^{*(1,2)}/Q^{*(1,1)}, F = Q^{*(2,3)}/Q^{*(2,2)}, G = Q^{*(3,3)}/Q^{*(1,1)}$$

$$H = Q^{*(4,4)}/Q^{*(2,2)}, I = T^{*2}Q^{*(1,4)}, J = T^{*2}Q^{*(1,5)}, K = T^{*2}Q^{*(2,4)} \ .$$

For additional cross-sections $Q^{(2,5)}$, $Q^{(2,6)}$, $Q^{(1,6)}$ and $Q^{(1,7)}$, the following equations for calculation of $Q^{*(l,s)}$ are to be used:

$$Q^{*(1,s)} = \frac{16(s-1)!}{(s+1)!} \frac{1}{\Lambda^2} \left[\ln \Lambda - 1.65443133 + \sum_{n=1}^{s-1} \frac{1}{n} \right]$$

$$Q^{*(2,s)} = \frac{48(s-1)!}{(s+1)!} \frac{1}{\Lambda^2} \left[\ln \Lambda - 2.15443133 + \sum_{n=1}^{s-1} \frac{1}{n} \right] \ . \tag{5.45}$$

5.3 Collision Frequency, Mean Free Path

For the present, the following simple assumptions are made: (1) there is a mixture of different species j, each having a translational temperature T_j; (2) the particles are rigid, non-attracting spheres with diameter d_j; (3) all particles of one species travel with the same speed in a reasonable choice of the mean kinetic speed

$$v_j = \left[\frac{8k_B T_j}{\pi M_j} \right]^{1/2} \tag{5.46}$$

where M_j is the mass of a single species with diameter d_j; and (4) all particles travel in a direction parallel to one of the coordinate axes, that is one-sixth of them travel at any given instant in the $+x$-direction, and so on.

Let us now consider a single particle of the j-th species moving in the $+x$-direction, and while doing so, it collides with the particles of the k-th species moving in one of the six coordinate directions. In actual practice the j-th species particle should move in a zigzag fashion between collisions, but it is now idealized by assuming that the zigzag path is lined up in the $+x$ direction. The relative velocities g_{jk} between the two colliding particles j and k are: $|\, v_j - v_k \,|$ if the particles are moving in the same direction, $|\, v_j + v_k \,|$, if they are moving in opposite directions, and $\sqrt{v_j^2 + v_k^2}$, if they approach each other from any of the four perpendicular directions. As in the previous section, we now consider all particles of the k-th species made motionless and the single j-th species approaching them with a relative speed g_{jk}. In time Δt, the moving particle covers a distance $g_{jk}\Delta t$. If there are particles of the k-th species in the volume element $Q_{jk}g_{jk}\Delta t$, where the collision cross-section $Q_{jk} = \pi(d_j^2 + d_k^2)/4$, then there will be collision. If there are n_k molecules of the k-th species in unit volume, then there are $(n_k Q_{jk}g_{jk}\Delta t)$ particles, which will collide.

Thus the collisional frequency is $(n_k Q_{jk}g_{jk})$. Taking collisions between particles moving in all six coordinate directions, the collision frequency is

$$\Gamma_{jk} = \frac{1}{6}n_k Q_{jk} \sum_6 g_{jk} = \xi_{jk} n_k Q_{jk}, \ \mathrm{s}^{-1} \tag{5.47}$$

where n_k = number density of the k-th species $[\mathrm{m}^{-3}]$, $Q_{jk} = \pi(d_j^2 + d_k^2)/4$ = collision cross-section $[\mathrm{m}^2]$ and

$$\xi_{jk} = \frac{1}{6}\left[|\, v_j + v_k \,| + |\, v_j - v_k \,| + \frac{2}{3}\sqrt{v_j^2 + v_k^2} \right] \tag{5.48}$$

is the average relative speed for a binary collision. If a particle collides with other particles of the same species (j = k), then

$$\xi_{jj} = v_j \left(\frac{1}{3} + \frac{2\sqrt{2}}{3} \right) = 1.279 v_j \ . \tag{5.49}$$

On the other hand, if $v_j \gg v_k$, for example, for collisions between electrons and heavy particles h, then $\xi_{eh} = v_e$.

Thus the total number of collisions per unit volume and time between all particles of the j-th species with all particles of the k-th species is

$$\Gamma'_{jk} = \frac{1}{2} n_j \Gamma_{jk} = \frac{1}{2} \xi_{jk} n_j n_k Q_{jk} \ . \tag{5.50}$$

The factor $(1/2)$ is put to take care of duplication of collisions between the same two particles. Since $\xi_{jk} = \xi_{kj}$, $Q_{jk} = Q_{kj}$, it is evident that. $\Gamma'_{jk} = \Gamma'_{kj}$. The mean free path of the species j for collision with species k is

$$\lambda_{jk} = v_j / \Gamma_{jk} = \frac{v_j}{\xi_{jk} n_k Q_{jk}} \sim \frac{1}{n_k Q_{jk}} \ . \tag{5.51}$$

It may be noted that λ_{jk} may not be the same as λ_{kj}. For example, in a quasi-neutral gas mixture of electrons e, singly charged ions i and atoms a, the number density of electrons $n_e =$ the number density of ions n_i, and $\xi_{ie} = \xi_{ei} = v_e$. Thus, $\lambda_{ie} = v_i/(v_i n_e Q_{ei}) = (v_i/v_e)\lambda_{ei}$, but $\lambda_{ie} = (n_e Q_{ei})^{-1} = \lambda_{ei}\sqrt{(T_i M_e)/(T_e M_i)}$. Similarly, $\lambda_{ea} = (n_a Q_{ea})^{-1}$, but

$$\lambda_{ae} = v_a/(v_e n_e Q_{ea}) = (n_a/n_e)\lambda_{ea}\sqrt{(T_h M_e)/(T_e M_h)} \ . \tag{5.52}$$

For determination of the mean free path of a single particle of j-th species in a gas mixture, one can find the following expression

$$\lambda_j = \frac{v_j}{\sum_k (\xi_{jk} n_k Q_{jk})} = \frac{v_j}{\sum_k \Gamma_{jk}} \sim \frac{1}{\sum_k (n_k Q_{jk})} \ . \tag{5.53}$$

The above expressions, derived under the assumption of very simple motions in directions parallel to one of the directions of the Cartesian coordinates, will now be derived from more rigorous analysis. From the kinetic theory of gases, the fraction of the number density of the j-th particle, dn_j/n_j in the velocity space between v_j and $v_j + dv_j$ is given by the distribution function given by (3.148),

$$f(v) = 4\pi v^2 \left(\frac{M}{2\pi k_B T} \right)^{3/2} \exp^{-Mv^2/(2k_B T)} \, dv \tag{5.54}$$

and the corresponding *volumetric collision frequency* is given by the relation (see, *Chapman* and *Cowling* [5])

$$\Gamma'_{jk} = Q_{jk} \int\!\!\int f_j f_k g^3 G^2 \mathrm{d}G \mathrm{d}g$$

$$= \frac{n_j n_k (M_j M_k)^{3/2} Q_{jk}}{\pi k_B^3 (T_j T_k)^{3/2}} \int_0^\infty \int_0^\infty \exp^{-\frac{1}{2k_B}\left(\frac{M_j v_j^2}{T_j} + \frac{M_k v_k^2}{T_k} \right)} g^3 G^2 \mathrm{d}G \mathrm{d}g \ . \tag{5.55}$$

Now,

$$\frac{1}{2k_B}\left(\frac{M_j v_j^2}{T_j} + \frac{M_k v_k^2}{T_k}\right)$$

$$= \frac{1}{2k_B}\left[\frac{M_j}{T_j}\left(G - \frac{M_k g}{M_j + M_k}\right)^2 + \frac{M_k}{T_k}\left(G + \frac{M_j g}{M_j + M_k}\right)^2\right]$$

$$\approx \frac{1}{2k_B}\left[G^2\left(\frac{M_j}{T_j} + \frac{M_k}{T_k}\right) + \frac{M_j M_k g^2}{(M_j + M_k)^2}\left(\frac{M_j}{T_j} + \frac{M_k}{T_k}\right)\right]$$

and thus,

$$\Gamma'_{jk} = \frac{n_j n_k (M_j M_k)^{3/2} Q_{jk}}{\pi k_B^3 (T_j T_k)^{3/2}}$$

$$\times \int_0^\infty \int_0^\infty \exp^{-\frac{1}{2k_B}\left[G^2\left(\frac{M_j}{T_j} + \frac{M_k}{T_k}\right) + \frac{M_j M_k g^2}{(M_j + M_k)^2}\left(\frac{M_j}{T_j} + \frac{M_k}{T_k}\right)\right]} g^3 G^2 \mathrm{d}G \mathrm{d}g \ .$$

Noting that

$$\int_0^\infty \exp^{-\frac{G^2}{2k_B}\left(\frac{M_j}{T_j} + \frac{M_k}{T_k}\right)} G^2 \mathrm{d}G = \frac{\sqrt{\pi k_B^3/2}}{\left(\dfrac{M_j}{T_j} + \dfrac{M_k}{T_k}\right)^{3/2}} \tag{5.56}$$

and

$$\int_0^\infty \exp^{-\frac{M_j M_k g^2}{2k_B(M_j + M_k)^2}\left(\frac{M_j}{T_j} + \frac{M_k}{T_k}\right)} g^3 \mathrm{d}g = \frac{2 k_B^2 (M_j + M_k)^4}{(M_j M_k)^2 \left(\dfrac{M_j}{T_k} + \dfrac{M_k}{T_j}\right)^2} \tag{5.57}$$

we get further the relation

$$\Gamma'_{jk} = \left(\frac{2k_B}{\pi}\right)^{1/2} \frac{Q_{jk} n_j n_k}{(T_j T_k)^{3/2}(M_j M_k)^{1/2}} \frac{(M_j + M_k)^4}{\left(\dfrac{M_j}{T_j} + \dfrac{M_k}{T_k}\right)^{3/2}\left(\dfrac{M_j}{T_k} + \dfrac{M_k}{T_j}\right)^2} \ . \tag{5.58}$$

Equating (5.58) with (5.50), we get

$$\xi_{jk} = \left(\frac{8k_B}{\pi}\right)^{1/2} \frac{(M_j + M_k)^4}{(T_j T_k)^{3/2}(M_j M_k)^{1/2}\left(\dfrac{M_j}{T_j} + \dfrac{M_k}{T_k}\right)^{3/2}\left(\dfrac{M_j}{T_k} + \dfrac{M_k}{T_j}\right)^2} \ . \tag{5.59}$$

Now two special cases are considered. As the first case we consider $T_j = T_k = T$ and in the second case we consider as the j-th species the electrons and as the k-th species the heavy particles ($M_e \ll M_k$). For the first case we get

$$\xi_{jk} = \left[\frac{8k_BT}{\pi}\left(\frac{M_j + M_k}{M_j M_k}\right)\right]^{1/2} \tag{5.60}$$

which for the further special case of $M_j = M_k$ becomes $\xi_{jj} = v_j\sqrt{2} = 1.414 v_j$ instead of (5.49). For the second case, we get $\xi_{eh} = \sqrt{8k_BT_e/(\pi M_e)} = \bar{v}_e$, the mean kinetic speed of the electrons.Further discussion about the collision frequency and the mean free path remain as before.

5.4 Reaction Rates and Vibrational and Temperature Nonequilibrium

The reaction rate (the number of particles being newly created or destroyed) has to depend on the number of collisions per unit time, as well as the value of the total kinetic energy of the two particles concerned. The fact that the rate at which molecules react is very much less than the rate at which they collide indicates that additional conditions are required for reaction beyond mere collision, especially that a collision that results in reaction must involve considerably more energy than is available in just one collision. In other words the total kinetic energy of the colliding particles must be larger than a minimum energy E_{ac}, the so-called *activation energy*, for the collision to result in the reaction. From the gas kinetic considerations it can be shown that the fraction of collisions that result in a reaction is of the order of

$$\exp^{-E_{ac}/(k_BT)} \, . \tag{5.61}$$

In addition to this, there is a factor which takes care of the relative geometrical positions during the collision, and is called the *steric factor* ϕ, whose determination poses an insoluble problem in quantum theory and must be found empirically. Thus we introduce this factor to the theoretically determined reaction rate, and write

$$\text{rate of reacting collisions} = \text{rate of collision}.\phi.\exp^{-E_{ac}/(k_BT)} \, . \tag{5.62}$$

For a reaction between the particles A and B in a bimolecular collision to form the particle AB, that is for the reaction of the type A + B $\rightarrow$ AB, the reaction rate of production of AB is given by the relation

$$\frac{dn_{AB}}{dt} = -\frac{dn_A}{dt} = -\frac{dn_B}{dt} = \phi\Gamma'_{AB}\exp^{-E_{ac}/(k_BT)} = kn_An_B \tag{5.63}$$

with the help of (5.50), we get the relation for the *reaction rate constant*

$$k = \phi\xi_{jk}Q_{jk}\exp^{-E_{ac}/(k_BT)} = CT^n\exp^{-E_{ac}/(k_BT)} \, . \tag{5.64}$$

For n_j, n_k in (5.63) we have considered the number density in unit m^{-3}, but the chemical engineers prefer use of concentration in the unit $\mathrm{kmol.m}^{-3}$; this is easily obtained by dividing the number density by the Avogadro number. Thus k is given in $\mathrm{m}^3\mathrm{kmol}^{-1}.\mathrm{s}^{-1}$ and the forward reaction rate has the unit $\mathrm{kmolm}^{-3}.\mathrm{s}^{-1}$. Accordingly C has the unit $\mathrm{m}^3\mathrm{kmol}^{-1}.\mathrm{s}^{-1}.\mathrm{K}^{-n}$. On the other hand for a three body reaction of type $\mathrm{A+B + M \to AB + M}$, the forward reaction rate has the same unit $\mathrm{kmol.m}^{-3}.\mathrm{s}^{-1}$, but then k is given in $\mathrm{m}^6.kmol^{-2}.\mathrm{s}{-}1$ and C has the unit $\mathrm{m}^6\mathrm{kmol}^{-2}.\mathrm{s}^{-1}.\mathrm{K}^{-n}$.

Noting from (5.48), that for binary collisions $\xi_{jk} \propto \sqrt{T}$ and hence, the exponent n in (5.64) is of the order of $(1/2)$. However, Q_{jk} may also be dependent on T and hence, n can be different from $(1/2)$. Equation (5.64) is commonly known as the *Arrhenius equation* (after the swedish physicist-chemist *Arrhenius*, who in 1889 first gave it a theoretical interpretation). Sample values of C, n and (E_{ac}/R^*) for different reacting gases have been given in Table 5.13; the resultant unit of k are expressed in cm^3, mole and seconds in combination appropriate for the given chemical equation.

A similar reaction is $\mathrm{AB + M \leftrightarrow A + B + M}$, in which the arrows in both directions mean that reactions in both directions are possible. While the forward reaction

$$\mathrm{AB + M} \overset{k_f}{\to} \mathrm{A + B + M} \qquad (5.65)$$

can be treated as done earlier, the reverse reaction rate

$$\mathrm{A + B + M} \overset{k_r}{\to} \mathrm{AB + M} \qquad (5.66)$$

is possible as a three-body recombination in which the third partner M must carry away the recombination energy between A and B. In case the reaction energy is not carried away by the third partner, and also if this excess energy after recombination is not given up as a radiative energy, both particles A and B must separate again. While k_f and k_r are the reaction rate constants of forward and backward reactions respectively, the overall rate is given by the expression

$$\frac{\mathrm{d}n_A}{\mathrm{d}t} = \frac{\mathrm{d}n_B}{\mathrm{d}t} = -\frac{\mathrm{d}n_{AB}}{\mathrm{d}t} = k_f n_{AB} n_M - k_r n_A n_B n_M \ . \qquad (5.67)$$

Now the reaction rate k can be written separately for the forward and backward reaction, for which the value of C and E_{ac} are considered separately for forward and backward reaction. At equilibrium, the reaction rate is zero, and as such the ratio

$$\frac{k_f}{k_r} = \frac{n_A n_B}{n_{AB}} = \frac{C_f}{C_r} \exp^{-(E_{ac,f}-E_{ac,r})/(R^*T)} = C \exp^{-\Delta H/(R^*T)} \qquad (5.68)$$

is equal to the *equilibrium constant* K_n if the number density is written as the number of particles per unit volume and all other variables are in compatible units. If, however, n denotes concentration in $\mathrm{kmol.m}^{-3}$, we write the

Table 5.13. Constants for typical reactions in air ([1])

Reaction	C	n	$Eac/k_B(K)$
$H_2 + O_2 = OH + OH$	1.70e13	0	24169
$OH+H_2 = H_2O + H$	2.20e13	0	2593
$H+O_2 = OH +O$	2.20e14	0	8459
$O + H_2 = OH +H$	1.80e10	1	4481
$OH +OH = H_2O + O$	6.30e12	0	549
$H + OH + M = H_2O + M$	2.20e22	-2	0
$H + O + M = OH + M$	6.00e16	-0.6	0
$H + H = H_2 + M$	6.40e17	-1	0
$H + O_2 + M = HO_2 + M$	1.70e15	0	-593
$HO_2 + H = H_2 + O_2$	1.30e13	0	0
$HO_2 + H = OH + OH$	1.40e14	0	544
$HO_2 + O = OH + O_2$	1.50e13	0	478
$HO_2 + OH = H_2O + O_2$	8.00e12	0	0
$HO_2 + HO_2 = H_2O_2 + O_2$	2.00e12	0	0
$H + H_2O_2 = H_2 + HO_2$	1.40e12	0	1813
$O + H_2O_2 = OH +HO_2$	1.40e13	0	3222
$OH + H_2O_2 = H_2O + HO_2$	6.10e12	0	720
$M + H_2O_2 = 2OH + M$	1.20e17	0	22910
$O + O + M = O_2 + M$	6.00e13	0	-593
$N + N + M = N_2 + M$	2.80e17	-0.75	0
$N + O_2 = NO + O$	6.40e9	1	3172
$N + NO = N_2 + O$	1.60e13	0	0
$N + OH = NO + H$	6.30e11	0.5	0
$H + NO + M = HNO + M$	5.40e15	0	-302
$H + HNO = NO + H_2$	4.80e12	0	0
$O + HNO = NO + OH$	5.00e11	0.5	0
$OH + HNO = NO + H_2O$	3.60e13	0	0
$HO_2 + HNO = NO + H_2O_2$	2.00e12	0	0
$HO_2 + NO = NO_2 + OH$	3.43e12	0	-131
$H + NO_2 = NO + OH$	3.50e14	0	755
$O + NO_2 = NO + O_2$	1.00e13	0	302
$NO_2 + M = NO + O + M$	1.16e16	0	33232

equilibrium constant as K_c. In the above equation ΔH is the heat release due to reaction. Thus knowing the value of K_n or K_c, and one of the reaction rate constants, it is possible to have the other reaction rate constant in compatible units.

A similar situation arises for the ionizing and recombining reactions for electrons. According to *Igra* and *Barcessat* [70], the following reactions may be considered:

$$\text{collisional:} \quad A^* + e \rightarrow A^+ + e + e \tag{5.69}$$

$$A + e \rightarrow A^* + e \tag{5.70}$$

$$\text{radiative:} \quad A^* \to A + h\nu \tag{5.71}$$

$$A^+ + e \to A + h\nu \ . \tag{5.72}$$

Herein A denotes an atom, A^* is an excited atom and A^+ denotes a charged ion. It is agreed, that for plasmas having densities up to $n_e = 10^{18}$ cm^{-3} and temperatures around 1 eV, the radiative processes can be ignored. Further it is conjectured, that the most probable recombination is the one in which a free electron is recaptured into one of the atomic excited states. Therefore, we can limit our interest to the reaction (5.69) only, for which the forward reaction rate constant is k_f and the reverse reaction rate constant is k_r. These reactions require an additional collision partner, so that during recombination the additional collision partner can take away the excess energy, else otherwise they will separate again. This additional collisional partner in reaction (5.69) is the electron. The rate of reaction of the electron number density is therefore,

$$\frac{\mathrm{d}n_e}{\mathrm{d}t} = k_f n_e n_a - k_r n_e^2 n_i = k_f n^2 x_e \left(x_a - x_e^2 n \frac{k_r}{k_f} \right) \tag{5.73}$$

where x is the *mole fraction* of the particular species. At equilibrium, denoted by superscript ($*$), the terms within the parentheses are equal to zero. In addition under quasi-neutrality condition, $n_i = n_e$ and we get the recombination rate constant in terms of the reverse reaction rate constant as follows:

$$k_r = (x_a^*/x_e^{*2})k_f/n^* \ . \tag{5.74}$$

Thus the rate of production of electron number density is given by the relation

$$\dot{n}_e = k_r n^3 x_e^3 \left[\left(\frac{x_e^*}{x_e} \right)^2 \left(\frac{x_a}{x_a^*} \right) \frac{n^*}{n} - 1 \right] \ . \tag{5.75}$$

For the general case of the electron temperature T_e larger than the heavy particle temperature T_h (temperature ratio $\theta = T_e/T_h > 1$) it is shown later in Chap. 6 (Sect. 6.7), that the relation

$$\frac{n}{n^*} = \frac{1 + x_e^*(\theta - 1)}{1 + x_e(\theta - 1)} \tag{5.76}$$

is valid. Under this condition we get

$$\dot{n}_e = k_r n^3 x_e^3 \left[\left(\frac{x_e^*}{x_e} \right)^2 \left(\frac{x_a}{x_a^*} \right) \left(\frac{1 + x_e(\theta - 1)}{1 + x_e^*(\theta - 1)} \right) - 1 \right] \ . \tag{5.77}$$

For the limiting case, $x_a = x_a^* \to 1$, $n \to n^*$, and we get

$$\dot{n}_e = k_r n_e(n_e^{*2} - n_e^2) \tag{5.78}$$

where k_r for a caesium seeded argon plasma is given by the relation

$$k_r = 1.0744 \times 10^{-21} T_e^{-4.5} , \text{ m}^6\text{s}^{-1} . \tag{5.79}$$

For the three-body recombination of potassium plasma the recombination rate constant is given by

$$k_r = 3.47 \times 10^{-20} T_e^{-4.765} , \text{ m}^6\text{s}^{-1} . \tag{5.80}$$

For argon plasma, *Igra* and *Barcessat* [70] have considered further the values given by different authors and decided to take

$$k_r = 6.2 \times 10^{-21} T_e^{-4.765} , \text{ m}^6\text{s}^{-1} \tag{5.81}$$

and *Hasted* [9] gave recombination coefficients for many gases.

We now consider the case of a vibrational non-equilibrium due to chemical processes. As a typical example, one can think of the flow behind shock in hypersonic flying bodies, or large expansion in a nozzle. For a system of harmonic oscillators which have permissible vibration energy levels

$$E_v = (v + 1/2)h\nu \tag{5.82}$$

where v is the vibrational quantum number, we begin by establishing the differential equation that governs the rate of change of the non-equilibrium values of the number of particles N_v as a function of time, for which it is assumed, consistent with detailed quantum mechanical studies of the transition probabilities, that the changes in energy of oscillators upon collision takes place only between adjacent energy levels, that is, $\Delta v = \pm 1$. Let $k_{v,v+1}$ be the rate constant for transition from the energy level v to $v + 1$, then the rate of change of the number of oscillations in any given energy level v is

$$\frac{\mathrm{d}N_v}{\mathrm{d}t} = -k_{v,v+1}N_v + k_{v+1,v}N_{v+1} - k_{v,v-1}N_v + k_{v-1,v}N_{v-1} . \tag{5.83}$$

For the special circumstances in which there is an equilibrium, it is obvious that $\mathrm{d}N_v/\mathrm{d}t = 0$. Further, assuming that the net interchange between any two adjacent states under equilibrium conditions is zero, one can write for the equilibrium case (denoted by the superscript asterisk),

$$-k_{v,v-1}N_v^* + k_{v-1,v}N_{v-1}^* = 0$$
$$-k_{v,v+1}N_v^* + k_{v+1,v}N_{v+1}^* = 0 .$$

Here the k's have the same value as when the system is out of equilibrium. Noting that under the equilibrium condition

$$\frac{N_v^*}{N_{v-1}^*} = \frac{k_{v-1,v}}{k_{v,v-1}} = \frac{\exp^{-[(v+1/2)h\nu/(k_BT)]}}{\exp^{-[(v-1/2)h\nu/(k_BT)]}} = \exp^{-h\nu/(k_BT)} \tag{5.84}$$

and

$$\frac{N^*_{v+1}}{N^*_v} = \frac{k_{v,v+1}}{k_{v+1,v}} = \frac{\exp^{-[(v+3/2)h\nu/(k_B T)]}}{\exp^{-[(v+1/2)h\nu/(k_B T)]}} = \exp^{-h\nu/(k_B T)} \tag{5.85}$$

and since from quantum mechanical study of the transition probabilities it is known that

$$k_{v,v-1} = v k_{1,0} \tag{5.86}$$

the other reaction rate constants can now be determined from the following relations:

$$k_{v+1,v} = (v+1)k_{1,0}$$
$$k_{v-1,v} = v k_{1,0} \exp^{-h\nu/(k_B T)}$$
$$k_{v,v+1} = (v+1)k_{1,0} \exp^{-h\nu/(k_B T)} \ .$$

Note that we can write alternative of (5.86) also as

$$k_{v,v+1} = (v+1)k_{0,1} \ . \tag{5.87}$$

By substituting the foregoing expression into (5.83), and with $\Theta_v = h\nu/k_B$, then

$$\frac{dN_v}{dt} = k_{1,0}\left[\{-vN_v + (v+1)N_{v+1}\} + \exp^{-\Theta_v/T}\{-(v+1)N_v + vN_{v-1}\}\right] \ . \tag{5.88}$$

Noting that the total vibration energy

$$E_{vib} = \sum_0^\infty N_v E_v = h\nu \sum_0^\infty \left(v + \frac{1}{2}\right)N_v \approx h\nu \sum_0^\infty vN_v \tag{5.89}$$

we get

$$\frac{dE_{vib}}{dt} = h\nu \sum_{v=0}^\infty v\frac{dN_v}{dt}$$

$$= h\nu k_{1,0} \sum_{v=0}^\infty v\left[-vN_v + (v+1)N_{v+1}\right.$$

$$\left. + \exp^{-\Theta_v/T}(-(v+1)N_v + vN_{v-1})\right] \ . \tag{5.90}$$

In equation (5.89), the energy quantum $(h\nu)$ has the dimension J. Thus if N_v is the number of particles per kmole (or per m^3) of gas in the vibrational level v, then E_{vib} has the dimension J.kmole^{-1} (or Jm^{-3}). Further, the series in (5.89) converges if $v \to \infty$, $N_v \to 0$. Otherwise we have to consider the anharmonic oscillator model.

Now adding separately for each of the two brackets,

$$\sum_{v=0}^{\infty} v\left[-vN_v + (v+1)N_{v+1}\right] = -\sum_{v=0}^{\infty} vN_v \tag{5.91}$$

and

$$\sum_{v=0}^{\infty} \left(-(v+1)N_v + vN_{v-1}\right) = \sum_{v=0}^{\infty} (v+1)N_v \tag{5.92}$$

we get from (5.90)

$$\frac{\mathrm{d}E_{vib}}{\mathrm{d}t} = k_{1,0}h\nu \sum_{v=0}^{\infty} N_v \left[-v + (v+1)\exp^{-\Theta_v/T}\right]$$

$$= k_{1,0}h\nu \left[\exp^{-\Theta_v/T} \sum_{v=0}^{\infty} N_v - \sum_{v=0}^{\infty} vN_v \left(1 - \exp^{-\Theta_v/T}\right)\right]. \tag{5.93}$$

Now since, $N = \sum N_v$ and $E_{vib}/(\sum vN_v) = h\nu$, (5.93) further becomes

$$\frac{\mathrm{d}E_{vib}}{\mathrm{d}t} = k_{1,0}h\nu \left[N\exp^{-\Theta_v/T} - \left(1 - \exp^{-\Theta_v/T}\right)\frac{E_{vib}}{k_B\Theta_v}\right]. \tag{5.94}$$

In the present approximation, by replacing $(v + 1/2)$ by v, and considering at equilibrium, the expression for vibrational internal energy (3.120), as

$$E_{vib}^* = \frac{R^*\Theta_v}{\exp^{\Theta_v/T} - 1} \tag{5.95}$$

in which R is the gas constant. Noting that $h\nu = k_B\Theta_v$, and introducing the notation

$$\tau = \frac{1}{k_{1,0}\left(1 - \exp^{-\Theta_v/T}\right)} \tag{5.96}$$

which has the dimension of time, (5.93) becomes

$$\frac{\mathrm{d}E_{vib}}{\mathrm{d}t} = \frac{E_{vib}^* - E_{vib}}{\tau}. \tag{5.97}$$

In (5.97) E_{vib}^* is the internal energy of the gas under the equilibrium condition and E_{vib} is the same under vibrational non-equilibrium condition. For the former, one could estimate it from the equilibrium condition at the translation temperature. Therefore, (5.97) gives the rate at which the vibrational energy relaxes to the (equilibrium) translation energy. However, the equation does not give explicitly a relation linking the rate of collision between molecules due to (relative) translational velocities of the molecules.

Equation (5.97) can be integrated easily to get the result

$$\frac{E_{vib} - E_{vib}^*}{(E_{vib})_{t=0} - E_{vib}^*} = \exp^{-t/\tau}. \tag{5.98}$$

Thus, in case of any departure from the equilibrium, energy is quickly brought (relaxed) to the equilibrium again in an interval of time of the order of the *relaxation time* τ. The most commonly cited theory for τ is given by *Landau* and *Teller* who obtained the result as

$$\tau = \frac{K_1 T^{5/6} \exp^{(K_2/T)^{1/3}}}{p\left(1 - \exp^{-\Theta_v/T}\right)} \tag{5.99}$$

where p is in bar, and K_1 and K_2 are constants depending on the excited species and also the species with which it collides to relax, and their values are given in Table 5.14.

Table 5.14. Constants K_1 and K_2 for vibration-relaxation of selected molecules

Species	Heat-bath molecule	K_1 (atm-ms)	K_2 (K)	approx.temp. range(K)
O_2	O_2	5.42e-5	2.95e6	800–1300
O_2	Ar	3.58e-4	2.95e6	1300–4300
N_2	N_2	7.12e-3	1.91e6	800–6000
NO	NO	4.86e-3	1.37e5	1500–3000
NO	Ar	6.16e-1	1.37e5	1500–4600

At sufficiently low temperatures $(T < \Theta_v)$, the terms under the parenthesis in denominator of (5.99) can be approximated to 1, and one may write

$$\tau = \frac{K_1 T^{5/6} \exp^{(K_2/T)^{1/3}}}{p} \, . \tag{5.100}$$

Taking the data from Table 5.14 for O_2-O_2 collision one can show, for example, that at $p = 1$ atm and $T = 2000K$, $\tau = 4.9$ ms. From (5.86), (5.96) and 5.100), one can get the following expression for the reaction rate constant for transition from the state v to $(v-1)$:

$$k_{v,v-1} = \frac{pv}{K_1 T^{5/6} \exp^{(K_2/T)^{1/3}}} \tag{5.101}$$

which can be used to evaluate the reaction rates between neighboring vibrational energy levels. It can be shown that both $k_{v,v-1}$ and $k_{v-1,v}$ increase proportional to v.

In connection with the flow problems of ionized gases it is necessary to study the ionization and dissociation rates on one hand, and on the other knowledge in the distribution of non-equilibrium vibration energy distribution. According to *Landrum* and *Candler* [76], "when a gas is heated the nonequilibrium vibrational distribution of diatomic molecules is formed in three identifiable stages: first, there is introduction of vibrational quanta over the lower molecular levels by the heating, next the introduced quanta are redistributed by collisional processes , primarily vibration-vibration (V-V) exchanges up the *vibrational ladder* of the molecule, and finally, the vibrational quanta are dissipated through gas heating by vibration-translation (V-T) relaxation or in chemical reactions such as dissociation. The ladder-climbing process occurs over a finite time and therefore significant dissociation cannot occur until an adequate number of highly vibrationally excited molecules are present. The net effect of this vibrational exchange is a reduction of the molecular dissociation rate".

For determination of V-T relaxation time the rate constants are obtained from the Landau-Teller theory. For chemical reactions with diatomic molecules like nitrogen and oxygen in air it is generally assumed that the rotational temperature is the same as the translational temperature, T, while there can be a separate vibrational temperature, T_v. A reaction model due to *Park* [97, 98, 99] predicts that the rate coefficients for dissociation are dictated by a geometrically effective temperature $\sqrt{TT_v}$ for the forward rate of dissociation. In addition we should also mention about the dissociation rate reactions by *Adamovich*, et al. [32, 33], and also a computational method given by *Capitelli* et al. [52]. For this purpose the latter have written down elaborate expressions for the rate coefficients of V-T and V-V reactions. Concerning the dissociation-recombination reactions, a pseudo level $(v+1)$ located just above the last bound level (v) is considered, through which the dissociation-recombination reaction passes. The last vibrational level (at dissociation) for nitrogen is taken at 45 and for oxygen at 33.

Finally, for high temperature gas plasma, there is another kind of nonequilibrium occurring in an electromagnetic field. By considering the energy transfer due to collision between the electrons and the heavy particles (*Gnoffo*, et al. [66]; *Park* [99]) it has been shown, that loss or gain in the kinetic energy is given by (5.23). Now in the case that there is an electric current flow, energy in the form of Joule heating j^2/σ is introduced, where j is the current density (Am^{-2}) and σ is the *electrical conductivity* $(\mathrm{A(Vm)}^{-1})$. It will be shown later that the electrical conductivity is given by the relation

$$\sigma = \frac{e^2 n_e^2}{2 M_e \Gamma'_{eh}} \tag{5.102}$$

where Γ'_{eh} is the volumetric collision frequency between the electrons and the heavies, and thus the temperature difference between the electrons and the heavy particles in case of the current flow is given by the relation

$$(T_e - T_h) = \frac{j^2 M_h}{3 e^2 n_e^2 k_B} \tag{5.103}$$

Thus at a low pressure and high current density, there can be an appreciable difference in temperature between the electrons and the heavy particles.

5.5 Exercise

5.5.1 Compute collision frequency, volumetric collision frequency and mean free path of air ($Q_{\mathrm{rigid}} = 11.0$ Å^2) at $T = 288$K and as a function of pressure.

5.5.2 Electron mole fraction for argon at 1 bar and 13,000K is about 0.3. Compute the number density of the electrons, ions, neutrals and all particles. Taking the various collision cross-sections for argon from Fig. 5.4, compute the mean free path and collision frequency for different specie of argon.

5.5.3 Estimate the vibration relaxation time for gas molecules given in Table 5.14.

6 Equilibrium Composition of a Reacting Gas Mixture

In the previous chapter we have discussed the question of reaction rate, and how the reaction rates for the forward and backward reactions may give the equilibrium composition at a given temperature and pressure. The discussion does not include the *pseudo equilibrium*, as it may happen at low temperatures, when both the forward and backward reactions rates are infinitesimal small. As an example, hydrogen and oxygen can be mixed at moderate pressures and room temperature without any reaction taking place. However, if a spark is introcuced then there will be explosive reaction to reach the equilibrium. Although, in principle, mere knowledge of the values of reaction rate constants for forward and backward reactions may be sufficient for calculation of the equilibrium composition, in practice, however, they are generally not known, and, therefore, an alternative method is required. Let us consider the following basic reaction,

$$\sum \alpha_j A_j \leftrightarrow \sum \beta_j B_j \tag{6.1}$$

for example,

$$\alpha_1 H_2 + \alpha_2 O_2 \leftrightarrow \beta_1 H_2 + \beta_2 O_2 + \beta_3 OH + \beta_4 H_2O + \beta_5 H + \beta_6 O + \dots . \tag{6.2}$$

In general, this overall reaction can be broken down into several series and parallel part reactions, in which for every reaction only two or three collision partners may be involved at any time, but as a result of all these reactions the total gas mixture attains a final composition at a given temperature and pressure, which does not change with time any further. Thus we can formally consider the reaction of the above type, and designate the left hand components arbitrarily as *reacting components*, and the right hand side components, except those which are in the left hand side also, as reacted components. We would now investigate on the basis of a mathematical model, which is called as *Vant' Hoff model*, the condition at a given temperature and pressure, that the reacting and reacted components are at equilibrium. This is not the only model used for the calculation of the equilibrium composition, but it is quite transparent and easily understandable. As a consequence of the model it is shown that one could use free enthalpy of various reacting components to determine the equilibrium composition. Further, the methodology has been extended in two directions, namely to multi-temperature plasmas and how to calculate the temperature derivative of composition, which can be used further to determine the reactive heat conductivity coefficient for a reactive gas mixture. Both these have been introduced in this book for the first time to the knowledge of this author. It need be mentioned, however, that the physical definition of multi-temperature gas itself is for the non-equilibrium thermal state and, therefore, computing equilibrium composition for multi-temperature model may be somewhat questionable.

6.1 Vant' Hoff Model of Chemical Reaction

The Vant' Hoff model of reaction uses semi-permeable membranes around a reactor, in which each membrane can allow only one pure component. Although such semi-permeable membranes are few and rare, for the study of very slow reactions, in which at every stage the temperature and pressure can be kept constant (reversible reaction!), concept of such membranes is of very great importance.

For a reversible reaction the type of equipment shown in Fig. 6.1 is used as a model, in which there are containers for different pure reacting and reacted components. In each of these containers a constant pressure p is maintained. The reacting components are now expanded at constant temperature in each of the cylinders to partial pressure p_j of that particular component to be maintained in the reactor for a reversible reaction, and thus, in each of the cylinders some work is gained and there is an exchange of heat with the reservoir. The reacting components are brought to the reactor in the exact ratio of a stoichiometric reaction, so that the mole ratio of each of them

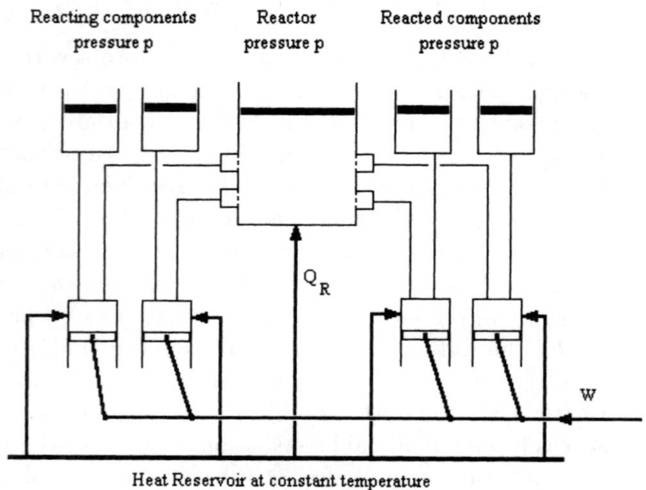

Fig. 6.1. Vant' Hoff model of reaction

corresponds to the ratio of the chemical valency between the components. Similarly the reacted components at partial pressure p_j are compressed in individual cylinders in the ratio of the chemical valency between the components to the container pressure p. To differentiate the chemical valency ν_j between the reacting and the reacted components, the former is given a negative sign and the latter a positive sign. After introduction of the components into the reservoir, they are allowed to react at a constant temperature T sufficiently slowly, so that the reaction can be assumed to be reversible, and a *heat of reaction* $Q_R = \Delta H = \sum \nu_j H_j$ (per unit mole of reacting components), which is negative in sign if heat is removed, is taken out from the reactor to the reservoir; H_j is the enthalpy of the components per mole. The total work gained from different cylinders (per mole) is then

$$W = R^* T \sum \nu_j \ln \frac{p_j}{p} = R^* T \sum \nu_j \ln x_j \ . \tag{6.3}$$

Herein R^* is the universal gas constant and $x_j = p_j/p$ is the *mole fraction*. Now for a reversible reaction, the total heat can be written in terms of the entropy change ΔS by the equation

$$Q = T \Delta S = W + Q_R = R^* T \sum \nu_j \ln x_j + \sum \nu_j H_j \ . \tag{6.4}$$

For a reversible reaction there is no change in the total entropy of the mixture, and one can write

$$\Delta S = \sum \nu_j S_j \tag{6.5}$$

and therefore,

$$W = R^* T \sum \nu_j \ln x_j = \sum \nu_j (T S_j - H_j) = \sum \nu_j G_j \tag{6.6}$$

where $G_j = T S_j - H_j$ is the *free-enthalpy* of the j-th component.
Defining an equilibrium constant as

$$K_x = \prod x_j^{\nu_j} \tag{6.7}$$

we get

$$\ln K_x = \sum \nu_j \ln x_j = \frac{1}{R^* T} \sum \nu_j G_j = \frac{1}{R^* T} \Delta G \tag{6.8}$$

where ΔG is the change in free enthalpy between reacting and reacted components.
By having an alternate definition of free-enthalpy

$$\bar{G}_j(T, p) = G_j(T, p) - R^* T \ln x_j \tag{6.9}$$

it can be shown that the condition for equilibrium can be written also as

$$\sum \nu_j \bar{G}_j = 0 \ . \tag{6.10}$$

Now the enthalpy of a species of a perfect gas is a function of temperature only, $H_j = H_j(T)$, whereas the entropy is a function of both the temperature and pressure, $S_j = S_j(T, p)$. However, the values of entropy for different gases are normally given in literatures at a standard pressure p_o, for example $p_o = 1$ bar. From thermodynamics we know that

$$S_j = S_j(T, p) = S_j(T, p_o) + R^* \ln(p_o/p) \tag{6.11}$$

which can be used further to evaluate the free-enthalpy of the species. From (3.59) and (6.8) further,

$$\ln K_x = \sum \nu_j \ln x_j = \frac{1}{R^*T} \sum \nu_j G_j = \ln \prod (Z_j/N_A)^{\nu_j} \tag{6.12}$$

where Z_j is the partition function of the j-th component [kmole^{-1}] and N_A is the Avogadro number [kmole^{-1}]. Thus,

$$K_x = \prod x_j^{\nu_j} = \prod (Z_j/N_A)^{\nu_j} \tag{6.13}$$

where $x_j = n_j/n$, n_j is the particle density of the j-th species [m^{-3}] and n is the particle density of all the specie [m^{-3}]. Partition function Z_j has the same dimension as N_A, and as such it is justified to replace N_A by n, and then Z_j will also have the dimension (m^{-3}). Hence we write

$$K_x = \prod (n_j/n)^{\nu_j} = \prod (Z_j/n)^{\nu_j} \tag{6.14}$$

and we get the *law of mass action* in terms of the partition function and particle density as

$$K_n = \prod n_j^{\nu_j} = \prod Z_j^{\nu_j} \; . \tag{6.15}$$

The above gives a new definition of equilibrium constant in terms of the number density. Further since $x_j = p_j/p$, where p_j is the partial pressure of the j-th species and p is the total pressure, we get another definition of the equilibrium constant in terms of the partial pressure

$$K_p = \prod p_j^{\nu_j} = \prod p^{\sum \nu_j} . K_x \; . \tag{6.16}$$

From (6.8) and (6.11) we write

$$\ln K_x = \frac{1}{R^*T} \sum \nu_j (TS_j - H_j)$$

$$= \frac{1}{R^*T} \sum \nu_j (TS_j(T, p_o) - H_j) + \ln(p_o/p)^{\sum \nu_j} \; .$$

Since the standard pressure is usually 1 bar or 1 atm, we can write a new definition of the equilibrium constant in terms of the standard free enthalpy

$$\ln K_p = \ln K_x + \sum \nu_j \ln p = \frac{1}{R^*T} \sum \nu_j (T S_j(T, p_o) - H_j)$$
$$= \frac{1}{R^*T} \sum \nu_j G_j(T, p_o)$$

which shows that K_p need be calculated at one standard pressure only when $K_p = K_x$. At any other pressure, K_x can then be computed with the help of (6.16).

6.2 Heat of Reaction

The term *heat of reaction* gives the energy that is absorbed or released in a chemical reaction. In order to explain the concept, we consider now the reaction of hydrogen and oxygen gas in the volume ratio 2:1 at the initial temperature T_1, and initiate reaction (combustion) by increasing the temperature of an infinitesimal small volume to start the combustion process. The mole mass of the mixture before reaction (') is

$$m' = (2 \times 2.016 + 32)/3 = 12.017 \ . \tag{6.17}$$

After reaction water is formed, let the maximum temperature after complete combustion and no heat loss be T_2, which is called the *adiabatic flame temperature*. In case the reaction is so complete that there are only water molecules, then after the reaction ($''$) the mole mass is $m'' = 18.016$. By the process of reaction (combustion) the total mole of mixture of hydrogen and oxygen is reduced to two-third the original number of moles.

Now we consider two models. In the first model, we put combustible gases (reactants) in a cylinder (closed by a symbolic weightless piston, which is movable if the pressure inside the cylinder is kept constant, or is kept fixed to keep the volume constant) and the entire cylinder is assumed to be thermally insulated (adiabatic wall). In the first model the heat of reaction is replaced by some heat from an external source. If we allow the piston to move in such a manner that the pressure is kept constant, it is evident from the *first law of thermodynamics* that the *heat of reaction* at constant pressure is equal to the change of enthalpy

$$Q_{Rp} = \Delta H = \sum \nu_j H_j \tag{6.18}$$

where the subscript j refers to the j-th species. On the other hand if the pressure is allowed to increase at constant volume (as in a *bomb calorimeter*), then again from the first law of thermodynamics, the heat of reaction is equal to the change of internal heat, that is

$$Q_{Rv} = \Delta E = \sum \nu_j E_j \ . \tag{6.19}$$

In case of an exothermic reaction the heat of reaction is positive, but it is negative in the case of an endothermic reaction. From such a definition, we can compute the temperature after reaction, if we assume an average specific heat and mole mass do not change much, that is, we model the heat of reaction as a source of heat from outside. Thus it is evident that the difference of the two heat of reaction (at constant pressure or volume) is

$$Q_{Rp} - Q_{Rv} = p(v'' - v') \ . \tag{6.20}$$

We consider now the second model, where the heat of reaction is included in the enthalpy of the reacting and the reacted components. In this model (Fig. 6.2), no heat is removed externally after the combustion process is over. Since no heat is exchanged (adiabatic wall), the specific (per unit mass or per unit mole of reacting components) enthalpy (for constant pressure process) or internal energy (for the constant volume process) before reaction (reacting gas components) is equal to the enthalpy after reaction (for reacted components). In this model, therefore, the temperature attained is the *adiabatic flame temperature*, T_2, which is larger than the temperature before reaction T_1. The heat of reaction can be found by cooling the reacted components to the initial temperature T_1. While the heat of reaction is somewhat different depending on whether it is a constant pressure or constant volume process, we consider further only one of them (for example, the constant pressure process, and the discussion on the other may follow on similar lines). According to the model, therefore, we write

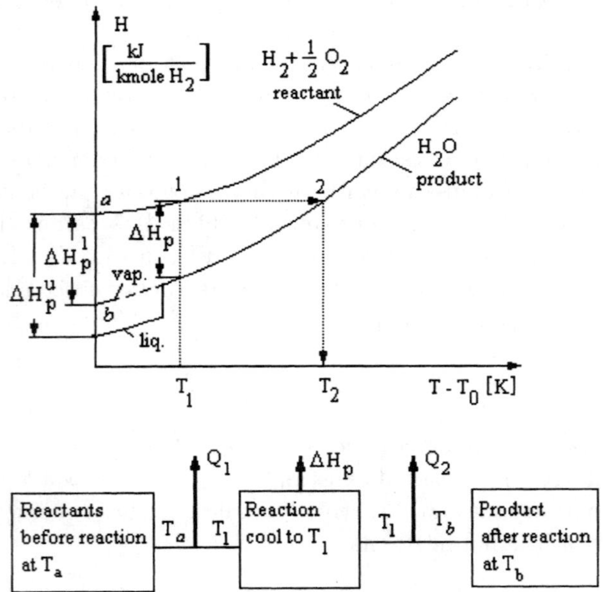

Fig. 6.2. Model to explain heat of reaction and adiabatic flame temperature

$$Q_{Rp}(T_1) = \int_{T_1}^{T_2} C_p'' \mathrm{d}T - \int_{T_1}^{T_1} C_p' \mathrm{d}T = \int_{T_1}^{T_2} C_p'' \mathrm{d}T \qquad (6.21)$$

where C_p' and C_p'' are specific heats (per unit mole or mass) of reacting or reacted components, respectively. The enthalpy of the reacted components at the temperature T_1, must, therefore, be lower (for exothermic reaction) and this information can be taken into consideration for the enthalpy values (*"absolute enthalpy"*). Thus for an exothermic reaction like hydrogen-oxygen reaction, the enthalpy of the reacting hydrogen and oxygen molecules are made zero at the reference (absolute) temperature, then the enthalpy of water vapor at the reference temperature is negative.

Very often, however, the heat of reaction in literatures is given as $Q_{Rp}(T_o)$. It can be shown that

$$Q_{Rp} = H'(T_1) - H''(T_1)$$

$$= H'(T_o) - H''(T_o) - \int_{T_o}^{T_1} (C_p'' - C_p') \mathrm{d}T$$

$$= Q_{Rp}(T_o) - \int_{T_o}^{T_1} (C_p'' - C_p') \mathrm{d}T \ . \qquad (6.22)$$

In literatures, the tabulated values of the enthalpy for various components are given in different ways. In this book, for example, the reference temperature is at the absolute temperature. In many books, however, this may be given relative to a standard temperature (288.15 K) and the heat of reaction at that standard temperature. These need be recalculated to the values at another temperature by the method discussed above.

In the present section, we consider the (thermodynamic) properties of various components before or after reaction by considering the mixture properties. How to obtain these mixture properties will now be discussed in the following section.

6.3 Properties of Mixture of Gases

Composition of a mixture of various gas specie can be described either in terms of the mole fraction of the j-th species $x_j = p_j/p = n_j/n$, or mass fraction $Y_j = \rho_j/\rho$, where p and ρ are the pressure and density, respectively and the quantities without any subscript are those for the mixture. Relation between these two can be obtained easily with the help of the equation of state and the definition of the mole mass m. Since,

$$p = \sum p_j = \sum \rho_j R_j T = R^* T \sum \rho_j/m_j = \rho R^* T/m \qquad (6.23)$$

where R^* is the *universal gas constant*. Therefore,

$$m = \sum x_j m_j = 1 / \sum (Y_j/m_j) \qquad (6.24)$$

and also

$$x_j = p_j/p = Y_j m/m_j \ . \tag{6.25}$$

In addition some of the calorific properties of gas specie like enthalpy, entropy, specific heat, etc. may be written either in terms of unit per mole (designated with capital letters) or per mass (designated with small letters) of the species. For example, the specific heat of the mixture at constant pressure can be written as

$$c_p = \sum Y_j c_{pj} = \sum Y_j C_{pj}/m_j = \frac{1}{m} \sum x_j C_{pj} = C_p/m \ . \tag{6.26}$$

Similarly, for specific heat at constant volume, enthalpy and internal energy we write

$$c_v = \sum Y_j c_{vj} = \sum Y_j C_{vj}/m_j = \frac{1}{m} \sum x_j C_{vj} = C_v/m \ . \tag{6.27}$$

$$h = \sum Y_j h_j = \sum Y_j H_j/m_j = \frac{1}{m} \sum x_j H_j = H/m \ . \tag{6.28}$$

$$e = \sum Y_j e_j = \sum Y_j E_j/m_j = \frac{1}{m} \sum x_j E_j = E/m \ . \tag{6.29}$$

The molar entropy of the gas changes with pressure. Hence if the rule is applied to the partial pressure, we get for the molar entropy

$$\begin{aligned} S(T,p) &= \sum x_j [S_j(T,p) - R^* \ln x_j] \\ &= \sum x_j [S_j(T,p_o) - R^* \ln x_j] - R^* \ln(p/p_o) \end{aligned} \tag{6.30}$$

where p_o is the *reference pressure*. Entropy per unit mass can of course be obtained by dividing the above expression by molemass m.

6.4 Equilibrium Composition of an Ideal Dissociating Diatomic Gas

In mid-fifties *Lighthill* proposed the following model for an ideal dissociating gas like H_2, O_2, N_2, etc. In all such cases we have a reaction of type

$$\mathrm{M} \leftrightarrow 2\mathrm{A} \ . \tag{6.31}$$

For such a reaction the molemass of the molecule is twice that of the atom, $m_M = 2m_A$, where the subscript M stands for the molecule and A for the atom and the molemass of the mixture is

$$m = \frac{2m_A}{(1+Y_A)} = \frac{m_M}{(2-Y_M)} = (2-x_A)m_A = \frac{1}{2}(1+x_M)m_M \tag{6.32}$$

where x is the mole fraction and Y is the mass fraction. Relationship between these are given by

$$x_A = \frac{2Y_A}{(1+Y_A)} = \frac{2(1-Y_M)}{(2-Y_M)} \tag{6.33}$$

and

$$x_M = \frac{Y_M}{(2-Y_M)} = \frac{1-Y_A}{1+Y_A} \,. \tag{6.34}$$

Therefore, for the equilibrium constant we write in terms of either the mole fraction x or the mass fraction Y as

$$K_x = \frac{x_A^2}{x_M} = \frac{x_A^2}{1-x_A} = \frac{4Y_A^2}{1-Y_A^2} = \frac{4(1-Y_M)^2}{Y_M(2-Y_M)}$$

$$= 2[TS_A(T,p_o) - H_A] - [TS_M(T,p_o) - H_M] - R^*T \ln\left(\frac{p}{p_o}\right) \tag{6.35}$$

or in terms of the partial pressure as

$$K_p = \frac{p_A^2}{p_M} = \frac{p_A^2}{(p-p_A)} = 2[TS_A(T,p_o) - H_A] - [TS_M(T,p_o) - H_M]$$

$$= \frac{1}{R^*T}[2G_A(T,p_o) - G_M(T,p_o)] \,. \tag{6.36}$$

One can write for K_x in terms of the partition function from (6.14) as

$$K_x = \frac{Z_A^2}{Z_M N_A} \tag{6.37}$$

where N_A is the *Avogadro number*; Z with subscripts A and M refer to partition function of atom or molecule per unit mole, respectively.

Now for evaluation of the partition function of atoms we have to consider the translation mode and the *energy of dissociation*, whereas for molecules the translation, rotation and vibration modes only, since the electrons, in the temperature range being considered, are to be in the ground state. From (3.67), (3.80) and (3.113) we write for the partition functions for atoms:

$$Z_A = \frac{R^*T}{p}\left(\frac{2\pi k_B M_A T}{h^2}\right)^{3/2}, \text{ kmole}^{-1} \tag{6.38}$$

and for molecules:

$$Z_M = \frac{R^*T}{p}\left(\frac{2\pi k_B M_M T}{h^2}\right)^{3/2}\frac{T}{2\Theta_r}\frac{1}{1-\exp^{-\Theta_v/T}}\exp^{E_D/(k_B T)}, \text{ kmole}^{-1} \tag{6.39}$$

where E_D is the energy of dissociation (per molecule), and Θ_r and Θ_v are the characteristic rotational and vibrational temperature of the molecule, respectively. Further since we are considering homopolar molecules, the value of the *symmetry factor* σ in (6.39) has been put equal to two. Therefore from (6.37–6.39), and noting that $M_M = 2M_A$, we get

$$K_x = \frac{Z_A^2}{Z_M N_A}$$

$$= \frac{1}{\sqrt{2}} \frac{k_B T}{p} \left(\frac{2\pi k_B M_A T}{h^2}\right)^{3/2} \frac{\Theta_r}{T} (1 - \exp^{-\Theta_v/T}) \exp^{-E_D/(k_B T)} \ . \quad (6.40)$$

Lighthill wrote (6.40) approximately in the form

$$K_x = \frac{4Y_A^2}{1 - Y_A^2} = \frac{\rho_d}{\rho} \exp^{-T_d/T} \quad (6.41)$$

where $T_d = E_D/k_B = H_D/R^*$ is the *characteristic dissociation temperature* and ρ_d is the *characteristic density of dissociation*; H_D is the *dissociation enthalpy* (per mole of molecules). Since the density

$$\rho = \frac{p}{RT} = \frac{pM}{k_B T} = \frac{2pM_A}{k_B T(1 + Y_A)} = \frac{2pm_A}{R^*T(1 + Y_A)} \quad (6.42)$$

we get for the *characteristic density of dissociation*

$$\rho_d = 2\sqrt{2} \frac{M_A \Theta_r}{1 + Y_A} \left(\frac{2\pi k_B M_A}{h^2}\right)^{3/2} \sqrt{T} \left(1 - \exp^{-\Theta_v/T}\right) \quad (6.43)$$

and for the *characteristic pressure of dissociation*

$$p_d = \rho_d/(RT_d) \ . \quad (6.44)$$

Noting that Y_A can have values only between 0 and 1, and for the other temperature dependent terms

$$\sqrt{T} \text{ and } (1 - \exp^{-\Theta_v/T}) \quad (6.45)$$

their temperature dependency in the range of interest (1000 to 7000 K) tend to cancel each other, Lighthill calculated the value of these for N_2 and O_2, which are slightly different from values given in Fig. 6.3, which contains also values for H_2 and Cl_2 computed by this author. Some of the characteristic values for selected diatomic molecular gases are given in Table 6.1.

Table 6.1. Characteristic values for selected diatomic molecules

Gas	m_A	Θ_r [K]	Θ_v [K]	E_D [eV]	p_d [bar]	ρ_d [kg.m$-$3]	T_d [K]
H_2	1.008	87.6	6333	4.476	9.14e6	2.13e3	51986
O_2	16.00	2.082	3400	5.080	9.89e5	3.23e4	59000
N_2	14.01	2.89	2280	7.373	1.98e7	3.90e4	85633
Cl_2	35.46	0.351	815	2.475	2.07e4	2.07e4	28745

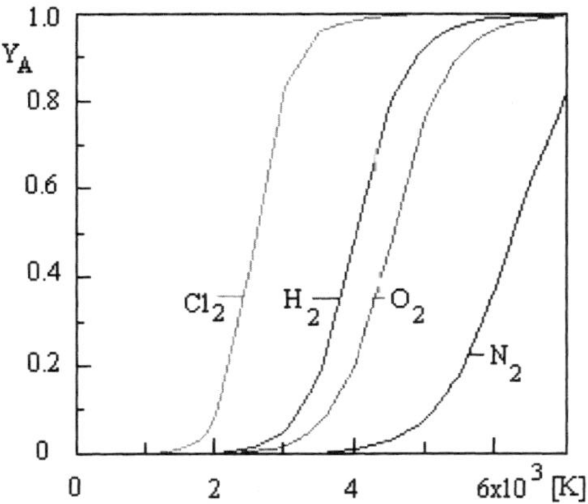

Fig. 6.3. Massfraction of atoms for four diatomic molecules at p - 1 bar

From numerical calculations at 1 bar, it is found that for the four gases under consideration, there is hardly any dissociation if at 1 bar pressure (T/T_d) is less than 0.05 and complete dissociation if it is larger than 0.08. Hence it is evident that for these gases dissociation occurs within a very narrow temperature range, where the atoms and molecules co-exist.

Since,

$$\rho = \frac{pm}{R^*T} = \frac{2pm_A}{R^*T(1+Y_A)} \tag{6.46}$$

and the equilibrium constant

$$K_x = \frac{4Y_A^2}{1-Y_A} = \frac{\rho_d R^*T(1+Y_A)}{2pm_A} \exp^{-T_d/T} \tag{6.47}$$

we get the relation for the mass fraction of atom

$$Y_A = \sqrt{C/(1+C)} \tag{6.48}$$

where

$$C = \frac{1}{8}\frac{\rho_d R^*T}{pm_A} \exp^{-T_d/T} \ . \tag{6.49}$$

As an example by keeping $p/p_d = 10^{-6}$ and in the temperature ratio T/T_d between 0.05 and 0.2, it can be seen that the atom mass fraction changes from 0.0036 to 0.997. Thus as a rule of thumb for starting dissociation, the temperature for a diatomic gas has to be higher than 0.02 T_d, which explains why the oxygen dissociates at a substantially lower temperature than nitrogen.

Following *Lighthill* we would now define a few other quantities. Noting that the mixture enthalpy per unit mass is

$$h = Y_A h_A + Y_M h_M \tag{6.50}$$

we write for the specific heat at constant pressure

$$c_{p,eff} = \left(\frac{\partial h}{\partial T}\right)_p = \left(Y_A\frac{\partial h_A}{\partial T} + Y_M\frac{\partial h_M}{\partial T}\right) + \left(h_A\frac{\partial Y_A}{\partial T} + h_M\frac{\partial Y_M}{\partial T}\right)$$

$$= c_{p,f} + \frac{\partial Y_A}{\partial T}(h_A - h_M) \ . \tag{6.51}$$

In above $c_{p,f}$ is the specific heat for the frozen composition. Now for the atom the specific heat at constant volume is

$$c_{vA} = \frac{3}{2}\frac{R^*}{m_A} \tag{6.52}$$

and for the molecule (without considering vibrational energy being effective)

$$c_{vM} = \frac{5}{2}\frac{R^*}{m_M} \approx \frac{6}{2}\frac{R^*}{2m_A} \approx \frac{3}{2}\frac{R^*}{m_A} = c_{vA} \tag{6.53}$$

and hence per unit of mass of the species they are approximately equal (more so if the vibrational energy component is zero). Hence for the total mixture enthalpy and mixture effective specific heat we write

$$h = \frac{3R^*T}{2m_A} + \frac{R^*T}{m} + \frac{H_DY_A}{2m_A} = \frac{3R^*T}{2m_A} + \frac{R^*T}{2m_A}(1 + Y_A) + \frac{H_DY_A}{2m_A}$$

where H_D is again the *dissociation enthalpy*. Hence the *effective specific heat* and the *frozen specific heat* at constant pressure per unit mass are

$$c_{p,\text{eff}} = \left(\frac{\partial h}{\partial T}\right)_p \frac{R^*}{2m_A}\left[(4 + Y_A) + (T + T_d)\frac{dY_A}{dT}\right] \tag{6.54}$$

and

$$c_{p,f} = \frac{R^*}{2m_A}(4 + Y_A) \ . \tag{6.55}$$

Thus the ratio of the two specific heats is

$$\frac{c_{p,\text{eff}}}{c_{p,f}} = 1 + \frac{T + T_d}{4 + Y_A}\frac{dY_A}{dT} \ . \tag{6.56}$$

The computed results of $(c_{p,\text{eff}}/c_{p,f})$ using (6.56), in which the necessary values (dY_A/dT) are computed by using (6.48) and the data of Table 6.1, are presented for four diatomic molecular gases in Fig. 6.4. The results show considerable increase in the value of specific heat in certain temperature range, the maximum being at about $0.08T/T_d$.

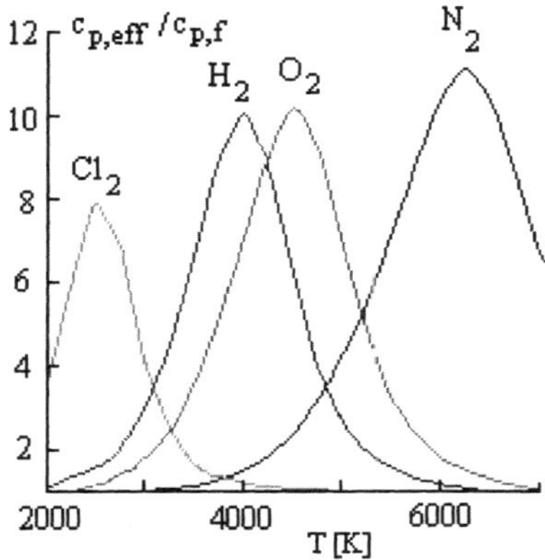

Fig. 6.4. Specific heat ratio for four diatomic molecular gases at $p = 1$ bar

6.5 Equilibrium Composition for a Multiple Component Gas

In the present Section a very powerful method, originally due to *Horn* and *Schueler* and used by *Neumann* and *Knoche* [113] is assumed to be given. The starting point is the state of the gas mixture (T, p), which is assumed as given. Let us now consider the reaction of the type given in (6.1), in which the quantities on the left hand side of the equation are the defined as *primary components* from which the reactions are started. The choice of the primary components is somewhat arbitrary and depends mainly on whether sufficient mole fraction of the primary components exist in the equilibrium gas mixture. Hence for (H_2, O_2) reaction, depending on the gas state, the selected primary components may either be (H_2, O_2) or (H, O). On the other hand for the reaction of the type

$$\alpha_1 C_m H_n O_p \leftrightarrow \beta_1 CO_2 + \beta_2 H_2 O + \beta_3 H + \beta_4 O + \beta_5 C_{\mathrm{gas}} + \beta_6 CO + \beta_7 OH + \dots$$
$$(6.57)$$

the right hand side of the equation can be obtained also by reacting C_{gas}, H_2, O_2 or C_{gas}, H and O or deducting H_2 and O_2 from H_2O and CO_2 molecules. Therefore, the selected primary components may be any of the gas components or something quite different depending on the ease of calculation of the equilibrium composition. Thus, we can now, in every chemical reacting gas mixture with n components, think of as being produced from r *primary components* and $(n - r)$ secondary components. We can also assume that every component is produced by combination of primary components as follows:

$$B_j \leftrightarrow \sum_{i=1}^{r} \nu_{i,j} B_i \qquad (6.58)$$

where

$\nu_{i,j}$ = valency relation between the primary components and all components;
i = index for primary components (i = 1, 2, 3, ... , r);
j = index for all components (j = 1, 2, 3, ... , n);
r = = number of primary components; and
n = total number of components.

For reaction between H_2 and O_2, and these being considered as primary components, it is, therefore, possible to write the following reactions:

$$H_2 \leftrightarrow 1.0H_2 + 0.0O_2$$
$$O_2 \leftrightarrow 0.0H_2 + 1.0O_2$$
$$OH \leftrightarrow 0.5H_2 + 0.5O_2$$
$$H \leftrightarrow 0.5H_2 + 0.0O_2$$
$$O \leftrightarrow 0.0H_2 + 0.5O_2$$

Therefore, for the above reaction $n = 5$, $r = 2$. Out of these reactions a valency matrix, the so-called ν-matrix are written, whose values are $\nu_{i,j}$ as follows:

$$\nu = \begin{vmatrix} 1.0 & 0.0 \\ 0.0 & 1.0 \\ 0.5 & 0.5 \\ 0.5 & 0.0 \\ 0.0 & 0.5 \\ 1.0 & 0.5 \end{vmatrix} \tag{6.59}$$

Now from (6.7), we write

$$K_{xj} = \frac{x_j}{\prod_i x_i^{\nu_{i,j}}} \tag{6.60}$$

for example,

$$K_{xOH} = \frac{x_{OH}}{x_{H_2}^{0.5} x_{O_2}^{0.5}}, \quad K_{xH} = \frac{x_H}{x_{H_2}^{0.5} x_{O_2}^{0.0}}, \text{eic.} \tag{6.61}$$

On the other hand, we get from (6.6) the relation

$$R^* \ln K_{x_j} = R^* \left[\ln x_j - \sum_{i=1}^{r} \nu_{i,j} x_i \right] = \frac{1}{T} \left[G_j - \sum_{i=1}^{r} \nu_{i,j} G_i \right] \tag{6.62}$$

where the free enthalpy of the j-th species is

$$G_j(T,p) = TS_j(T,p) - H_j(T) = T[S_j(T,p_o) + R^* \ln(p_o/p)] - H_j(T) . \tag{6.63}$$

We can now determine the mole fraction of all the components in the gas mixture at a given temperature and pressure as follows. First, we compute G_j and the equilibrium constant from (6.62). We now assume or guess the value of x_i for the primary components and we compute the value of x_j for secondary components from (6.62). We have therefore a mole matrix vector $\mathbf{X}(n, 1)$, which has the first r guessed mole fraction values of the primary components, followed by $(n - r)$ computed molefraction of the secondary components. Now the following two compatibility relations must be satisfied.

The first compatibility condition comes from the fact that the sum of mole fractions of all the components must be equal to one, that is

$$\sum_{j=1}^{n} x_j = 1 \ . \tag{6.64}$$

In practical terms, the first compatibility becomes

$$x_{H_2} + x_{O_2} + x_{OH} + x_H + x_O + x_{H_2O} = 1 \ . \tag{6.65}$$

Secondly, if all the components are reconverted (by backward reaction) to their (original) primary components, then the ratio of these are prescribed by the equation

$$q_k = \frac{\alpha_k}{\alpha_1} = \frac{\sum_{j=1}^{n} x_j \nu_{j,k}}{\sum_{j=1}^{n} x_j \nu_{j,1}}, \ k = 2,3, \ldots, r \ . \tag{6.66}$$

By cross-multiplication, we get now

$$\alpha_1 \sum_{j=1}^{n} x_j \nu_{j,k} - \alpha_k \sum_{j=1}^{n} x_j \nu_{j,1} = \sum_{j=1}^{n} x_j (\nu_{j,k} - q_k \nu_{j,1}) = \sum_{j=1}^{n} x_j \psi_{k,j} = 0 \ \tag{6.67}$$

$$(q_k = \alpha_k / \alpha_1, q_1 = 1.0) \ .$$

As an example for H_2 and O_2 being reacted stoichiometrically (two moles H_2 reacting with one mole O_2), the value of $q_2 = 1/2$, and thus the second compatibility relation is

$$q_2 = \frac{1}{2} = \frac{0.x_{H_2} + 1.0x_{O_2} + 0.5x_{OH} + 0.x_H + 0.5x_O + 0.5x_{H_2O}}{1.x_{H_2} + 0.x_{O_2} + 0.5x_{OH} + 0.5x_H + 0.x_O + 1.0x_{H_2O}} \ . \tag{6.68}$$

By cross-multiplication of (6.68) now becomes

$$-1.0x_{H_2} + 2.0x_{O_2} + 0.5x_{OH} - 0.5x_H + 1.0x_O + 0.0x_{H_2O} = 0 \ . \tag{6.69}$$

From (6.65, 6.69), we can now write a matrix equation

$$\psi \cdot \mathbf{X} = \mathbf{b} \ . \tag{6.70}$$

Thus from the two compatibility equations, we can define a ψ-matrix, whose coefficients are

$$\psi_{k,j} = \begin{cases} = 1 & \text{for } k = 1 \\ = (\nu_{j,k} - q_k \nu_{j,1}) & \text{for } k = 2,3, \ldots, r \end{cases} \ . \tag{6.71}$$

In general terms, a ψ-matrix of dimension (r, n) will have only 1 in the first row and in subsequent $(r\text{-}1)$ rows, the other elements, which are obtained by cross-multiplication.

The equivalent ψ-matrix for the H_2-O_2 reaction will thus be

$$\psi = \begin{vmatrix} 1.0 & 1.0 & 1.0 & 1.0 & 1.0 & 1.0 \\ -1.0 & 2.0 & 0.5 & -0.5 & 1.0 & 0.0 \end{vmatrix}. \tag{6.72}$$

Now by matrix multiplication of $\psi(r,n)$ with the $\mathbf{X}(n,1)$ vector (containing x_j-values), the compatibility conditions to be satisfied are

$$b_k = \begin{cases} 1 & \text{for } k = 1 \\ 0 & \text{for } k = 2,3,\dots,r \end{cases}. \tag{6.73}$$

The above compatibility condition however, in general, may not be satisfied since mole fractions are not of desirable quantity. In order to get the desired mole fraction, we would undertake a variation of the mole fraction of the primary components, which will automatically change the mole fraction of the secondary components also. While the formal expression of (6.70) is

$$b_k = \sum_{j=1}^{n} \psi_{k,j} x_j \tag{6.74}$$

the variation formulation of the molefraction is

$$\delta b_k = \sum_{j=1}^{n} \psi_{k,j} \delta x_j = \sum_{j=1}^{n} \psi_{k,j} x_j \delta \ln x_j . \tag{6.75}$$

Now from (6.62),

$$R^* \delta \ln K_{xj} = 0 = R^* \left[\delta \ln x_j - \sum_{i=1}^{n} \nu_{j,i} \delta \ln x_i \right] . \tag{6.76}$$

Thus, the variation formulation of the mole fraction can be written as

$$\delta b_k = \sum_{j=1}^{n} \left[\sum_{i=1}^{r} \psi_{k,j} x_j \nu_{j,i} \right] \delta \ln x_i = \sum_{i=1}^{r} g_{i,k} \delta \ln x_i \tag{6.77}$$

where

$$g_{i,k} = \sum_{j=1}^{n} \psi_{k,j} x_j \nu_{j,i} . \tag{6.78}$$

It may be noted that (6.78) is obtained by multiplying $[\psi]$ with a diagonal matrix having diagonal elements x_j and then finally by multiplying with ν-matrix. Thus the square g-matrix (with elements $g_{i,k}$) will have the dimension (r, r).

Now Equation (6.77) is written in a difference form. Noting that Δb_k is the difference between the actual b'_k and desired b_k, we write

$$\Delta b_k = b'_k - b_k = \sum_{i=1}^{r} g_{i,k} \Delta \ln x_i \qquad (6.79)$$

and we get the desired change in the primary component mole fraction by solving (6.79). Multiple iteration is possible, according to the Flow Chart (Fig. 6.5), to get the desired result. The described method in general is very powerful and gives results with a high degree of accuracy within a few iterations. However, the method has the following limitations: (1) the guessed values of x_i should be good approximation (a quick hand calculation of the composition of the primary components on the basis of the calculated equilibrium constant may help!), and (2) the values of any primary component should not be too small, which can be understood by noting from (6.60) that, in such a case, the numerator also will automatically be zero (it is better to change the primary components!). This is explained with the help of calculations of air plasma given in Fig. 6.6, which is given at $p = 1$ bar and temperatures up to 50,000 K. It is seen that the oxygen dissociates first and in fact there are practically no oxygen molecules beyond 5,000 K. However, single-charged ionization of the oxygen takes place at a temperature higher than that for the nitrogen. Hence it is alright to consider oxygen molecule as a primary component till about 4,500 K, then we could take oxygen atom as a primary component, and so on.

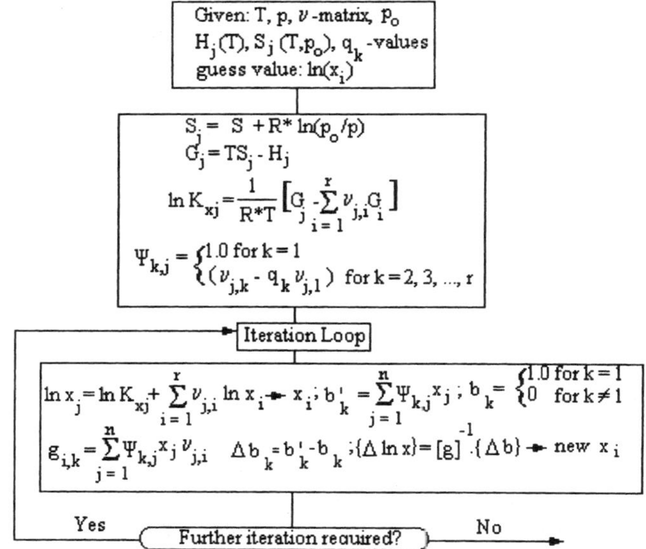

Fig. 6.5. Chart for equilibrium calculations

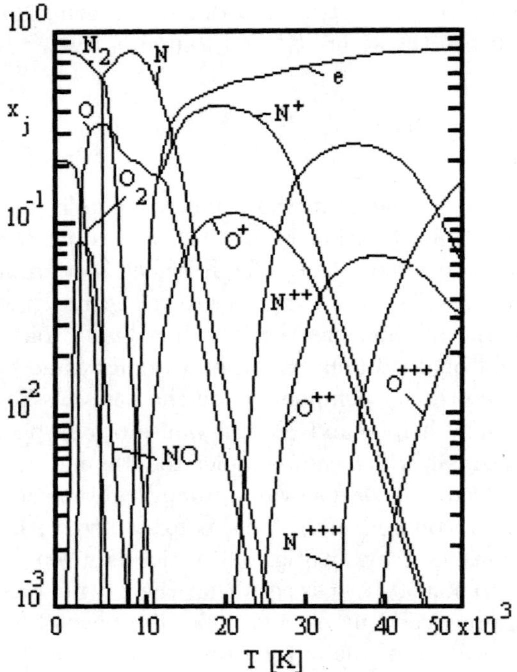

Fig. 6.6. Composition of air plasma at $p = 1$ bar

While the general procedure to compute equilibrium composition of a reacting gas mixture has been discussed in this section, in the subsequent sections we discuss the procedures to compute the equilibrium gas composition for ionized gases for single or two-temperature cases.We would also discuss in Sect. 6.7 an extension of the present method for the purpose of computation of temperature derivative of the mole fraction, which is required for calculation of the equilibrium reactive heat conductivity coefficient. However, we would first discuss the method to compute mole fraction of components of ionized gases, by using a slightly different method and not through the free-energy.

6.6 Equilibrium Composition
for a Pure Monatomic Gas Plasma

Let us now assume a gas consisting of ions, neutrals and electrons. As a special case, neutrals can be considered as zero-charged ions. Thus we may introduce an index i whose value gives the number of charges for the particle: neutrals have the value of index i = 0, for singly-charged ions i = 1, for doubly-charged ions i = 2, etc. A mixture of such neutrals, ions and electrons is called, according to I. Langmuir, a plasma, but for the present we consider a plasma of a pure monatomic gas only. However, even for a plasma made out of a pure gas calculation would have been very complicated, if a wide range of charges were present at a time. Fortunately, as the calculations show, we have only the electrons and the ions with i and (i+1) charges, and only in a limited overlapping temperature range there are electrons and ions with i, (i+1) and (i+2) charges. Therefore, we have to consider, mostly three specie in a pure gas plasma, although in a limited temperature range we would consider four specie case also. We consider, therefore, a plasma with only the electrons, and the ions with i and (i+1) charges for the present, for which the reaction considered is

$$A_i \leftrightarrow A_{i+1} + e \; . \tag{6.80}$$

Equilibrium constant can now be written with the help of (6.15) as

$$K_n = \frac{n_e n_{i+1}}{n_i} = \frac{Z_e Z_{i+1}}{Z_i} \; . \tag{6.81}$$

Now among the partition functions, Z_e has only the translation component, and for Z_i and Z_{i+1} have translational and excitational components. These can be written as follows:

$$Z_e = 2 \left(\frac{2\pi M_e k_B T}{h^2} \right)^{3/2} \tag{6.82}$$

$$Z_i = \left(\frac{2\pi M_i k_B T}{h^2} \right)^{3/2} Z_{i,\text{exc}} \tag{6.83}$$

and

$$Z_{i+1} = \left(\frac{2\pi M_{i+1} k_B T}{h^2} \right)^{3/2} Z_{i,\text{exc}} \exp^{-I_i/(k_B T)} \tag{6.84}$$

where

$$Z_{i,\text{exc}} = \sum g_{i,r} \exp^{-E'_{i,r}/(k_B T)} \approx g_{i,o} \tag{6.85}$$

and

$$Z_{i+1,\text{exc}} = \sum g_{i+1,r} \exp^{-E'_{i+1,r}/(k_B T)} \approx g_{i+1,o} \; . \tag{6.86}$$

In above g_r is the statistical weight for the r-th energy level, E'_r is the corresponding excitation energy of the electrons in bound state, and I_i and I_{i+1} are the ionization energy (potential) for transition between the ground energy levels of the i-th and (i+1)-th ionization. Substituting (6.82–6.86) into (6.81) and noting that the mass of ions at different ionization states are approximately equal, $M_i \approx M_{i+1}$, we get the expression for the equilibrium constant

$$K_n = \frac{n_e n_{i+1}}{n_i} = n \frac{x_e x_{i+1}}{x_i} = 2 \left(\frac{2\pi M_e k_B T}{h^2} \right)^{3/2} \frac{Z_{i+1,\text{exc}}}{Z_{i,\text{exc}}} \exp^{-I_i/(k_B T)} \; . \tag{6.87}$$

Equation (6.87) can be simplified considerably , if we assume that the first energy level is large in comparison to the ground level. For such a case, the excitation partition function can be replaced by the statistical weight at the ground level, and we get from (6.87) also

$$K_n = \frac{n_e n_{i+1}}{n_i} = n\frac{x_e x_{i+1}}{x_i} = 2\left(\frac{2\pi M_e k_B T}{h^2}\right)^{3/2} \frac{g_{i+1,0}}{g_{i,0}} \exp^{-I_i/(k_B T)} \quad .$$

(6.88)

Equation (6.87) or (6.88) can be solved under the condition of quasi-neutrality

$$n_e = i n_i + (i+1) n_{i+1}$$

(6.89)

and the total particle density prescribed by the temperature and pressure is given by the relation

$$n = n_e + n_i + n_{i+1} = p/(k_B T) \quad .$$

(6.90)

Equation (6.87) can also be written (and similarly, (6.88) in terms of the molefraction of the j-th species, $x_j = n_j/n$, (j = e,i and i+1)), which, by convention, is written as Saha Function, S, after *Meghnad Saha*, the indian physicist who developed it, as follows:

$$S \equiv K_x = \frac{K_n}{n} = \frac{x_e x_{i+1}}{x_i}$$

$$= \frac{2}{p}\left(\frac{2\pi M_e}{h^2}\right)^{3/2} (k_B T)^{5/2} \frac{Z_{i+1,\mathrm{exc}}}{Z_{i,\mathrm{exc}}} \exp^{-I_i/(k_B T)}$$

$$= 6.6 \times 10^{-7} \frac{T^{5/2}}{p} \frac{g_{i+1,0}}{g_{i,0}} \exp^{-I_i/(k_B T)}$$

(6.91)

where T is in K (Kelvin) and p is in bars. Value of $g_{i,o}$ and I_i for a few elements are given in Table 6.2 for a quick estimate of the plasma composition. Also given are m, the mole mass, and N_c, the charge number of the atom, which is equal to the number of electrons in the atom if fully ionized. Now Equations (6.89) and (6.90), in terms of the mole fraction becomes

$$x_e = i x_i + (i+1) x_{i+1}$$

(6.92)

and

$$x_e + x_i + x_{i+1} = 1 \quad .$$

(6.93)

Accordingly,

$$x_{i+1} = -i + (i+1)x_e \quad \text{and} \quad x_i = (i+1) - (i+2)x_e$$

(6.94)

with the limit

$$\frac{i}{i+1} \leq x_e \leq \frac{i+1}{i+2} \quad .$$

(6.95)

Table 6.2. Statistical weights and ionization potential [eV] of selected atoms

N_c	Elem.	m	$g_{0,0}$	$I_{1,0}$	$g_{1,0}$	$I_{2,0}$	$g_{2,0}$	$I_{3,0}$	$g_{3,0}$	$I_{4,0}$	$g_{4,0}$
1	H	1.008	2	13.59							
2	He	4.003	1	24.58	2	54.40					
3	Li	6.94	2	5.39	1	75.62	2	122.42			
6	C	12.01	9	11.26	6	24.38	1	47.86	2	64.48	1
7	N	14.01	4	14.54	9	29.60	6	47.43	1	77.45	2
8	O	16.00	9	13.61	4	35.15	9	54.93	6	77.39	1
10	Ne	20.18	1	21.56	6	41.07	9	63.50	4	97.16	9
11	Na	23.00	2	5.14	1	47.29	6	71.65	9	98.88	4
18	A	39.94	1	15.75	6	27.62	9	40.90	4	59.79	9
19	K	39.10	2	4.34	1	31.81	6	46.00	9	60.90	4
36	Kr	83.70	1	14.00	6	24.56	9	36.90			

Thus, when there are only neutral atoms ($i = 0$), singly charged ions and
electrons, and $x_e = x_i$ is between 0 and 0.5. and $x_a = 1 - 2x_e$ is between 1
and 0.

From Equations (6.91–6.93), we can write now a quadratic equation of
x_e,

$$x_e^2 - \frac{i - (i+1)S}{i+1}x_e - S = 0 \tag{6.96}$$

which can be solved easily. From the solution for $i = 0$, it can shown, that
for $S \ll 1$, $x_e \approx \sqrt{S}$ and for $S \gg 1$, $x_e \approx 0.5$.

A quick estimate of the mole fraction of the electrons x_e with the help of
(6.91) at $p = 1$ bar and $T = 20{,}000$ K for argon gas gives a value of $x_e =
0.45$ if for the ratio of statistical weights at ground state $(g_{i+1,0}/g_{i,0})$ a value
of 6.0 is taken, whereas $x_e = 0.4953$, if the partition functions are evaluated
in a more elaborate manner (the values of partition function ratios for argon
plasma are given in Table 6.3).

While in all these calculations the ionization potential of argon for first
ionization ($I_o = 15.75$ eV) was taken, it has been found experimentally that,
as the temperature is increased, the *ionization potential* of particles is lowered
due to a polarizing effect of the charged particles. This indicates that there is

Table 6.3. Partition function ratio for an excited ion to atom for argon

T[K]	5,000	6,000	7,000	8,000	9,000	10,000	11,000
$Z_{1,\text{exc}}/Z_{0,\text{exc}}$	5.325	5.419	5.490	5.591	5.591	5.628	5.658

T[K]	13,000	14,000	15,000	16,000	17,000	18,000	19,000
$Z_{1,\text{exc}}/Z_{0,\text{exc}}$	5.707	5.726	5.744	5.759	5.773	5.736	5.799

a reduction in the amount of energy to ionize, that is, the ionization potential is reduced. Several criteria have been suggested by many authors for this, and *Veis* [108], comparing the results of these authors suggested for *lowering the potential* the formula

$$\Delta I_i = \frac{q_{\mathrm{eff}}^2 e^2}{4\pi\epsilon_o \lambda_D} \tag{6.97}$$

where

$\epsilon_o = $ *dielectric constant in vacuum* $= 8.855\times10^{-12}$, $\mathrm{As(Vm)}^{-1}$
$e = $ *elementary charge* $= 1.602\times10^{-19}$, As
$q_{\mathrm{eff}} = $ *effective charge number* $= \mathrm{i} + 1$

and λ_D is the *Debye shielding distance* (discussed in later Chap. 8, Sect. 8.2), which for the general case of different temperature for the electrons and heavy particles (temperature ratio $\theta = T_e/T_h > 1$) is given by the relation

$$\lambda = \left[\frac{2\epsilon_o k_B T_e}{e^2 n_e(1+\theta)}\right]^{0.5}. \tag{6.98}$$

Further in the calculation of the partition function of excitation the question arises as how far the energy levels should be counted. It is known that the statistical weight g is proportional to the square of the principal quantum number. Thus as $n \to \infty$, $g \to \infty$ and the partition function of excitation $Z_{\mathrm{exc}} \to \infty$, unless a limit is imposed in considering the energy levels. Suggestions were made by some authors that the maximum energy for calculating the partition function should be such that the distance between the electrons and the nucleus should be less than the mean distance between the particles. It is shown in Chap. 2 that the maximum energy is inversely proportional to the maximum possible quantum number, whereas the nucleus-electron distance is directly proportional to the square of the principal quantum number. If this distance is to be less than the mean distance between the particles of the order of $n^{1/3}$, the *lowering the potential* should be proportional to $n^{1/3}$. A criterion suggested by *Unsöld* [25] for the limiting maximum energy is, therefore

$$E_{i,r_{\max}} \leq I_i - \frac{3e^2}{\epsilon_o}(q_{\mathrm{eff}} n_e)^{1/3} = I_i - 5.4273 \times 10^{-8}(q_{\mathrm{eff}} n_e)^{1/3}. \tag{6.99}$$

The results herein is in electron-volts, if the electron particle density n_e is in m^{-3}. It is obvious that the cut-off energy can not be less than the ground state energy level $E_{i,o}$.

Now we consider the overlapping temperature region where n_i is not zero, but also n_{i+2} is not zero. For such a case we write the Saha function for two part equilibrium

$$S_i = \frac{x_e x_{i+1}}{x_i} = 6.6 \times 10^{-7} \frac{T^{5/2}}{p} \frac{Z_{i+1,\mathrm{exc}}}{Z_{i,\mathrm{exc}}} \exp^{-I_i/(k_B T)} \qquad (6.100)$$

$$S_{i+1} = \frac{x_e x_{i+2}}{x_{i+1}} = 6.6 \times 10^{-7} \frac{T^{5/2}}{p} \frac{Z_{i+2,\mathrm{exc}}}{Z_{i+1,\mathrm{exc}}} \exp^{-I_{i+1}/(k_B T)} \qquad (6.101)$$

with the two auxiliary equations, sum of mole fraction equal to one,

$$x_e + x_i + x_{i+1} + x_{i+2} = 1 \qquad (6.102)$$

and quasi-neutrality condition,

$$x_e - i x_i - (i+1)x_{i+1} - (i+2)x_{i+2} = 0 \ . \qquad (6.103)$$

Equations (6.100–6.103) can now be solved to determine four unknown mole-fractions. While the method of calculation is a straight forward solution of a polynomial equation, the situation becomes complicated because of the non-linearity, if there are more than three specie in the mixture. Because of this reason, we discuss in the following an alternate general iterative method, which can be applied for any number of specie. Accordingly, we consider a system of equation for reaction

$$A_e = 1A_e + 0A_{i+1}$$
$$A_{i+1} = 0A_e + 1A_{i+1}$$
$$A_i = 1A_e + 1A_{i+1}$$
$$A_{i+2} = -1A_e + 1A_{i+1} \qquad (6.104)$$

where the right hand side components may be considered to be the *primary components* (out of which all other components including each of them are produced) and the other components are *secondary components*. We have, therefore, two primary components and two *secondary components*. These can now be written in the form of a valency matrix (ν-matrix)

$$\nu = \begin{vmatrix} 1 & 0 \\ 0 & 1 \\ 1 & 1 \\ -1 & 1 \end{vmatrix} \ . \qquad (6.105)$$

Further, in analogy to (6.70) we write the matrix equation

$$\psi \cdot \mathbf{X} = \mathbf{b} \qquad (6.106)$$

where

$$\psi = \begin{vmatrix} 1 & 1 & 1 & 1 \\ 1 & -(i+1) & -i & -(i+2) \end{vmatrix} , \quad \mathbf{X} = \begin{vmatrix} x_e \\ x_{i+1} \\ x_i \\ x_{i+2} \end{vmatrix} , \quad \mathbf{b} = \begin{vmatrix} 1 \\ 0 \end{vmatrix} \ . \qquad (6.107)$$

In actual calculation, the values of mole fraction for the primary components, x_e and x_{i+1}, are guessed at first and x_i and x_{i+2} are calculated. When $\mathbf{X}$ vector is multiplied with ψ one gets, however, a vector, $\mathbf{b}'$, which is, generally different from $\mathbf{b}$. Therefore, we must modify the mole fraction of the primary components and consequently also the secondary components to modify $\mathbf{X}$, done by the method already discussed in Sect. 6.4 after Equation (6.70).

6.7 Equilibrium Composition of a Multiple Temperature Gas Plasma

Already in the beginning of a new section, the question may be raised, whether equilibrium does not mean equal temperature for all species in all energy modes. Unfortunately this is not so, especially when transfer of energy in one mode is quite different (much smaller) than another mode. On the basis of experimental evidence, the electrons are known to have higher translational temperature than the heavy particles (atoms, ions) in electric discharges, near cold walls, etc. In a study of the magneto-gas-dynamic generators using combustion gas products it was realized (*Kerrebrock* [28, 73]), that the recombination rates in molecular gases were too high to permit electron densities to develop, but concluded that non-equilibrium ionization might be feasible in atomic working gases. In experiments with potassium seeded argon and helium plasma (*Zukowski*, et al. [111, 112]) the electrical conductivity determined from the measured values of total current and the electric field strength seemed to agree well with the values calculated with the help of an equilibrium composition model, for which the temperature in the Saha equation was replaced by the electron temperature [28]. A two-temperature model was also established from the measured heat transfer to a cooled anode (*Shih*, et al. [105]). It was found that in electric discharges the rotational and vibrational temperatures often deviated from the plasma temperature (*Shahin* [31]), since perturbation in the populations of the rotational and vibrational levels occur in collisions with metastable atoms, and in collisions with chemical reaction, including those leading to the dissociation of polyatomic molecules or the recombination of molecular ions. For example in electric arcs where the electron density is high, the rotational temperatures, which were observed, were close to the arc temperature (the arc temperature, measured by the spectrometry, is the excited electron temperature). Further in reaction zones for flames, abnormally high rotational and vibrational temperatures resulted from excitations due to elementary chemical reactions (*Emmons* [62]). On the other hand if the electronic collisions were less, the inelastic collisions with electrons provided the electronic excitation of the molecules, but did not perceptibly alter the angular momentum with the result that the rotational temperature represented the translational temperature of the molecules. Because of difficulties in delineating exactly the range in which the vibrational, rotational and excitational temperatures deviate from the translational temperature, the multi-temperature model described in this section is restricted to the temperature range in which no molecules are present. Since the method by *Kerrebrock* to modify the Saha equation by replacing the temperature by the electron temperature gives too high value of electron density at lower temperature range, another method due to *Veis* [108] has been used. For completeness, we should mention also the two-temperature model of *Morro* and *Romeo* [89], which is obtained from the condition of the vanishing reaction rates. The problem in this method is that the equilibrium composition is dependent on the reaction rates (which may be very inaccurate) and hence the results may be quite different from the values one can obtain from the more rigorous Saha equation. Another two-temperature model, similar in structure like that of *Morro* and *Romeo*, but derived on the grounds of more thermodynamic laws, was reported by *Sanders*, et al. [107].

Obviously a two- or multi-temperature plasma is not at equilibrium, and left to itself for sufficiently long time, a common "*single*" temperature will be found. However from the examples discussed in the previous paragraph, it is evident that a "*quasi-steady*" model will be properly represented by a two-temperature or a multi-temperature model, the former being discussed in this section. Therefore, the state of the plasma being considered is assumed to be characterized by the pressure p, the electron temperature T_e and the heavy particles translational temperature T_h. The temperature ratio $\theta = T_e/T_h$ is assumed to be greater than or equal to one. The excitation temperature of the electrons in the bound state of the heavy particles is T_e, and we do not consider presence of molecules. The gas mixture may consist of a number of components of various specie designated with subscript j. The equation of state for such a plasma is

$$p = k_B \sum n_j T_j = k_B n \sum x_j T_j = n k_B T_m \qquad (6.108)$$

where the translational temperature of a specie is T_j, the total number density $n = \sum n_j$ and the mole-fraction $x_j = n_j/n$. We can, therefore evaluate an average temperature T_m from the relation

$$T_m = \sum x_j T_j \ . \qquad (6.109)$$

Since we are considering only a two temperature model, and we have to consider also the quasi-neutrality condition (the sum of opposite charges cancel each other globally), (6.109) becomes

$$T_m = x_e T_e + T_h \sum_h x_j = x_e T_e + (1 - x_e)T_h = T_e \left[x_e + \frac{1 - x_e}{\theta} \right] \ . \qquad (6.110)$$

Thus for $x_e \to 0$, $T_m = T_h$ and for $x_e \to 1$, $T_m = T_e$. In addition to the state defined in terms of the total pressure and two temperatures we need to know the mole fraction of the electrons, which can be guessed initially.The starting point of calculation of equilibrium composition for multi-temperature plasma is the definition of free enthalpy of the j-th species, which is given by the relation for the single temperature gas as

$$G_j = R^* T_j \ln \left(\frac{Z_j}{N_A} \right) \qquad (6.111)$$

where N_A is the *Avogadro number*, R^* is the *universal gas constant* and Z_j is the partition function given by the relation

$$Z_j = Z_{j,tr} \cdot Z_{j,r} \cdot Z_{j,v} \cdot Z_{j,\text{exc}} \cdots . \qquad (6.112)$$

In the above equation the second subscript denotes the mode of energy. For example, "tr" denotes translation, "r" denotes rotation, "v" denotes vibration, "exc" denotes electronic excitation, and so on. Some of the additional contributions, shown mathematically with dots, are due to the interaction between different energy types like rotation and vibration, etc.

In case there are different temperatures for different modes of energy, the above equation may be written as follows:

$$G_j = R^* T_{j,tr} \ln Z_{j,tr} + R^* T_{j,r} \ln Z_{j,r}$$
$$+ R^* T_{j,v} \ln Z_{j,v} + R^* T_{j,\text{exc}} \ln Z_{j,\text{exc}} \cdots \qquad (6.113)$$

where for each mode of energy the partition function relation are as follows:

$$Z_{j,tr} = \frac{1}{n_j} \left(\frac{2\pi k_B m_j T_{j,tr}}{N_A h^2} \right)^{3/2} ; Z_{j,r} = \frac{T_{j,r}}{\Theta_{j,r}}$$

$$Z_{j,v} = \frac{1}{2 \sinh[\Theta_{j,v}/(2T_{j,v})]}$$

$$Z_{j,\text{exc}} = \sum g_{j,m} \exp^{-E_{j,m}/(k_B T_{j,\text{exc}})} = Z'_{j,\text{exc}} \exp^{-I_j/(k_B T_{j,\text{exc}})} .$$

Herein k_B is the *Boltzmann constant*, m_j is the mole mass of the j-th species, n_j is the *particle number density*, h is the *Planck constant*, $\Theta_{j,r}$ is the characteristic rotational temperature of the species and $\Theta_{j,v}$ is the characteristic vibrational temperature of the species. Further, the expression for the vibrational partition function is written for a diatomic molecule in which the atoms vibrate as a harmonic oscillator and the expression for the excitation contains the subscript "m" to refer to a particular energy level for a species like neutrals, singly charged ions, doubly charged ions, etc. Taken from the ground excitation level of the neutral atom, the excitation energy E_m is the sum of the ionization energy up to the ground level of the species from the ground level of the neutral atom of the same species, $I_{j,o}$, and the relative excitation energy E'_{jm} from the ground level of the particular ionization level. Thus, the modified partition function, $Z'_{j,\text{exc}}$ accounts from the ground state of the particular specie only. In addition, the partition function expressions for the contribution of vibration and rotation of diatomic molecules have to be appropriately altered for the case of more complicated molecules.

There appears to be experimental evidence that for non-thermal-equilibrium plasma the excitation temperature of the bound electrons are at equilibrium with the translational temperature of the free electrons. There is, however, not much of information for such evidence for rotational and vibrational temperatures, which need be considered in the case of slightly conducting molecular gas, as in the case of seeded combustion plasma for magneto-gas-dynamic applications. In view of this, further analysis in this section is done only for the mixture of the particles like monatomic neutrals (atoms), ions and electrons. For the first two the translational temperature is T_h, but the excitation temperature is the same as the electron translational temperature, T_e. We consider now, therefore, the general case of a gas mixture consisting of atoms (these are considered to be a special case of ions with charge index zero) and ions of different elements designated with superscripts k, and the electrons with e, and the state is given by (T_e, T_h, p). Thus at any given state the gas mixture, for each element in the mixture, ions and neutrals are designated by charge index i, i+1 and i+2 (i = 0 for neutrals). Thus we consider the reactions

$$A_i^k \leftrightarrow A_{i+1}^k + e \; ; \; A_{i+2}^k \leftrightarrow A_{i+1}^k - e \; . \tag{6.114}$$

From (6.63), the equivalent Saha equation for the two reactions will now be

$$S_i^k = x_e \left(\frac{x_{i+1}^k}{x_i^k} \right)^{1/\theta}$$

$$= \frac{2Z_{i+1,\text{exc}}^{\prime k}}{Z_{i,\text{exc}}^{\prime k}} \frac{1}{n} \left(\frac{2\pi m_e k_B T_e}{N_A h^2} \right)^{3/2} \exp^{-I_i^k/(k_B T_e)} \tag{6.115}$$

$$S_{i+1}^k = x_e \left(\frac{x_{i+2}^k}{x_{i+1}^k} \right)^{1/\theta}$$

$$= \frac{2Z_{i+2,\text{exc}}^{\prime k}}{Z_{i+1,\text{exc}}^{\prime k}} \frac{1}{n} \left(\frac{2\pi m_e k_B T_e}{N_A h^2} \right)^{3/2} \exp^{-I_{i+1}^k/(k_B T_e)} \; . \tag{6.116}$$

Calculation is done by first guessing the mole fractions x_e and x_{i+1}^k after which x_i^k and x_{i+2}^k are computed with the help of (6.115, 6.116). Further calculations are done by the procedure given in the previous section and the result of calculation for a two-temperature argon plasma at 1 bar is given in Fig. 6.7 (*Bose* [45]). The method is quite straight forward.

The method is further simplified, if at any time, three species only are considered ($x_{i+2} = 0$), and we consider only (6.115). We get now the following

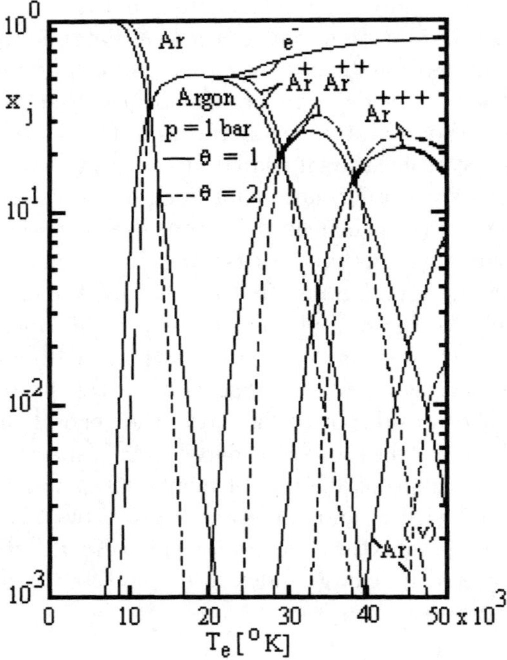

Fig. 6.7. Composition of two-temperature argon plasma at 1 bar

polynomial equation

$$x_{i+1}^{\theta+1} + i x_{i+1}^{\theta} + (i+2)x_{i+1} - S_i^{\theta} = 0 \qquad (6.117)$$

which for $\theta = 1$ reduces to a quadratic equation. Otherwise for $\theta \neq 1$, (6.117) can be solved by a method such as *Newton-Raphson method*.

6.8 Temperature Derivatives of Equilibrium Gas Mixtures

When a hot gas is confined in an enclosure with comparatively cold wall, at the center of the enclosure there is hot dissociated or ionized gas, while near the cold wall there are mainly neutral particles. Therefore gradient of the various species are created, which cause diffusion of particles with corresponding transfer of energy transfer. This energy transfer is in addition to that will occur due to "*pure*" conduction only. One way we can compute the diffusive mass and energy transfer if we can determine the mole fraction derivative with respect to temperature for the single-temperature case, or partial derivatives with respect to multiple temperatures, assuming, of course, that there is always local equilibrium.

In the previous two sections we discussed methods for determination of mole-fraction of multicomponent gas mixtures. The method requires, in the first instance, determination of the *equilibrium constant*, which is required further in an iterative fashion for determination of the mole fraction of various components, x_j, in the mixture. In principle, one could use the method further for determination of the value of (dx_j/dT), the temperature derivative of the mole fraction of the j-th species, which we will require later in our discussion about the *reactive heat conductivity coefficient*. However, determination of this temperature derivative of the mole fraction by the above iterative method is time consuming and hence, therefore, a simple method is given for this purpose in this section first for a *single temperature* (temperature of all specie is the same) reacting gas mixture. Starting point in our discussion is (6.74), the temperature derivative of which is

$$\frac{db_i}{dT} = 0 = \sum_{j=1}^{n} \psi_{kj} x_j \frac{d\ln x_j}{dT} = \sum_{j=1}^{n} \psi_{k,j} x_j \left[\frac{d\ln K_{xj}}{dT} + \sum_{i=1}^{r} \nu_{ji} \frac{d\ln x_i}{dT} \right]$$

$$= \sum_{j=1}^{n} \psi_{kj} x_j \frac{d\ln K_{xj}}{dT} + \sum_{j=1}^{n} \psi_{kj} x_j \sum_{i=1}^{r} \nu_{ji} \frac{d\ln x_i}{dT} \ . \tag{6.118}$$

From the definition of **g**-matrix in (6.78), we can now write the above equation as

$$\sum_{i=1}^{r} g_{i,k} \frac{d\ln x_i}{dT} = - \sum_{j=1}^{n} \psi_{kj} x_j \frac{d\ln K_{xj}}{dT} \ . \tag{6.119}$$

Since, from the equilibrium composition calculation of mole fraction described in previous two sections, the mole fraction x_j is known, the coefficient matrix g_{ik} can be determined easily. In the right hand side the temperature derivative of $\ln K_{xj}$ can be determined numerically or analytically (if $\ln K_x$ is given, for example, as a polynomial of T), and thus a simple solution of a system of simultaneous linear equation is required to obtain the temperature derivative of all other components, since from (6.60) we can write

$$\frac{d\ln x_j}{dT} = \frac{d\ln K_{xj}}{dT} + \sum_{i=1}^{r} \nu_{ji} \frac{d\ln x_i}{dT} \ . \tag{6.120}$$

While we discussed in section (6.4) the method to compute $(c_{p,\text{eff}}/c_{p,f})$ for ideal diatomic gases, we would now determine similar properties for a single species gas plasma, like for noble gases or other monatomic gases. This is one of the simplest cases of investigation. We consider only three components: electrons, i-th and (i+1)-th ions (i = 0 for neutral atoms). First we discuss for single temperature plasma. Now from (6.94) we write

$$\frac{d\ln x_{i+1}}{dT} = \frac{1}{x_{i+1}} \frac{dx_{i+1}}{dT} = \frac{d\ln x_e}{dT} \tag{6.121}$$

and

$$\frac{\mathrm{d}\ln x_i}{\mathrm{d}T} = \frac{1}{x_i}\frac{\mathrm{d}x_i}{\mathrm{d}T} = \frac{-(2+i)x_e}{(1+i)-(2+i)x_e}\frac{\mathrm{d}\ln x_e}{\mathrm{d}T} \ . \tag{6.122}$$

Further from (6.91)

$$\begin{aligned}
\frac{\mathrm{d}\ln K_x}{\mathrm{d}T} &= \frac{\mathrm{d}\ln x_e}{\mathrm{d}T} + \frac{\mathrm{d}\ln x_{i+1}}{\mathrm{d}T} - \frac{\mathrm{d}\ln x_i}{\mathrm{d}T} \\
&= \left[\frac{2(1+i)-(2+i)x_e}{(i+1)-(i+2)x_e}\right]\frac{\mathrm{d}\ln x_e}{\mathrm{d}T} = \frac{1}{T}\left(\frac{5}{2}+\frac{I_i}{k_BT}\right)
\end{aligned} \tag{6.123}$$

and hence,

$$\frac{\mathrm{d}x_e}{\mathrm{d}T} = \frac{x_e}{T}\left[\frac{(i+1)-(i+2)x_e}{2(1+i)-(2+i)x_e}\right]\left(\frac{5}{2}+\frac{I_i}{k_BT}\right) \ . \tag{6.124}$$

Since,

$$\frac{i}{i+1} \le x_e \le \frac{i+1}{i+2}, \ (i = 0,\,1,\,2,\,\cdots) \tag{6.125}$$

it can be shown easily that at the higher limit of x_e for a given ionization index i, $\mathrm{d}x_e/\mathrm{d}T \to 0$. On the other hand only for $x_e \to 0$ (that is, i $= 0$), $\mathrm{d}x_e/\mathrm{d}T = 0$.

We would now derive similar expressions for temperature gradient for multi-temperature plasma of a single element gas. Noting from the equilibrium two temperature model with three components of a single species, as described by the modified Saha equation

$$S = nx_e\left(\frac{x_{i+1}}{x_i}\right)^{1/\theta} \tag{6.126}$$

it can be easily shown that

$$\frac{\partial x_e}{\partial T_h} = -\frac{g}{fT_h} + \frac{5}{2f}\left[\frac{\theta}{T_e}+\frac{I_i}{k_BT_e^2}\right] \tag{6.127}$$

and

$$\frac{\partial x_e}{\partial T_e} = \frac{g\theta}{fT_h} \tag{6.128}$$

where

$$\begin{aligned}
f &= \frac{\theta(x_e+a)+x_e}{(x_e+a)x_e} - \frac{\theta^2(1-bx_e)-b(1-x_e)-\theta}{(1-bx_e)[x_e(\theta-1)+1]} \\
g &= \frac{1-x_e}{x_e(\theta-1)+1} - \ln\left(\frac{S}{x_e}\right) \\
a &= -\frac{i}{i+1}, b = \frac{i+2}{i+1} \ .
\end{aligned} \tag{6.129}$$

The above expressions can now be used easily to get the two *"reactive heat conductivity coefficients"* for the electrons and the heavy particles.

6.9 Effect of Radiation

The calculation of the equilibrium composition of ionized gases was described so far without taking into account the role of the radiation. This lacuna will now be removed by trying to estimate the effect of radiation. Effects of radiation on the general composition of the plasma are two-fold. Firstly, in a volume of radiating gas of a dimension so large that the absorption is significant and the gas radiates an equilibrium radiation, there are associated radiation quantities,

$$\text{pressure}: p_R = \frac{4\sigma}{3c}T^4 = 0.252 \left(\frac{T}{10^5}\right)^4 , \text{ bar} \tag{6.130}$$

$$\text{enthalpy}: h_R = \frac{16\sigma}{3\rho c}T^4 = \frac{4}{\rho}p_R = \frac{100.8}{\rho}\left(\frac{T}{10^5}\right)^4 , \text{ kJ.kg}^{-1} \tag{6.131}$$

$$\text{entropy}: s_R = \frac{16\sigma}{3\rho c}T^3 = \frac{100.8}{\rho}\left(\frac{T}{10^5}\right)^3 , \text{ kJ.kg}^{-1}\text{K}^{-1} \tag{6.132}$$

where ρ is the *gas density* (kg.m^{-3}), $\sigma = 5.672 \times 10^{-11}$ kW.m^{-2}.K^{-4}, and $c = $ *speed of light* $= 2.998 \times 10^8$ m.s^{-1}.

It is noted that at temperatures around 10–20,000 K, the partial pressure of the equilibrium radiation is sufficiently small to warrant special consideration. Even at temperatures, in which this partial pressure of equilibrium radiation may be significant, the gas may not actually be radiating as a black body for most cases, since the local radiation equilibrium is only allowed if the *mean free path of radiation* is so small that the radiation can be *trapped* between the particles, which is an impossible condition for laboratory plasmas. At the outset, the total pressure, the enthalpy and the entropy should include contribution from individual species, as well as due to radiation and thus,

$$p_{\text{total}} = p + p_R, h_{\text{total}} = h + h_R \quad \text{and} \quad s_{\text{total}} = s + s_R . \tag{6.133}$$

Thus from the given total pressure one may subtract the radiation pressure, which is zero for laboratory plasmas, to obtain the gas total pressure p, which in turn may be used to calculate the equilibrium composition given by methods given in earlier sections.

In many cases, however, one can assume equilibrium among different ionization and recombination mechanisms to determine the overall equilibrium. Thereby it is assumed, for simplification, that all atoms and ions are in their respective ground states only, and the interaction with the radiative energy is to ionize further or deionize. Thus, excitation and de-excitation are not considered, and only the following reactions are taken into account.

Ionization by collision by fast electrons and three-body recombination:

$$A_e + e + \frac{1}{2}M_e v_e^2 \leftrightarrow A_{i+} + e + e . \tag{6.134}$$

The basic requirement in such a case is that the kinetic energy of the colliding electron must be larger than the ionization energy. On the other hand, a mere coming in contact of an electron with the ion A_{i+1} is not sufficient to recombine to get A_i, unless the ionization energy that is freed is taken away to avoid re-separation. This excess energy is given up to the third colliding electron, or as will just be discussed, also by emitting radiation. The number of ionizations per unit volume and time by collision with fast electrons is $N_1 = k_1 n_i n_e$, where k_1 represents the transition probability for ionization by collision $(\mathrm{m^3 s^{-1}})$, and n_i, n_e are the number densities $(\mathrm{m^{-3}})$ for the i-th ion and electrons, respectively. Similarly for the three-body recombination, the number of ionization per unit volume and time is $N_2 = k_2 n_{i+1} n_e^2$, where k_2 is in $(\mathrm{m^6 s^{-1}})$

$$A_i + h\nu \leftrightarrow A_{i+1} + e \ . \tag{6.135}$$

For this the number of ionizations per unit volume and time by radiation is $N_3 = k_3 n_i$, where k_3 is in $(\mathrm{s^{-1}})$. It may be pointed out that the number of transitions should also strictly depend on the intensity of radiation about whose influence we would discuss later. The number of photo-recombinations per unit volume and time is $N_4 = k_4 n_{i+1} n_e$. Now since, $N_1 + N_3 = N_2 + N_4$, the relation for the equilibrium constant is

$$K_n = \frac{n_{i+1} n_e}{n_i} = \frac{k_1 n_e + k_3}{k_2 n_e + k_4} \ . \tag{6.136}$$

While the above equation is very simple and could formally be used for calculation of the equilibrium composition, there are uncertainties in knowing the values from k_1 to k_4. A limiting case may be considered, if the produced radiation may leave the gas volume freely without absorption, in which case $N_3 = k_3 = 0$. Thus specially under this condition

$$K_n = \frac{n_{i+1} n_e}{n_i} = \frac{k_1 n_e}{k_2 n_e + k_4} \ . \tag{6.137}$$

Although uncertainties exist in the values of k_1, k_2 and k_4, *Elwert's formula* (*Knoche* [113]; *Elwert* [61]) gives an order of magnitude of these quantities.

6.10 Exercise

6.10.1 Compute at a given pressure and temperature the molefraction of all components in the dissociation of (a) CO_2; (b) N_2H_4; and (c) H_2O.

6.10.2 Compute at a given pressure and temperature the mole fraction of all components in the ionization of all the five noble gases.

6.10.3 For the diatomic molecules, for which the characteristic temperature and dissociative energy are given in Table 2.1, derive a close-form equation for the equilibrium constant. [Hint: write down the expressions for partition functions for molecules and atoms; how you will take care of the dissociation energy?].

7 Transport Properties
of High Temperature Gases

In last three chapters we have discussed several aspects of equilibrium state. In Chap. 3, it was the equilibrium energy distribution of the particles, in Chap. 4, it was radiation with special emphasis to equilibrium or blackbody radiation and in Chap. 6 we have talked about chemical equilibrium by considering the extremum of the free enthalpy. Even the multi-temperature model implies a quasi-equilibrium, because changes do not take place very slowly. All these require a uniform distribution of the particle number distribution or the state, and a departure from the equilibrium from point to point can cause the flow of mass, moment and energy.

While the transport of particles, momentum and energy for gases at moderate temperatures have been dealt suitably by *Hirschfelder et al.* [10] and *Chapman* and *Cowling* [5], these are now being extended for gases in which there are charged particles, and with or without magnetic fields. In the presence of electromagnetic fields, singly-charged particles are subjected to additional forces, for which the Maxwell's equations are discussed first in Sect. 7.1, followed by a discussion on the motion of such particles. Section 7.2 deals next with the transport properties of a collision-dominated gas, followed by a discussion in Sect. 7.3 on the special features of a collision-dominated ionized gas plasma in electromagnetic fields. Thus the special features like the mobility coefficient, the ambi-polar diffusion coefficient and the electrical conductivity coefficient are considered.

7.1 Motion of a Singly-Charged Particle
in Electromagnetic Fields

We begin our study of the motion of charged particles in electromagnetic fields by a discussion on the convention of signs used. The electric charge q_j of a single particle of the j-th species has values $-e$, 0 and $+ie$, for an electron, a neutral and ion, which has been formed by removing i electrons (i = 0 for neutrals) from neutral atoms. It may be noted that $e = 1.602 \times 10^{-19}$ As is the elementary charge. In case two charged particles j and k are placed in a vacuum at a distance r apart, then the force acting on each other in the direction of the line connecting the position of the two particles is given by the *Coulomb's law*

$$\mathbf{F} = \frac{1}{4\pi\epsilon_o}\frac{q_j q_k}{r^2} \tag{7.1}$$

where ϵ_o is the *dielectric constant* in vacuum. In case the right hand side of
(7.1) is positive, then the particles repel each other, and if they are negative,
then they attract each other. If one of the particles is held stationary (for
example, q_k), then the force divided by q_j is called the *electric field* given by
the relation

$$\mathbf{E} = \frac{\mathbf{F}}{q_j} = \frac{1}{4\pi\epsilon_o}\frac{q_k}{r^2} \ . \tag{7.2}$$

Thus, the electric field, $\mathbf{E}$ is the force acting on a unit charge by another
charge in vacuum. However, if there is an intervening medium (not vacuum)
between the two charges, then a shift in the position of charges in the intervening medium can give rise to a distribution of dipole sources affecting the
original electric field $\mathbf{E}$ to change to $\mathbf{D}$, the relation between the two being

$$\mathbf{D} = \epsilon\mathbf{E} \tag{7.3}$$

where ϵ is the *permittivity* (dielectric) of the medium and $\mathbf{D}$ is the *electric
displacement*.

In (7.2), the electric field in a direction away from the charged particle k
is positive if q_k is positive and gives the direction in which the positive unit
charge must move. In case there are several charges q_k, which are distributed,
then the electric field at the position of the particle j is given by the vector
addition of fields due to each particle k,

$$\mathbf{E} = \sum \mathbf{E}_k = \frac{1}{4\pi\epsilon_o}\sum \frac{q_k}{r_{jk}^2} \ . \tag{7.4}$$

The total electric field is dependent only on the position with respect to the
fixed k particles, and thus no work is done, if the single j-th particle is moved
around and brought back to the original position. Thus,

$$\int \mathbf{F}\cdot\mathbf{ds} = q_j \int \mathbf{E}\cdot\mathbf{ds} = 0 \tag{7.5}$$

from which it immediately follows from the Stokes theorem and the assumption of a steady electric field that

$$\nabla \times \mathbf{E} = 0 \ . \tag{7.6}$$

It follows further that one can define a *potential U*, such that

$$\mathbf{E} = -\nabla U \ . \tag{7.7}$$

Generally speaking, we are not interested in individual particles, and therefore, we consider the charge density

$$n_c = \sum n_k q_k. \ \mathrm{Asm}^{-3} \tag{7.8}$$

where k represents a species type (ion, electron, atom, etc.) and n_k is the particle number density of the k-th species $[\mathrm{m}^{-3}]$. Use of the *Gauss theorem* in converting from the volume integral to the surface integral yields

$$\nabla \cdot \mathbf{E} = n_c/\epsilon_o \tag{7.9}$$

and then finally to the *Poisson's equation*

$$\nabla^2 U = -n_c/\epsilon_o \ . \tag{7.10}$$

If there is an intervening medium, (7.9) now gets modified to

$$\nabla \cdot \mathbf{E} = \frac{n_c}{\epsilon} = \frac{n_c}{\epsilon_o} \cdot \frac{\epsilon_o}{\epsilon} \ . \tag{7.11}$$

The ratio ϵ/ϵ_o is the relative dielectric constant for the medium. The effect of different dielectric constants for two different media is shown from (7.7) that at the interface, the tangential component of the electric field in the two media are the same ($E_{t1} = E_{t2}$), but for normal components $D_{n1} = D_{n2}$, and thus the ratio of the normal components of the electric field is $E_{n1}/E_{n2} = \epsilon_2/\epsilon_1$.

Similar to the Coulomb's law giving the force acting between two charged particles, it is an experimentally observed fact that when electric currents I and I' flow through two parallel conductors at distance r apart, each having length l and l', respectively, then a mutual force in vacuum exerted between them is given by the relation

$$\mathbf{F} = \frac{\mu_o}{4\pi} \frac{II'll'}{r^2} \tag{7.12}$$

where μ_o is the *magnetic permeability* in vacuum. Just like the definition of the electric field, the *magnetic induction* $\mathbf{B}$ is defined in the manner to give the force per unit of the product of the current and length (Il), when there is no intervening medium (vacuum), and we get the relation

$$\mathbf{B} = \frac{\mu_o}{4\pi} \frac{I'l'}{r^2} \ . \tag{7.13}$$

Once again we may not be interested in a single current-carrying wire, but there may be a complete distribution of the current density $\mathbf{j}$. By using the *Stokes theorem* it can be shown again, that in vacuum

$$\nabla \times \mathbf{B} = \mu_o \mathbf{j} \ . \tag{7.14}$$

While the electric current flowing through a wire may induce by (7.13) a magnetic induction $\mathbf{B}$ in the surrounding vacuum, this equation is modified if there is an intervening medium. If the intervening medium is a gas with charged particles around, there may be other motions in the gas (for example, gyrating motion) giving rise to magnetic moments. Thus, (7.14) is changed to

$$\nabla \times \mathbf{H} = \mathbf{j} \tag{7.15}$$

where $\mathbf{j}$ is now the current density of the guiding centers and $\mathbf{H}$ is the magnetic field. Between $\mathbf{B}$ and $\mathbf{H}$ the relation is

$$\mathbf{B} = \mu \mathbf{H} \tag{7.16}$$

and the ratio (μ/μ_o) is a measure of other effects of the current flow on magnetization and μ is the *permittivity* of the gas through which the current is flowing. While these factors are for a steady state case, in an unsteady case these are modified. For example, as a consequence of *Faraday's law of induction*, the expression we get is

$$\nabla \times \mathbf{E} = -\frac{\partial \mathbf{B}}{\partial t} \tag{7.17}$$

which becomes the same as (7.6) for the steady state case only. Similarly, as a consequence of free charge density, the principle of conservation of electric current may not be valid and (7.15) is suitably modified to include the effect of the rate of change of the displacement current vector $\mathbf{D}$. As a consequence, one gets the following electro-magnetic equations, known as *Maxwell's equations*, in their usual form as follows:

$$\nabla \cdot \mathbf{D} = n_c \tag{7.18}$$

$$\nabla \cdot \mathbf{B} = 0 \tag{7.19}$$

$$\mathbf{D} = \epsilon \mathbf{E} \tag{7.20}$$

$$\mathbf{B} = \mu \mathbf{H} \tag{7.21}$$

$$\nabla \times \mathbf{E} = -\frac{\partial \mathbf{B}}{\partial t} \tag{7.22}$$

$$\nabla \times \mathbf{H} = \mathbf{j} + \frac{\partial \mathbf{D}}{\partial t} \tag{7.23}$$

From (7.19) the auxiliary condition at the interface of two media is that B_n, the normal component of $\mathbf{B}$ across the two media, does not change ($B_{n1} = B_{n2}$). This has important consequences in the study of magneto-gas-dynamic shocks.

For collisionless plasmas (ionized gases), the equation of motion of a single particle is given by the relation

$$M_j \frac{d\mathbf{w}_j}{dt} = q_j(\mathbf{E} + \mathbf{w}_j \times \mathbf{B}) \tag{7.24}$$

where M_j is the mass of the particle and $\mathbf{w}_j$ is the velocity of a single particle. Now the following special cases may be considered.

a) *No magnetic induction*, $\mathbf{B} = 0$, but $\mathbf{E}$ constant over space and time. Thus the constant acceleration in the direction of the electric field is

$$\frac{d\mathbf{w}_j}{dt} = \frac{q_j}{M_j}\mathbf{E} \ . \tag{7.25}$$

(b) *No electric field,* $\mathbf{E} = 0$, but $\mathbf{B}$ constant over space and time. Thus the force acting on the particle will be at right angle to both $\mathbf{w}_j$ and $\mathbf{B}$ and the particle must move around the magnetic induction lines $\mathbf{B}$ with a radius of curvature R_{cj}. Since $d\mathbf{w}_j/dt = w_j^2/R_{cj}$, from (7.24), we may write

$$\frac{w_j^2}{R_{cj}} = \frac{q_j}{M_j} w_j B \tag{7.26}$$

and one gets the equation for the radius of the circular path ($=$ *Larmor* or *cyclotron radius*)

$$R_{cj} = \frac{w_j M_j}{q_j B} \tag{7.27}$$

and the (*radian*) *cyclotron frequency*

$$\omega_{cj} = \frac{w_j}{R_{cj}} = \frac{q_j B}{M_j} \ . \tag{7.28}$$

While Equation (7.27) does not give an explicit expression for Larmor radius since w_j must be known; an order of magnitude estimation can be done if the latter is replaced by the mean or most probable kinetic speed. It can then be shown, that the Larmor radius has to be much smaller for the electrons than the one for the ions.

Looking in the direction of the magnetic induction $\mathbf{B}$, it may be noted that the positive charges gyrate around the magnetic induction line counter-clockwise, and the electrons clockwise. However, because of the very small mass of the electrons with respect to that for the ions, the cyclotron frequency for the electrons is much larger than that for the ions. Further, the ratio of the radian cyclotron frequency to the collision frequency is given by the relation

$$\xi_j = \omega_{cj}/\Gamma_j \tag{7.29}$$

where Γ_j is the collision frequency, which may be calculated with the method discussed in Chap. 5, which has thus important consequences for the transport properties.

(c) Let $\mathbf{E}$ and $\mathbf{B}$ be constant in space and time, and let $\mathbf{E}$ be perpendicular to $\mathbf{B}$. Further let

$$\mathbf{w}_j^* = \mathbf{w}_j - \frac{\mathbf{E} \times \mathbf{B}}{B^2} \ . \tag{7.30}$$

Thus from (7.24)

$$\begin{aligned}
M_j \frac{d\mathbf{w}_j}{dt} &= q_j \left[\mathbf{E} + \left\{ \mathbf{w}_j^* + \frac{\mathbf{E} \times \mathbf{B}}{B^2} \right\} \times \mathbf{B} \right] \\
&= q_j \left[\mathbf{E} + \mathbf{w}_j^* \times \mathbf{B} + \frac{1}{B^2} (\mathbf{E} \times \mathbf{B}) \times \mathbf{B} \right] \\
&= q_j \left[\mathbf{E} + \mathbf{w}_j^* \times \mathbf{B} - \mathbf{E} \right. = q_j (\mathbf{w}_j^* \times \mathbf{B}) \tag{7.31}
\end{aligned}$$

which is the same equation, as if the particles, having the velocity $\mathbf{w}_j^*$, were gyrating around the magnetic induction lines and without any effect of the electric fields. Thus in crossed electric and magnetic field, there is a drift of the charged particles

$$\mathbf{w}_{Dj} = \mathbf{w}_j - \mathbf{w}_j^* = \frac{\mathbf{E} \times \mathbf{B}}{B^2} \tag{7.32}$$

which is in a perpendicular direction to both electric and magnetic fields.

(d) Charged particles in crossed magnetic and gravitational fields, $\mathbf{B}$, is perpendicular to $\mathbf{g}$. Equation (7.24) is modified to

$$M_j \frac{d\mathbf{w}_j}{dt} = q_j(\mathbf{w}_j \times \mathbf{B}) + M_j \mathbf{g} \ . \tag{7.33}$$

Let

$$\mathbf{w}_j^* = \mathbf{w}_j - \frac{M_j}{q_j} \frac{\mathbf{g} \times \mathbf{B}}{B^2} \tag{7.34}$$

which follows

$$M_j \frac{d\mathbf{w}_j^*}{dt} = q_j \left[\left(\mathbf{w}_j^* + \frac{M_j}{q_j} \frac{\mathbf{g} \times \mathbf{B}}{B^2} \right) \times \mathbf{B} \right] + M_j \mathbf{g} = q_j(\mathbf{w}_j^* \times \mathbf{B}) \ . \tag{7.35}$$

There may be a drift of the particle

$$\mathbf{w}_{Dj} = \mathbf{w}_j - \mathbf{w}_j^* = \frac{M_j}{q_j} \frac{\mathbf{g} \times \mathbf{B}}{B^2} = \frac{\mathbf{g} \times \mathbf{B}}{\omega_c B} \ . \tag{7.36}$$

It is of interest to note that the mechanism just described plays a significant role in trapping charged particles near the earth due to the interaction of the earth's magnetic and gravitational fields.

7.2 Collision-Dominated Ionized Gas

While the thermal motion of gas 'particles' are random in nature, and is dependent on the local temperature of the gas, T_j, there are other velocities, which are directional in nature. The foremost among the latter is the gas dynamic velocity, which is due to the pressure gradient, that is imposed on the gas externally as boundary condition in the domain. The other directed velocities are more diffusive in nature such as those due to concentration gradient or electric potential gradient (electric field). Different particles diffuse in different directions simultaneously and hence it is convenient to have an average of velocities of all particles. This average can, either be done as a mass-averaged velocity $\mathbf{V}$ or number-averaged (or molar-averaged) velocity $\mathbf{V}^*$. The thermal velocities, for which while the directions are random, the magnitude can be averaged again, and the different ways of averaging (either simple mean or root mean square) was discussed already in Chap. 3. For the moment, let us now designate the random thermal velocity of all species as v, and the mass-averaged velocity of the j-th species, V_j, is superimposed on it, and thus the velocity of a single particle of the j-th species at any given instance is given by the relation

$$\mathbf{w}_j = \mathbf{v} + \mathbf{V}_j \ . \tag{7.37}$$

On the other hand, if we add the thermal velocity to the mass averaged velocity of *all* species, V, then we write a similar relation

$$\mathbf{w} = \mathbf{v} + \mathbf{V} \ . \tag{7.38}$$

Subtracting one from the other, the difference $\mathbf{w}_j - \mathbf{w} = \mathbf{V}_j - \mathbf{V}$ is the mass-averaged drift velocity of the j-th species due the concentration gradient (*diffusion*), the temperature gradient (*thermo-diffusion*), the electromagnetic fields (conduction of electric current), etc. One could, of course, define similarly molar-averaged drift velocity. While a more precise solution to the Boltzmann integro-differential equation is necessary not only for the exact calculation of the effects due to the gradients mentioned as above and the fields, the method is highly mathematical, and in the present case a simple approach will be taken so as not to lose track of the essential physical concepts (the method is called by *Hirschfelder*, et al. [10] as *ultra simplified theory*). To begin with, we do not consider any electromagnetic field and discuss the diffusion, heat conduction and viscosity coefficients, for a gas in general, but an ionized gas in particular.

In order to estimate these coefficients we consider global motion of a species moving parallel to the three planes and in the direction of the x-coordinate (Fig. 7.1) with a net velocity $V_j = V_j(z)$. The distance between the two plates is of the order of magnitude of a mean free path, λ_j, so that the particles do not undergo any collision within this distance from the wall. Thus the particles which leave the plane A has a momentum $(M_j V_j)_A$, where M_j is the mass of a single particle. Assuming that the random velocity of the species is such that at any instant one-sixth of the total number of particles move along one of the coordinate directions with mean kinetic speed, $\bar{v}_j$, then the flux of the particles from the plane A to the plane O is $\bar{v}_j n_j / 6$. Thus the flux of the momentum of the x-component per unit area in the z-direction from A to O plane is $\bar{v}_j n_j (M_j V_j)_A / 6$. Similarly the *momentum flux* from the B-plane to O-plane is $-\bar{v}_j n_j (M_j V_j)_B / 6$. Thus the net flow of the momentum flux from A and B planes to the O plane is

$$\tau_{xzj} = \frac{1}{6} \bar{v}_j n_j M_j (V_{jA} - V_{jB}), \ \text{Nm}^{-2} \ . \tag{7.39}$$

Provided we assume a linear distribution of the macrospeed V_j,

$$V_{jA} = V_{jO} - \lambda_j \frac{\mathrm{d}V_j}{\mathrm{d}z} \ , \ V_{jB} = V_{jO} + \lambda_j \frac{\mathrm{d}V_j}{\mathrm{d}z} \tag{7.40}$$

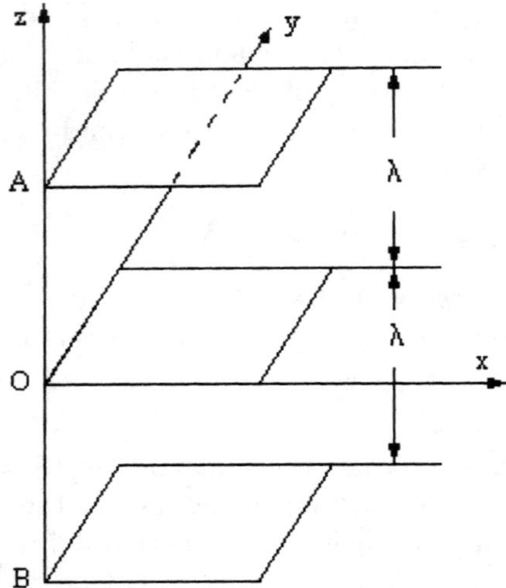

Fig. 7.1. Schematic diagram to explain transport properties

and thus, $(V_{jA} - V_{jB}) = -2\lambda_j dV_j/dz$ and

$$\tau_{xzj} = -\frac{1}{3}\bar{v}_j n_j M_j \lambda_j \left(\frac{dV_z}{dz}\right) . \tag{7.41}$$

By *Newton's law*, however

$$\tau_{xzj} = \mu_j \left(\frac{dV_z}{dz}\right) \tag{7.42}$$

where μ_j is the *dynamic viscosity coefficient* of the j-th species. Thus,

$$\mu_j = \frac{1}{3}\bar{v}_j n_j M_j \lambda_j, \ \mathrm{kgm^{-1}s^{-1}} . \tag{7.43}$$

Similarly at A, the particles have the near-microscopic kinetic energy $(M_j \bar{v}_j^2/2)_A = (3k_B T_{jA}/2)$ and the net flux of energy in z-direction at O for the j-th species is

$$q_j = \frac{1}{6} n_j k_B \bar{v}_j \frac{3}{2}(T_{jA} - T_{jB}) . \tag{7.44}$$

Assuming again a linear distribution of the temperature, we may write $(T_{jA} - T_{jB}) = -2\lambda_j dT_j/dz$, we write for the heat flux as

$$q_j = \frac{1}{2}\bar{v}_j n_j k_B \lambda j \frac{dT_j}{dz} \tag{7.45}$$

Now from *Fourier's law*

$$\mathbf{q}_j = -k'_{cj}\left(\frac{\mathrm{d}T_z}{\mathrm{d}z}\right)$$

(7.46)

where k'_{cj} is the *heat conductivity coefficient* of the j-th species considering translational motion alone. By taking the translational component of the energy transfer alone, one gets therefore the relation,

$$k'_{cj} = \frac{1}{3}\bar{v}_j n_j k_B \lambda_j = \frac{3}{2}R_j \mu_j \ , \ \mathrm{Wm}^{-1}\mathrm{K}^{-1}$$

(7.47)

in which R_j is the gas constant for the j-th species. This equation is adequate to predict the heat conductivity coefficient for pure monatomic gases. The theory, however, does not consider the transfer of other modes of internal energy within the molecules. *Eucken's method* gives the other contributions to the *heat conductivity coefficient*

$$k''_{cj} = 0.88\left[0.4\left(\frac{c_{pj}}{R_j}\right) - 1\right]k'_{cj} \ .$$

(7.48)

and it may be noted that for a pure gas of the j-th species (no gas mixture),

$$k_{cj} = k'_{cj} + k''_{cj}$$

(7.49)

Finally, the net flux of molecules at O in z-direction is

$$\dot{n}_j = \frac{1}{6}\bar{v}_j(n_{jA} - n_{jB}), \ \mathrm{m}^{-2}\mathrm{s}^{-1} \ .$$

(7.50)

Once again by assuming a linear distribution of the number density, $(n_{jA} - n_{jB}) = -2\lambda_j \mathrm{d}n_j/\mathrm{d}z$, and noting the *Fick's law*,

$$\dot{n}_j = -D_j\left(\frac{\mathrm{d}n_z}{\mathrm{d}z}\right)$$

(7.51)

we get the expression of the *diffusion coefficient*

$$D_j = \frac{1}{3}\bar{v}_j \lambda_j \ , \ \mathrm{m}^2\mathrm{s}^{-1} \ .$$

(7.52)

We would now like to discuss the motion of charged particles with charge e_j in a collision-dominated gas in presence of an externally applied electric field, $\mathbf{E}'$. We assume now that $\mathbf{E}'$ is in the positive z-direction and the charge of the particles is such that it accelerates in the direction of $\mathbf{E}'$. Now let us assume one charged particle comes from B to O, for which it has the free-flight time of $\tau_j = \lambda_j/v_j$, v_j being the kinetic speed of the molecule. Change in the velocity will now be

$$\Delta v_j = \frac{e_j \lambda_j}{M_j v_j}\mathbf{E}' \ .$$

(7.53)

Similarly for a charged particle coming from A to O, there is a retardation in the velocity. However, if the kinetic speed of the particle, v_j, is deducted from the speed of the above two particles, it gives an average *field drift velocity* given by the relation

$$\mathbf{w}_j = \frac{e_j \lambda_j}{M_j v_j}\mathbf{E}' = b_j \mathbf{E}' \tag{7.54}$$

where b_j is the *mobility coefficient* given by the relation

$$b_j = \frac{e_j \lambda_j}{M_j v_j} = \frac{3 e_j D_j}{M_j v_j^2} = \frac{e_j D_j}{k_B T_j} \ . \tag{7.55}$$

In the above equation the relation for the average kinetic energy (translational)

$$\frac{1}{2}M_j v_j^2 = \frac{3}{2}k_B T_j \tag{7.56}$$

is used.

We now use for mean kinetic speed (3.159), for mean free path (5.53) and for n_j from the equation of state, and the relations for the three *transport properties equations* for the j-th species become

(a) *dynamic viscosity coefficient*:

$$\mu_j = 8.385853 \times 10^{-6}\sqrt{T_j m_j}/Q_{jj}^{(2,2)} \ , \ \mathrm{kgm^{-1}s^{-1}} \tag{7.57}$$

where m_j is the mole mass of the species and $Q_{jj}^{(2,2)}$ is the collision cross-section for momentum and energy transfer in $\mathrm{\AA}^2$.

(b) *Translation energy contribution* of the *heat conductivity coefficient*:

$$k_{cj}' = 2.61644 \times 10^{-4}\sqrt{T_j/m_j}/Q_{jj}^{(2,2)} \ , \ \mathrm{kWm^{-1}K^{-1}} \tag{7.58}$$

and for the contribution of other forms of internal energy it is (7.48).

(c) *Self-diffusion coefficient*:

$$D_j = 2.62823 \times 10^{-7}\sqrt{T_j^3/m_j}/(pQ_{jj}^{(1,1)}) \ , \ \mathrm{m^2 s^{-1}} \tag{7.59}$$

where p is the pressure in bar and $Q^{(1,1)}$ is the collision cross-section for diffusion transport in $\mathrm{\AA}^2$.

Explanation of the diffusion coefficient for a single species gas may be somewhat difficult to visualize, but when we use a radioactive gas isotope of almost the same molecular weight as that of non-radioactive isotope, it is possible to explain it in terms of the diffusion of one or the other isotopes. For binary or multicomponent gas mixtures, the diffusion coefficient can be described with the help of the mean free path for binary collisions or multiple collisions, as has been described in Chap. 5. The latter is discussed also in Sect. 7.3.

It need to be explained further, that the collision cross-section to be used in (7.57–7.59) is the same for the rigid sphere model only, but otherwise they could be different. In the Lennard-Jones 6-12 model $Q_{jj} = Q_{jj}^{(2,2)}$ for viscosity and heat conductivity coefficients, and $Q_{jj} = Q_{jj}^{(1,1)}$ for diffusion coefficient.

7.3 Diffusion, Ambi-polar Diffusion and Mobility

In the previous section the diffusion coefficient is explained with respect to a single component gas. Even for a multi-component gas the diffusion coefficient relation (7.52) gives the value of only one component gas mixture with the corresponding particle flux relation (7.51). These relations are written once again as follows:

Particle flux:

$$\dot{n}_j = n_j \mathbf{V}_j = -D_j \nabla n_j \tag{7.60}$$

Diffusion coefficient:

$$D_j = \frac{1}{3} \bar{v}_j \lambda_j = \frac{\bar{v}_j^2}{3 \sum \xi_{jk} n_k Q_{jk}} \ . \tag{7.61}$$

By considering the state equation

$$p = n k_B T \ . \tag{7.62}$$

Equation (7.60) can now be written as

$$\dot{n}_j = n_j \mathbf{V}_j = -D_j \nabla n_j = -D_j \nabla (n x_j)$$

$$= -D_j \left[n \nabla x_j + \frac{x_j}{k_B} \nabla \left(\frac{p}{T} \right) \right] \tag{7.63}$$

where $x_j = n_j / n$ is the mole fraction of the j-th species. Further, we get the relation of the particle average speed as

$$\mathbf{V}_j = -D_j \left[\frac{1}{x_j} \nabla x_j + \frac{1}{p} \nabla p - \frac{1}{T} \nabla T \right] \ . \tag{7.64}$$

In the above equation the first term is due to the gradient in the mole fraction of the j-th species, the second term is due to the pressure gradient (*pressure diffusion*) and the third term is due to the (*thermo-diffusion*). For the present we consider the first term only, which is equivalent of stating that the total number density of all specie, n, is constant, and we write

$$\mathbf{V}_j = -\frac{D_j}{x_j} \nabla x_j \ . \tag{7.65}$$

For many cases it is convenient to work with the mass-density flux, rather than the number density flux. The relation for this can be derived easily by multiplying (7.63) with the mass of the species, M_j to get

$$\dot{\rho}_j = \rho_j \mathbf{V}_j = -D_j \nabla \rho_j \ . \tag{7.66}$$

In the definition of the diffusion coefficient we have so far assumed the particle flux dependent basically on its own gradient only, except considering the collision with other particles. However the diffusion process has to satisfy certain compatibility conditions. For example in reacting gases, some gases, which are in abundant in hot regions, diffuse to the colder wall, during which they may also undergo collisions with reacting partners, and consequently the reacted particles diffuse out of the cold regions and move to hotter regions; this, of course, depends on the reaction rates. For very fast reactions, there can be local chemical equilibrium, while for very slow reactions the composition may not change in the gas phase. However if there is catalytic surface reaction at the wall, then there can still be gradients of individual species in the gas phase.

Since the diffusion coefficient and the corresponding flux relation depend on the temperature of the species, its mole mass and the collision cross-section, it is evident that even for a bi-molecular gas mixture at a common temperature and common binary cross-section (neglecting collisions between particles of its own kind), but with different mole mass, the flux of each species is different. Hence in a gas container at a given pressure and temperature (constant volumetric number density), the gas particles with lighter molecules will have higher particles flux rate causing an imbalance of pressure and a non-zero particle-average or mass-average velocity. Thus the above relations for the diffusion coefficients can not be applied in a straight forward manner without satisfying certain compatibility conditions regarding average velocity of the particles, and for this purpose we define now first the global particle-average and followed by the mass-average velocity as follows:

Particle average velocity:

$$\mathbf{V}^* = \frac{1}{n} \sum n_j \mathbf{V}_j = -\frac{1}{n} \sum D_j \nabla n_j = -\sum x_j \mathbf{V}_j = -\sum D_j \nabla x_j \tag{7.67}$$

where V^* is the number-average velocity. By defining a diffusion velocity of the j-th species with respect to the particle-average velocity, $\mathbf{V}_j^{*\prime} = \mathbf{V}_j - \mathbf{V}_j^*$, obviously $\sum n_j V_j^{*\prime} = 0$. We introduce now an effective diffusion coefficient, D_{jm}^*, which represents diffusion with respect to the global average velocity, and for such the diffusive flux is

$$n_j V_j^{*\prime} = n_j(V_j - V^*) = -D_{jm}^* \nabla n_j = -\left(D_j \nabla n_j - \frac{n_j}{n} \sum_k D_k \nabla n_k \right) \tag{7.68}$$

from which we get the relation for the diffusion coefficient of the j-th species against the gas mixture as

$$D^*_{jm} = D_j - \frac{n_j}{n}\sum_k D_k \frac{\partial n_k}{\partial n_j} \approx (1 - x_j)D_j - x_j \sum_{k \neq j} D_k \frac{\partial x_k}{\partial x_j} \tag{7.69}$$

where x_j is the mole fraction of the j-th species.

mass-average velocity:

$$\mathbf{V} = \frac{1}{\rho}\sum \rho_j \mathbf{V}_j = -\frac{1}{\rho}\sum D_j \nabla \rho_j \tag{7.70}$$

where $\mathbf{V}$ is the mass-average velocity. By a similar method as above, we define now a diffusion velocity of the j-th species with respect to the particle-average velocity, and we write for the diffusion coefficient

$$D_{jm} = D_j - Y_j \sum_k D_k \frac{\partial \rho_k}{\partial \rho_j} \approx (1 - Y_j)D_j - Y_j \sum_{k \neq j} D_k \frac{\partial Y_k}{\partial Y_j} \tag{7.71}$$

where Y_j is the mass fraction of the j-th species. The analysis may be simpler, for a binary mixture, however.

For binary mixture of two gas components A and B, where the partial derivative is equal to -1, Eqs. (7.69, 7.71) become first binary diffusion coefficient for particle-averaged velocity

$$D^*_{Am} = D^*_{Bm} = x_A D_B + x_B D_A \tag{7.72}$$

and similarly the binary diffusion coefficient with mass-averaged velocity as

$$D_{Am} = D_{Bm} = Y_A D_B + Y_B D_A \ . \tag{7.73}$$

We now evaluate the binary diffusion coefficient, for the particular case when the collision between identical particles of each species are neglected. Although this is a very sweeping assumption but it is used by most authors. As explained above the molar binary diffusion coefficient between the j-th and k-th specie is

$$D^*_{jk} = x_k D_j + x_j D_k = \frac{\bar{v}_j^2 + \bar{v}_k^2}{3n\xi_{jk}Q_{jk}} \tag{7.74}$$

where the mean relative velocity is

$$\xi_{jk} = \frac{1}{6}\left[\mid \bar{v}_j + \bar{v}_k \mid + \mid \bar{v}_j - \bar{v}_k \mid + \sqrt{\bar{v}_j^2 + \bar{v}_k^2} \right] \tag{7.75}$$

and the non-dimensional mean relative velocity is

$$\xi'_{jk} = \xi_{jk}\sqrt{\frac{2}{\bar{v}_j^2 + \bar{v}_k^2}} \ . \tag{7.76}$$

For j = k, we get, of course, the value

$$\xi'_{jj} = (1 + 2\sqrt{2})/3 \tag{7.77}$$

which is double the value for collisions between identical particles, when calculated by the more rigorous kinetic theory (*Chapman* and *Cowling* [5]). Hence a correction is introduced as explained in the next paragraph.

Now in the general case of binary mixture of gas particles with different temperature (the electrons and the heavy particles in a high temperature gas in an electric field may have different temperatures), and the mean and the square of the mean kinetic speeds can be written in terms of the respective temperature. Thus the binary molar diffusion coefficient is

$$D^*_{jk} = C\frac{(x_j T_j + x_k T_k)}{p Q_{jk}}\sqrt{\frac{1}{2}\left(\frac{T_j}{m_j} + \frac{T_k}{m_k}\right)} \tag{7.78}$$

where the expression for C, after the above correction is brought in, is

$$C = \frac{1.051}{\xi'_{jk}}\sqrt{\frac{8k_B^3 N_A}{\pi}}\,. \tag{7.79}$$

In multiple species gas mixture, determination of the diffusion coefficient is difficult, and it is found convenient to define an *effective diffusion coefficient*, D_{jm}, for the diffusion of the i-th component in a mixture (Bird, et al. [2]), for which the following formulas are given:

a) For trace component of $j \neq i$ in nearly pure species i, $D_{im} \approx D_{ii}$,
b) For system in which all the binary diffusion coefficients, D_{ij} are the same, that is, $D_{im} = D_{ij}$, and
c) All j-th species ($j \neq i$) move with the same velocity (or are stationary),

$$\frac{1 - x_i}{D_{im}} = \sum_{j \neq i} \frac{x_j}{D_{ij}}\,. \tag{7.80}$$

The last expression is for the diffusion of molar flux.

While the above discussion is for a gas-mixture, it is evident that the diffusion coefficient can be calculated depending on the requirement of particle-averaged diffusion velocity or the mass-averaged diffusion velocity. While for gas dynamic flow problems, written in terms of the mass flow velocity is ideal for mass-averaged diffusion, there can be other situations where the particle number-averaged diffusion is more appropriate, especially if there are charged particles, where the Coulomb forces play a dominant role. In order to simplify the problem we do not consider the diffusion between ions with same charge (neutrals are considered as having charge zero). Even if there are more number of specie, all particles of same charge are considered together.

We thus consider the diffusion coefficient of an ionized gas mixture, in which there are the electrons, the i-th ions and the (i+1)-th ions (note that i = 0 is for neutrals). Initially we will not consider the effect of the electric field that is induced because of the light electrons trying to diffuse out faster from the gas mixture than the heavier ions and neutrals. We can, therefore, write the two compatibility conditions

$$x_e + x_i + x_{i+1} = 1 \tag{7.81}$$

$$x_e - ix_i - (i+1)x_{i+1} = 0 \ . \tag{7.82}$$

While (7.81) states that the sum of the mole fraction of all specie is equal to zero, (7.82) gives the quasi-neutrality condition. As a result we get mole fractions of the two positive ions in terms of the mole fraction of the electrons, and as such we get two equations for the gradients of x_i and x_{i+1} in terms of gradients of the mole fraction of electrons, x_e, as follows:

$$\nabla x_i = -(i+2)\nabla x_e \ , \ \nabla x_{i+1} = (i+1)\nabla x_e \ . \tag{7.83}$$

Thus it can be conjectured, that without an electric field and due to mole fraction gradient alone, (i+1)-th ions accompanied by (i+1) electrons move in the direction of the lower mole fraction of these ions and i-th ions accompanied by i number of electrons move in the opposite direction (note, that i = 0 for neutrals). In fact the situation for (i+1)-th ions is like a very devoted husband, who will always be accompanied by (i+1) number of wives and not allow the ions to go alone. However, if there is an electric field $\mathbf{E}$, the motion of the charged particles are guided by the effect of the electric field, through the mobility coefficient b_j, (7.55), and the number density gradient; for the present case we assume that the total number density is uniform or else additional terms have to be considered. For such a case we write down expressions for the current density of the three components as follows:

$$\mathbf{j}_e = -en_e\mathbf{V}'_e = en_eb_e\mathbf{E} + eD_en\nabla x_e \tag{7.84}$$

$$\mathbf{j}_i = ien_i\mathbf{V}'_i = ien_ib_i\mathbf{E} - eD_in\nabla x_i \tag{7.85}$$

$$\mathbf{j}_{i+1} = (i+1)en_{i+1}\mathbf{V}'_{i+1} = (i+1)en_{i+1}b_{i+1}\mathbf{E} - eD_in\nabla x_{i+1} \tag{7.86}$$

Now without an externally applied electric field and due to the mole fraction gradient alone, the quasi-neutrality condition, as per the above conjecture, requires that

$$\mathbf{j}_i = i\mathbf{j}_e \text{ and } \mathbf{j}_{i+1} = -(i+1)\mathbf{j}_e \tag{7.87}$$

and hence,

$$\mathbf{j}_e + \mathbf{j}_i + \mathbf{j}_{i+1} = 0 \ . \tag{7.88}$$

We note further from (7.55) that for the three components the mobility coefficient are given by the relations

$$b_e = \frac{eD_e}{k_BT_e} \ , \ b_i = \frac{ieD_i}{k_BT_h} \text{ and } b_{i+1} = \frac{(i+1)eD_{i+1}}{k_BT_h} \tag{7.89}$$

where it is assumed that the positive ions are at a common translational temperature, T_h.

Multiplying (7.84) with $(i+1)n_{i+1}b_{i+1}$ and (7.86) with $n_e b_e$, subtracting one from the other so as to eliminate the term containing the electric field, and noting (7.87), we get

$$\frac{\mathbf{j}_e}{en_e} = \frac{D_e x_{i+1} b_{i+1} + D_{i+1} x_e b_e}{x_{i+1} b_{i+1} + x_e b_e} \nabla x_e$$

$$\approx \left(D_e \frac{x_{i+1} b_{i+1}}{x_e b_e} + D_{i+1} \right) \nabla x_e \; . \tag{7.90}$$

Introducing a new diffusion coefficient in such a way that the electron current density due to the electron number density gradient and without any electric field is the same as due to the electric field and the regular diffusion due to number density gradient, we write

$$\frac{\mathbf{j}_e}{en_e} = D_{\mathrm{amb}} \nabla x_e \tag{7.91}$$

where

$$D_{\mathrm{amb}} = D_e \frac{x_{i+1} b_{i+1}}{x_e b_e} + D_{i+1} \; . \tag{7.92}$$

Noting further that

$$\frac{b_e}{D_e} = \frac{e}{k_B T_e} \quad \text{and} \quad \frac{b_{i+1}}{D_{i+1}} = (i+1)\frac{e}{k_B T_h} \; . \tag{7.93}$$

Thus,

$$D_{\mathrm{amb}} = D_{i+1} \left[(i+1)\theta \frac{x_{i+1}}{x_e} + 1 \right] \tag{7.94}$$

is the *"ambi-polar diffusion coefficient"* and $\theta = T_e/T_h$ is the temperature ratio. Substituting (7.94) back into (7.84), we get now the relation for the electric field that would develop inside the plasma as

$$\mathbf{E}_{\mathrm{amb}} = \frac{b_e}{x_e}(D_e - D_{\mathrm{amb}})\nabla x_e \; . \tag{7.95}$$

The above mechanism is responsible for making an electrically insulated wall at an electric negative potential so that the electrons are retarded and the ions are accelerated, so that they reach the surface together. Since, $M_i = M_{i+1} + M_e$ and the requirement that the sum total of the mass flux due to diffusion must be equal to zero, is given as

$$\sum \rho_j \mathbf{V}'_j = n\left(-M_e + M_i - M_{i+1}\right) D_{\mathrm{amb}} \nabla x_e = 0 \; . \tag{7.96}$$

As a consequence, the i-th ions with charge i (i=0 for neutrals) move in opposite direction with the (common) effective diffusion coefficient D_{amb}.

7.4 Viscosity, Heat Conductivity and Electrical Conductivity

In Sect. 7.2 we have discussed the relations for viscosity, heat conductivity
and self-diffusion coefficients for a pure gas. We would now discuss derivation
of these relations in case the gas is a mixture of several specie or components.
One way will be to define these properties in terms of the mixed gas properties like number density, mass, mean free path and mean kinetic speed. For
example, one could write from (7.43), the relation for the viscosity coefficient

$$\mu = \frac{1}{3}\bar{v}nM\lambda = \frac{1}{3}\bar{v}M/Q \tag{7.97}$$

where Q is the average collision cross-section, M is the mixture average mass
of the particles and λ is the mixture *mean free path*. Obviously, μ depends
on the mole fraction of the various specie through the mixture particle mass.
However the actual relations for a gas mixture are somewhat more elaborate,
which we will now discuss.

We would now consider the following two cases. Firstly we consider various
chargeless particles at a common temperature T, and secondly we consider
the case where at least one of the colliding partners is electrons. It has been
shown already that in the latter case for the electron-electron collisions and
the electron-heavy particles, the effective temperature, because of very high
speed of the electrons in comparison to that for heavy particles, is mainly
due to the translation temperature of the electrons, T_e. However, for collision
between heavy particles the important temperature is the heavy particles
temperature, T_h. Thus, for a plasma consisting of the electrons and the heavy
particles it is possible to have separate calculation of the transport properties
for collisions between the heavy particles only, and for collisions in which at
least one of the colliding partners is an electron.

For the heavy particles of the j-th species, the viscosity coefficient is obtained from the relation

$$\mu_j = 8.385853 \times 10^{-6}\sqrt{T_h m_j}/Q_{jj}^{(2,2)}, \ \text{kgm}^{-1}\text{s}^{-1} \tag{7.98}$$

where m_j is the mole mass of the j-th species and $Q_{jj}^{(2,2)}$ is the collision cross-
section in Å^2. The collision cross-section is represented in the above equation
by $Q_{jk}^{(2,2)}$ [Å^2], which is called the *bracket integral* (*Hirschfelder*, et al. [10]),
but for collision between rigid spherical molecules and $(l,s) = (1,1)$ it is equal
to the collision cross-section for rigid molecules

$$Q_{jk}^{(1,1)} = \pi d_{jk}^2 = \pi(d_j + d_k)^2/4 \ . \tag{7.99}$$

Similarly for collisions between the j-th and k-th species (index j,k for any of
the combinations between i-th and (i+1)-th ions), the viscosity coefficient is

$$\mu_{jk} = 8.385853 \times 10^{-6} \sqrt{T_h \frac{2m_j m_k}{m_j + m_k}} / Q_{jk}^{(2,2)}, \ \mathrm{kgm^{-1}s^{-1}} \ . \tag{7.100}$$

The viscosity coefficient of the mixture can now be obtained from the relation with a ratio of determinants

$$\mu = \frac{\begin{vmatrix} H_1 1 & \cdots & H_{1N} & x_1 \\ \cdots & \cdots & \cdots & \cdots \\ H_{N1} & \cdots & H_{NN} & x_N \\ x_1 & \cdots & x_N & 0 \end{vmatrix}}{\begin{vmatrix} H_1 1 & \cdots & H_{1N} \\ \cdots & \cdots & \cdots \\ H_{N1} & \cdots & H_{NN} \end{vmatrix}} , \ \mathrm{kgm^{-1}s^{-1}} \tag{7.101}$$

where

$$H_{jj} = \frac{x_j^2}{\mu_j} + \sum_{k=1, j \neq k}^{N} \frac{2x_j x_k m_j m_k}{\mu_{jk}(m_j + m_k)^2} \left[\frac{5}{3A_{jk}^*} + \frac{m_k}{m_j} \right]$$

$$H_{jk} = - \frac{2x_j x_k m_j m_k}{\mu_{jk}(m_j + m_k)^2} \left[\frac{5}{3A_{jk}^*} - 1 \right] , \ j \neq k$$

$$A_{jk}^* = Q_{jk}^{(2,2)} / Q_{jk}^{(1,1)} \ .$$

In addition, x_j is the mole fraction of the j-th species and N is the total number of the heavy particles. It may be noted that in the above relations the subscripts j and k refer to the heavy particles specie only. This is because, the role of the electrons in the momentum transfer in an ionized gas is neglected.

For the heat conductivity coefficients of the heavy particles similar expressions are used. The equivalent expressions of (7.98, 7.101) for the pure conduction due to the particles translation energy k' are now

$$k_j' = 2.61644 \times 10^{-4} \sqrt{T_h / m_j} / Q_{jj}^{(2,2)} \ , \ \mathrm{kWm^{-1}K^{-1}} \tag{7.102}$$

and for collision between heavy particles (index j,k for any of the combinations between i-th and (i+1)-th ions), the equation for *"pure"* conduction is

$$k_{jk}' = 2.61644 \times 10^{-4} \sqrt{T_h(m_j + m_k)^2 / (2m_j m_k)} / Q_{jk}^{(2,2)} \ , \ \mathrm{kWm^{-1}K^{-1}} \ . \tag{7.103}$$

The mixture heat conduction due to *"pure"* conduction can now be computed from the following relation consisting of ratio of two determinants:

$$k'_{ch} = \frac{\begin{vmatrix} L_{11}^{00} & \cdots & L_{1N}^{00} & L_{11}^{01} & \cdots & L_{1N}^{01} & 0 \\ \cdots & \cdots & \cdots & \cdots & \cdots & \cdots & \cdots \\ L_{N1}^{00} & \cdots & L_{NN}^{00} & L_{N1}^{01} & \cdots & L_{NN}^{01} & 0 \\ L_{10}^{1N} & \cdots & L_{1N}^{10} & L_{11}^{11} & \cdots & L_{1N}^{11} & x_1 \\ \cdots & \cdots & \cdots & \cdots & \cdots & \cdots & \cdots \\ L_{N1}^{10} & \cdots & L_{NN}^{10} & L_{N1}^{11} & \cdots & L_{NN}^{11} & x_N \\ 0 & \cdots & 0 & x_1 & \cdots & x_N & 0 \end{vmatrix}}{\begin{vmatrix} L_{11}^{00} & \cdots & L_{1N}^{00} & L_{11}^{01} & \cdots & L_{1N}^{01} \\ \cdots & \cdots & \cdots & \cdots & \cdots & \cdots \\ L_{N1}^{00} & \cdots & L_{NN}^{00} & L_{N1}^{01} & \cdots & L_{NN}^{01} \\ L_{11}^{10} & \cdots & L_{1N}^{10} & L_{11}^{11} & \cdots & L_{1N}^{11} \\ \cdots & \cdots & \cdots & \cdots & \cdots & \cdots \\ L_{N1}^{10} & \cdots & L_{NN}^{10} & L_{N1}^{11} & \cdots & L_{NN}^{11} \end{vmatrix}} , \quad \mathrm{kWm^{-1}K^{-1}} \qquad (7.104)$$

where subscripts 1 to N refer to i-th and (i+1)-th ion and for $j = k \neq 1$:

$$L_{jj}^{00} = 0$$

$$L_{jj}^{01} = 5\frac{x_j x_l m_L (1.2C_{jl}^* - 1)}{(m_j + m_l)A_{jl}^* k_{jl}} = L_{jj}^{10}$$

$$L_{jj}^{11} = -\frac{4x_j^2}{k_{jj}} + \frac{2x_j x_l [7.5m_j^2 + 6.25m_l^2 - 3m_l^2 B_{jl}^* + 4m_j m_l A_{jl}^*]}{(m_j + m_l)^2 A_{jl}^* k_{jl}}$$

and for $j \neq k$:

$$L_{jk}^{00} = \frac{2x_j x_k}{A_{jk}^* k_{jk}}$$

$$L_{jk}^{01} = -5\frac{x_j x_k m_j [1.2C_{jk}^* - 1]}{(m_j + m_k)A_{jk}^* k_{jk}} = \frac{m_j}{m_k}L_{jk}^{10}$$

$$L_{jk}^{11} = \frac{2x_j x_k m_j m_k [13.75 - 3B_{jk}^* - 4A_{jk}^*]}{(m_j + m_k)^2 A_{jk}^* k_{jk}} .$$

In the above,

$$A_{jk}^* = Q_{jk}^{(2,2)}/Q_{jk}^{(1,1)} , \quad B_{jk}^* = (5Q_{jk}^{(1,2)} - 4Q_{jk}^{(1,3)})/Q_{jk}^{(1,1)}$$

$$C_{jk}^* = Q_{jk}^{(1,2)}/Q_{jk}^{(1,1)} .$$

For the contribution of the internal degrees of freedom of molecules with two or more atoms to the heat conductivity coefficient k_j'', *Eucken's semi-empirical formula* gives

$$k_{cj}'' = 0.88 \left[0.4\frac{C_{pj}}{R^*} - 1 \right] \qquad (7.105)$$

where C_{pj} is the *molar specific heat* of the j-th species at constant pressure and R^* is the *universal gas constant*. The expression is valid for $C_{pj}/R^* > 2.5$, and k''_{cj} is put equal to zero if this condition is not satisfied. The contribution of the internal energy to the heat conductivity coefficient for a mixture is obtained from semi-empirical formula, like that due to Brokaw, discussed later, and can be added to the mixture heat conductivity coefficient due to translation only.

The contribution of the electrons for the heat and electrical conductivities are obtained by evaluating first the determinant elements q^{rt}, for which the following relations are given:

$$q^{00} = 8 \sum_h x_e x_j Q_{ej}^{(1,1)} \tag{7.106}$$

$$q^{01} = 8 \sum_h x_e x_j [2.5 Q_{ej}^{(1,1)} - 3 Q_{ej}^{(1,2)}] \tag{7.107}$$

$$q^{11} = 8\sqrt{2} x_e^2 Q_{ee}^{(2,2)} + 8 \sum_h x_e x_j [6.25 Q_{ej}^{(1,1)} - 15 Q_{ej}^{(1,2)} + 12 Q_{ej}^{(1,3)}] \tag{7.108}$$

$$q^{02} = 8 \sum_h x_e x_j [4.75 Q_{ej}^{(1,1)} - 10.5 Q_{ej}^{(1,2)} + 6 Q_{ej}^{(1,3)}] \tag{7.109}$$

$$\begin{aligned}
q^{12} = {} & 8\sqrt{2} x_e^2 [1.75 Q_{ee}^{(2,2)} - 2 Q_{ee}^{(2,3)}] \\
& + 8 \sum_h x_e x_j [(175/16) Q_{ej}^{(1,1)} - (315/8) Q_{ej}^{(1,2)} + 57 Q_{ej}^{(1,3)} \\
& - 30 Q_{ej}^{(1,4)}]
\end{aligned} \tag{7.110}$$

$$\begin{aligned}
q^{22} = {} & 8\sqrt{2} x_e^2 [(77/16) Q_{ee}^{(2,2)} - 7 Q_{ee}^{(2,3)} + 5 Q_{ee}^{(2,4)}] \\
& + 8 \sum_h x_e x_j [(1225/64) Q_{ej}^{(1,1)} - (735/8) Q_{ej}^{(1,2)} + 199.5 Q_{ej}^{(1,4)} \\
& + 90 Q_{ej}^{(1,5)}]
\end{aligned} \tag{7.111}$$

$$q^{03} = 8 \sum_h x_e x_j [(105/16) Q_{ej}^{(1,1)} - (189/8) Q_{ej}^{(1,2)} + 27 Q_{ej}^{(1,3)} - 10 Q_{ej}^{(1,4)}] \tag{7.112}$$

$$\begin{aligned}
q^{13} = {} & 8\sqrt{2} x_e^2 [(63/12) Q_{ee}^{(2,2)} - 4.5 Q_{ee}^{(2,3)} + 2.5 Q_{ee}^{(2,4)}] \\
& + 8 \sum_h x_e x_j [(525/32) Q_{ej}^{(1,1)} - (315/4) Q_{ej}^{(1,2)} + 162 Q_{ej}^{(1,3)} \\
& - 160 Q_{ej}^{(1,4)} + 60 Q_{ej}^{(1,5)}]
\end{aligned} \tag{7.113}$$

$$\begin{aligned}
q^{23} = {} & 8\sqrt{2} x_e^2 [(945/128) Q_{ee}^{(2,2)} - (261/16) Q_{ee}^{(2,3)} + (125/8) Q_{ee}^{(2,4)} - 7.5 Q_{ee}^{(2,5)}] \\
& + 8 \sum_h x_e x_j [(3675/128) Q_{ej}^{(1,1)} - (11025/64) Q_{ej}^{(1,2)} \\
& + (1953/4) Q_{ej}^{(1,3)} - 752.5 Q_{ej}^{(1,4)} + 615 Q_{ej}^{(1,5)} - 210 Q_{ej}^{(1,6)}]
\end{aligned} \tag{7.114}$$

$$q^{33} = 8\sqrt{2}x_e^2[(14553/1024)Q_{ee}^{(2,2)} - (1215/32)Q_{ee}^{(2,3)} + (1565/32)Q_{ee}^{(2,4)}$$

$$- (135/4)Q_{ee}^{(2,5)} + 15Q_{ee}^{(2,6)} + Q_{ee}^{(4,4)}] + 8\sum_h x_e x_j[(11025/256)Q_{ej}^{(1,1)}$$

$$- (19845/64)Q_{ej}^{(1,2)} + (17577/16)Q_{ej}^{(1,3)} - 2257.5Q_{ej}^{(1,4)}$$

$$+ 2767.5Q_{ej}^{(1,5)} - 1890Q_{ej}^{(1,6)} + 560Q_{ej}^{(1,7)}] \tag{7.115}$$

The fourth order diffusion coefficient D_{ee}, *thermodiffusion coefficient* D_e^T, and the corresponding *electrical* and *thermal conductivities* due to *"pure"* conduction, σ and k_{ce} are now obtained from the relations which contain the ratio of determinants as follows:

$$[D_{ee}]_4 = \frac{3x_e}{2n}\left(\frac{2\pi k_B T_e}{M_e}\right)^{1/2} \frac{\begin{vmatrix} q^{11} & q^{12} & q^{13} \\ q^{21} & q^{22} & q^{23} \\ q^{31} & q^{32} & q^{33} \end{vmatrix}}{|q|} \tag{7.116}$$

$$[D_e^T]_4 = \frac{15x_e^2}{4}\sqrt{2\pi M_e k_B T_e}\,\frac{\begin{vmatrix} q^{01} & q^{02} & q^{03} \\ q^{21} & q^{22} & q^{23} \\ q^{31} & q^{32} & q^{33} \end{vmatrix}}{|q|} \tag{7.117}$$

$$\sigma = \frac{e^2 n_e [D_{ee}]_4}{k_B T_e} \tag{7.118}$$

$$k_{ce}' = \frac{75x_e^2 k_B}{8}\sqrt{2\pi k_B T_e/M_e}\,\frac{\begin{vmatrix} q^{00} & q^{02} & q^{03} \\ q^{20} & q^{22} & q^{23} \\ q^{30} & q^{32} & q^{33} \end{vmatrix}}{|q|} \tag{7.119}$$

$$k_{ce} = k_{ce}' - \frac{k_B[D_e^T]^2}{nx_e M_e^2 [D_{ee}]_4} \tag{7.120}$$

where n is the number density (the number of particles per unit volume, m^{-3}), M_e is the mass of an electron, e is the elementary charge, k_B is the Boltzmann constant, x_e is the mole fraction of the electrons, and $|q|$ is the determinant of the matrix

$$|q| = \begin{vmatrix} q^{00} & q^{01} & q^{02} & q^{03} \\ q^{10} & q^{11} & q^{12} & q^{13} \\ q^{20} & q^{21} & q^{22} & q^{23} \\ q^{30} & q^{31} & q^{32} & q^{33} \end{vmatrix}. \tag{7.121}$$

While the above procedure by rigorous theory is applicable even for multi-temperature plasma, a much simpler but approximate method has been given by *Brokaw* [51] (these are inadequate for ionized gases however) and for which the resulting equations are as follows:

Viscosity coefficient of a mixture:

$$\mu_{\text{mix}} = \sum_{i=1}^{n} \frac{\mu_i x_i}{\sum_{j=1}^{n} \phi_{i,j} x_j} \tag{7.122}$$

Heat conductivity coefficient of a mixture:

$$k'_{c,\text{mix}} = \sum_{i=1}^{n} \frac{k'_{c,i} x_i}{\sum_{j=1}^{n} \psi_{i,j} x_j} \;,\; k''_{c,\text{mix}} = \sum_{i=1}^{n} \frac{k''_{c,i} x_i}{\sum_{j=1}^{n} \phi_{i,j} x_j} \tag{7.123}$$

$$k_{c,\text{mix}} = k'_{c,\text{mix}} + k''_{c,\text{mix}} \tag{7.124}$$

where

$$\phi_{i,j} = \frac{[1 + (\mu_i/\mu_j)^{1/2}(m_j/m_i)^{1/4}]^2}{2\sqrt{2}[1 + (m_i/m_j)]^{1/2}} \tag{7.125}$$

$$\psi_{i,j} = \phi_{i,j} \left[1 + \frac{2.42(m_i - m_j)(m_i - 0.142 m_j)}{(m_i + m_j)^2}\right] \tag{7.126}$$

m_i, m_j = mole mass of the i-th or j-th species, respectively.

We would now discuss the transport of the thermal energy due to diffusion of particles from the region of higher temperature to lower temperature. In the process, the composition changes with a reaction rate that is dependent on the local pressure and temperature. If the reaction rate is very slow, then there will not be much of energy transfer due to diffusion, a special case will be the catalytic reaction on the wall surface at the boundary. In that case the energy release will be only on the surface enhancing the heat flux at the surface. On the other hand, if the reaction rate is very fast, then there will be local equilibrium condition and there will be energy release everywhere due to diffusion and reaction. This diffusion and shifting equilibrium is the target of our immediate discussion.

7.5 Diffusion and Radiative Heat Conduction

While the heat conduction due to transfer of translational and other internal energies by collision (so called, *"pure"* conduction) is significant, there are other forms of transfer of energy of equal significance. The first of these is the *diffusive-reactive heat conduction.*

For the general gas mixture, the transport of energy due to diffusion is $\sum \rho \mathbf{V}'_j h_j$, which we would derive in Chap. 11. For the present, however, we derive an expression for the *diffusive-reactive heat conduction coefficient* for a reactive gas mixture.

From section (7.4), we write for the mass-flux of the j-th species as

$$\rho_j \mathbf{V}'_j = -\rho D_{jm} \nabla Y_j \tag{7.127}$$

where D_{jm} is the effective mass diffusive coefficient of the j-th species in a gas mixture and $Y_j = \rho_j/\rho$ is the mass-fraction. Therefore the energy flux is

$$\sum \rho_j \mathbf{V}'_j h_j = -\rho \sum_j D_{jm} h_j \nabla Y_j = -\left(\rho \sum D_{jm} h_j \frac{\partial Y_j}{\partial T} \right) \nabla T = -k_d \nabla T$$

(7.128)

where k_d is the diffusive-reactive heat conduction coefficient. In order to evaluate this we require to evaluate $\partial Y_j/\partial T$, which can be easily computed from $\partial x_j/\partial T$ discussed in Sect. 6.8. For the ionized gases fortunately, we have the relation for a (common) *ambipolar diffusion coefficient*, and $\partial x_j/\partial T$ can be evaluated with the help of close-form expressions in Sect. 6.8.

Now the diffusive heat-flux due to recombination of ionized particles is given by the relation

$$\sum \rho_j \mathbf{V}'_j h_j = n \sum x_j \mathbf{V}'_j H_j = n(-H_e + H_{i+1} - H_i) D_{\text{amb}} \nabla x_e$$

$$= n(-H_e + H_{i+1} - H_i) D_{\text{amb}} \left(\frac{\partial x_e}{\partial T_e} \nabla T_e + \frac{\partial x_e}{\partial T_h} \nabla T_h \right) \quad (7.129)$$

where n is the number density, D_{amb} is the ambi-polar diffusion coefficient, h is the mass specific enthalpy, and H is the molar specific enthalpy. Hence we get the two *"reactive-diffusive heat conductivity coefficients"* for electrons and heavy particles

$$k_{re} = n(-H_e + H_{i+1} - H_i) D_{\text{amb}} \frac{\partial x_e}{\partial T_e}$$

(7.130)

and

$$k_{rh} = n(-H_e + H_{i+1} - H_i) D_{\text{amb}} \frac{\partial x_e}{\partial T_h}$$

(7.131)

which are added to the respective *"pure"* heat conductivity coefficients. Derivative of electron mole fraction with respect to T_e and T_h can be derived easily in a close-form. The ambi-polar diffusion coefficient, assuming that i-th ion accompanied with i electrons move in one direction and (i+1)-th ion accompanied with (i+1) electrons move in the opposite direction, is given by the relation (7.92)

$$D_{\text{amb}} = D_{i+1} \left[(i+1)\theta \frac{x_{i+1}}{x_e} + 1 \right]$$

(7.132)

where D_{i+1} is the diffusion coefficient of (i+1)-th ion due to gradient in number density of the particles. D_{i+1} is computed from a complicated relationship including binary and self-diffusion coefficients of individual specie in such a manner that the sum of diffusive flux is zero. The total heat conductivity coefficient for heavy particles and electrons, k_h and k_e, are obtained by adding the *"pure"* and *"diffusive"* parts in the heat conductivity. The volumetric collision cross-section between electrons and heavy particles are given by the relation

$$\Gamma'_{eh} = v_e n^2 x_e \sum x_j Q_{ej} \ , \ \mathrm{m^{-3}s^{-1}} \tag{7.133}$$

where for collision cross-section $Q_{ej} = Q_{ej}^{(1,1)}$ is taken and v_e is the mean kinetic speed given by the relation $v_e = \sqrt{8k_B T_e/(\pi M_e)}$.

Finally, we present the results of calculation of transport properties of two-temperature argon plasma at 1 bar (Fig. 7.2). It can be seen that at temperatures around 15,000 K and above for argon plasma at 1 bar pressure the viscosity coefficient and heavy particles heat conductivity coefficient are much smaller than at temperatures around 10,000 K; however since the mass density decreases with temperature much faster than the dynamic viscosity coefficient, the kinematic viscosity coefficient μ/ρ increases with temperature. Further at high temperatures the heat conductivity coefficient (thermal energy transport) and the electrical conductivity are mainly due to transport of electrons. Such transport properties results for air and other noble gas plasma have been presented by (Bose [45]), and also such calculations have been done for diatomic gases, water vapor, etc.

For the reacting and radiating gas mixture, there can be, under certain circumstances heat flux due to diffusive radiation transport. In case there

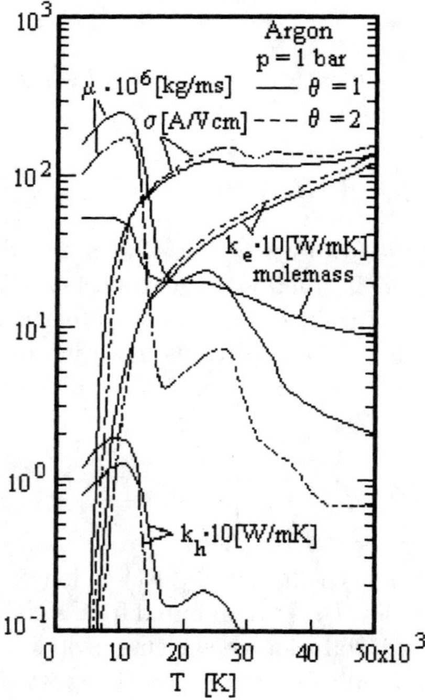

Fig. 7.2. Transport properties of two-temperature plasma

exists a local equilibrium of radiation, the heat flux is given by the relation (written for simplicity as one-dimensional radiative transport)

$$d\mathbf{q}^R = -\frac{D_R}{c}\frac{\partial I^*}{\partial x},\ \mathrm{Wm}^{-3} \tag{7.134}$$

where the radiative diffusion coefficient D_R is given by the relation

$$D_R = \frac{4\pi c}{3\kappa_R} \tag{7.135}$$

$c = $ *velocity of light*, and $\kappa_R = $ the *average Rosseland absorptions coefficient* which depends both on the gas density and temperature, m^{-1}.

The radiation energy intensity for *black-body radiation* I^* is given by (4.36), from which and (7.134), one can write

$$d\mathbf{q}^R = -\frac{4\sigma}{\pi}\frac{D_R}{c}T^3\frac{\partial T}{\partial x} = -k_R\frac{\partial T}{\partial x}\ . \tag{7.136}$$

Thus one gets the relation for the so-called *"radiative heat conductivity coefficient"* k_R as

$$k_R = \frac{4\sigma}{\pi}\frac{D_R}{c}T^3\ . \tag{7.137}$$

It may be recalled that in presence of a temperature gradient, the radiative heat flux is only one mechanism of energy transport (valid strictly for optically thick or diffusive radiation, for a discussion of which please see Chap. 4), the other two being due to pure conduction and due to transport of energy by diffusion of particles and subsequent recombination. Thus for the total heat flux, the total heat conductivity coefficient is the sum of all the three *"heat conductivity coefficients"* due to pure conduction, diffusion and diffusive radiation. However for most of the terrestrial applications the transmitting medium radiative flux can be considered as *"optically thin"*, and hence the contribution of radiation in the total heat conductivity coefficient may be neglected.

7.6 Effect of Magnetic Field on the Transport Properties of Ionized Gases

We have not considered so far the effect of the magnetic field on the transport properties of ionized gases. For this purpose, we derive first for an ideal gas the relation for the electrical conductivity. It has already been noted that, in presence of an externally applied electric field, $\mathbf{E}'$, the corresponding *field drift velocity* is given by the relation

$$\mathbf{w}'_j = \frac{e_j\lambda_j}{M_jv_j}\mathbf{E}' = b_j\mathbf{E}' \tag{7.138}$$

where the *mobility coefficient*

$$b_j = \frac{e_j \lambda_j}{M_j v_j} = \frac{3 e_j D_j}{M_j v_j^2} = \frac{e_j D_j}{k_B T_j} \ , \ \mathrm{m^2 V^{-1} s^{-1}} \ . \tag{7.139}$$

For multiple charged particles in a gas mixture (plasma), and in presence of the externally applied electric field, the current density is now given by the expression

$$\mathbf{j} = \sum \mathbf{j}_j = \sigma_o \mathbf{E} = \mathbf{E} \sum e_j n_j \mathbf{V}'_{fj} = \mathbf{E} \sum e_j b_j n_j \ , \ \mathrm{Am^{-2}} \tag{7.140}$$

and thus the *electrical conductivity* is given by the relation

$$\sigma_o = \sum e_j b_j n_j \ , \ \mathrm{AV^{-1} m^{-1}} \ . \tag{7.141}$$

The above electric conductivity σ_o is a scalar quantity, and the relation is valid if there is no strong magnetic field. This requires that there is no strong electric current also, because an electric current induces a magnetic field. Since, in a strong magnetic induction $\mathbf{B}$, there is a charge-drift in all directions except in the direction of $\mathbf{B}$, therefore, parallel to the magnetic induction, the electric field is given by the relation

$$\sigma_3 = \sigma_o = \sum \frac{n_j q_j^2}{M_j \Gamma_j} \approx \frac{e^2 n_e}{M_e \Gamma_e} \ , \ \mathrm{AV^{-1} m^{-1}} \tag{7.142}$$

where Γ_e is the electron collision frequency $[\mathrm{s^{-1}}]$. If an electric field $\mathbf{E}$ is applied perpendicular to the direction of $\mathbf{B}$, there are effects on the electrical conductivity in the two directions perpendicular to $\mathbf{B}$. First, in the direction of the electric field, the conductivity is

$$\sigma_1 = \sum \frac{n_j q_j^2}{M_j \Gamma_j [1 + \xi_j^2]} \approx \frac{\sigma_o}{[1 + \xi_j^2]} \tag{7.143}$$

where $\xi = \omega_{cj} / \Gamma_j$ gives the ratio of the radian cyclotron frequency to the collision frequency. Equation (7.143) reduces to (7.142) for small values of ξ_j. However, for large values of ξ_j,

$$\sigma_1 = \sum \frac{n_j q_j^2}{M_j \Gamma_j \xi_j^2} = \sum \frac{n_j M_j \Gamma_j}{B^2} \ , \ \mathrm{AV^{-1} m^{-1}} \ . \tag{7.144}$$

Since the mass of ions M_i is much larger than the mass of electrons M_e, the contribution of these ions at high magnetic induction is larger than that of electrons. Thus, the ion current across a strong magnetic induction is several order of magnitude larger than that of electrons, which is the reverse of that in the direction parallel to the magnetic field.

The other conductivity arises from a current flowing perpendicular to both $\mathbf{E}$ and $\mathbf{B}$, and is given by the relation

$$\sigma_2 = \sum \frac{n_j q_j^2 \xi_j}{M_j \Gamma_j [1 + \xi_j^2]} \approx \sigma_o \frac{\xi_e}{1 + \xi_e^2} \ . \tag{7.145}$$

These expressions for the electrical conductivity are independent of any modification of the electric field due to the magnetic field and the mass-average velocity field. The directions of these conductivities have been explained in Fig. 7.3.

We would now discuss the derivation of the Ohm's law in presence of magnetic field. For this purpose we consider the flow field in the x-direction, an externally applied magnetic induction, $\mathbf{B} = \{0, B_y, 0\}$ in the y-direction and an externally applied electric field $\mathbf{E}' = \{0, 0, E_z'\}$ in the z-direction (Fig. 7.4). If $\mathbf{w}_j$ is the velocity of a single charged particle, then the force acting on the single particle is

$$\mathbf{F}_j^1 = q_j(\mathbf{E}' + \mathbf{w}_j \times \mathbf{B}) \ . \tag{7.146}$$

Now the mass-averaged velocity of the j-th species, $\mathbf{V}_j$, may be obtained from the distribution of the individual particles of the j-th species, $\mathbf{w}_j$. Further, $\mathbf{V}_j$ is the sum of the mass-averaged velocity of all species, $\mathbf{V}$, and the diffusive

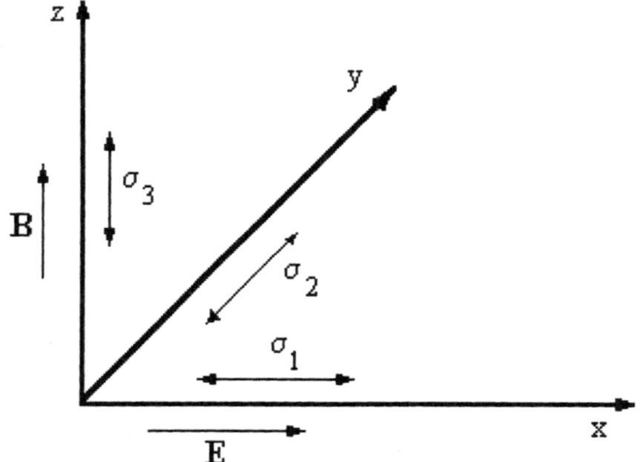

Fig. 7.3. Explaining electrical conductivity as vector

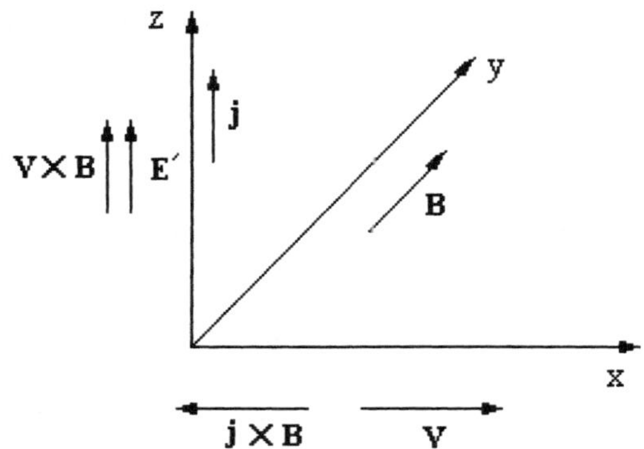

Fig. 7.4. Flow and electromagnetic fields

velocity (mass-diffusion due to particle number density gradient and due to motion in electro magnetic field), $\mathbf{V}'_j$. Hence by multiplying (7.146) by the number density of the j-th species, n_j, we get the relation for force as

$$\mathbf{F}_j = n_j \mathbf{F}^1_j = n_j q_j (\mathbf{E}' + \mathbf{V}_j \times \mathbf{B}) = n_j q_j (\mathbf{E}' + \mathbf{V} \times \mathbf{B} + \mathbf{V}'_j \times \mathbf{B}) \ . \quad (7.147)$$

At the outset one can add (7.147) over all specie to get the total volumetric force. Since the quasi-neutrality condition and the definition of the current density require that

$$\sum n_j q_j = 0 \quad \text{and} \quad \mathbf{j} = \sum n_j q_j \mathbf{V}'_j \qquad (7.148)$$

and hence,

$$\mathbf{F} = \sum n_j \mathbf{F}_j = \mathbf{j} \times \mathbf{B} \qquad (7.149)$$

is in the x-direction.

   Further from (7.147) and Fig. 7.4, it can be seen that there is a total electric field $\mathbf{E} = \mathbf{E}' + \mathbf{V} \times \mathbf{B}$ in the z-direction with the consequent current density $\mathbf{j} = \sigma_1 \mathbf{E}$. If assumed that this current density is carried mainly by the electrons, which is a very good assumption, then there is an associated field velocity of the electrons as

$$\mathbf{V}'_{fe} = -\frac{\mathbf{j}}{e n_e} \ . \qquad (7.150)$$

An electric field, therefore, is induced in the x-direction and one gets the relation for the *generalized Ohm's law* as

$$\mathbf{j} = \sigma_1 (\mathbf{E}' + \mathbf{V} \times \mathbf{B}) - \sigma_2 (\mathbf{j} \times \mathbf{B})/(e n_e) \qquad (7.151)$$

the last term being the Hall current density in the negative x-direction. It is explained more in detail in Chap. 11.

7.7 Transport Properties of an Ideal Dissociating Gas

In Sect. 6.4 we have already discussed the thermophysical properties of an ideal dissociating diatomic molecular gas (*Lighthill gas*). Here we would now discuss some simple relations regarding diffusion of such a gas and the related issue of the diffusive-reactive heat conduction. It may now be recalled, that we have now a reaction of type

$$M \leftrightarrow 2A \qquad (7.152)$$

for which the heat-flux in the direction normal to a wall (y-direction) due to conduction (both "*pure*" conduction and the diffusive-reactive conduction are included, but not the effect of the radiation) is given by the relation

$$\mathbf{q} = -k_c \frac{\mathrm{d}T}{\mathrm{d}y} - \rho D_{MA} \frac{\mathrm{d}Y_M}{\mathrm{d}y}(h_M - h_A)$$

$$= -k_c \frac{\mathrm{d}T}{\mathrm{d}y} + \rho D_{MA} \frac{\mathrm{d}Y_A}{\mathrm{d}y}(h_M - h_A) \ . \tag{7.153}$$

In above, k_c is the heat conductivity coefficient due to *"pure"* conduction, D_{MA} is the diffusion coefficient based on the mass density gradient, Y_M and Y_A are the mass-fraction of the molecule and the atom, respectively, and h_M and h_A are the corresponding specific (with respect to unit mass of the gas component) enthalpy. Noting from Sect. 6.4 again that

$$c_{p,\mathrm{eff}} = \left(\frac{\partial h}{\partial T}\right)_p = \left[Y_A \frac{\partial h_A}{\partial T} + Y_M \frac{\partial h_M}{\partial T} + \left(h_A \frac{\partial Y_A}{\partial T} + h_M \frac{\partial Y_M}{\partial T}\right)\right]$$

$$= c_{p,f} + \frac{\partial Y_A}{\partial T}(h_A - h_M) \ . \tag{7.154}$$

we can re-write (7.153) as

$$\mathbf{q} = -\left[k_c + \rho D_{MA}(c_{p,\mathrm{eff}} - c_f)\right]\frac{\mathrm{d}T}{\mathrm{d}y} = -k_\mathrm{eff}\frac{\mathrm{d}T}{\mathrm{d}y} \ . \tag{7.155}$$

By defining a *Lewis number* for the *chemically frozen gas* from the relation

$$\mathrm{Le}_f = \frac{\rho c_{p,f} D_{MA}}{k_c} \tag{7.156}$$

we get the relation for the ratio of the effective conductivity to the frozen conductivity as

$$\frac{k_\mathrm{eff}}{k_c} = 1 + \mathrm{Le}_f\left[\frac{c_{p\ \mathrm{eff}}}{c_{p,f}} - 1\right]$$

$$= (1 - \mathrm{Le}_f) + \mathrm{Le}_f\frac{c_{p,\mathrm{eff}}}{c_{p,f}} \ . \tag{7.157}$$

It is seen that the ratio of heat conductivity is linearly dependent on the ratio of the specific heat, and they are specially proportional if Le_f is equal to one. Now from Sect. 6.4 it is seen that the latter changes from the value one in certain temperature change only. It is also known that the frozen Lewis number for a gas is very near one, it is possible to estimate the effective heat conductivity coefficient for the ideal dissociating gases.

From the knowledge of properties for the monatomic and diatomic gases, we can now write for frozen dimensionless properties the following relations:

$$\textit{Prandtl number}\ \mathrm{Pr}_f = \frac{\mu c_{p,f}}{k_{cf}} = \frac{2}{3}\frac{(4 + Y_B)}{(3.75 + 1.25 Y_B)} \tag{7.158}$$

$$\textit{Schmidt number}\ \mathrm{Sc}_f = \frac{\mu}{\rho D_{MA}} = \frac{1 + Y_B}{2} \tag{7.159}$$

$$\textit{Lewis number}\ \mathrm{Le}_f = \frac{\mathrm{Pr}_f}{\mathrm{Sc}_f} \tag{7.160}$$

Hence, from (6.51, 7.157) we get the relation for the *heat conductivity ratio* as

$$\frac{k_{\text{eff}}}{k_c} = 1 + \frac{2}{3}\left[\frac{Y_B(1-Y_B)}{3.75 + 1.25Y_B}\left(1 + \frac{T_d}{T}\right)^2\right] \qquad (7.161)$$

These results are shown in Fig. 7.5. The results show that the total heat conductivity coefficient by including "*pure*" conduction and diffusion but without radiation to be up to about four times the value due to "*pure*" conduction alone.

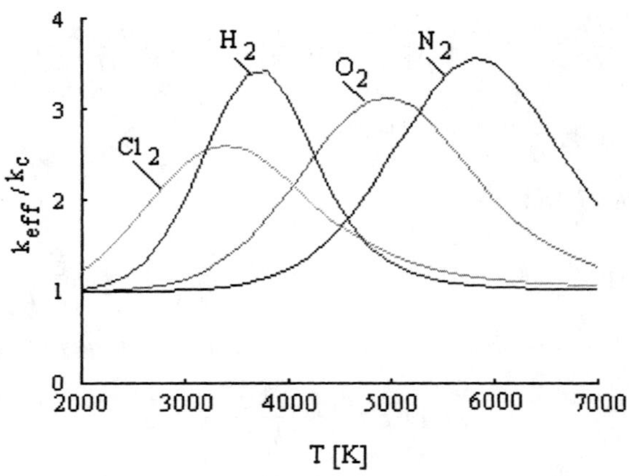

Fig. 7.5. Heat conductivity coefficient ratio for ideal diatomic gas

7.8 Exercise

7.8.1 For a given temperature and pressure, compute the values of the frozen and effective heat conductivity coefficients for H_2, O_2, N_2 and Cl_2.

7.8.2 Compute, for given pressure and temperature(s), the transport properties of a plasma of a known gas.

7.8.3 From the values of collision cross-section computed under problem (Sect. 5.5.2), compute the relative change in the value of the electric conductivity in different directions if the magnetic induction in a direction is 10,000 Gauss.

7.8.4 Examine the statement that at a given temperature, a lighter molecule will have smaller viscosity coefficient but larger heat conductivity coefficient.

7.8.5 Compute electron and ion collision frequency with magnetic induction 10,000 Gauss. [Ans: 1.759e11; 2.4168e6 s^{-1}]

8 Boundary Effects
for High Temperature Gases

Phenomena closely related to the transport of mass, momentum and energy in gas mixtures are those which take place at the boundary of gas volumes. Firstly, because of the temperature gradient, there is a transport of dissociated and ionized particles to cooler regions giving rise to considerable increase in the heat transfer rate, and for ionized gases a loss in the number of the charged particles also. To keep the electric current flowing in the ionized gases, new charged particles have to be created. This starts with the generation of the charged particles like the electrons on the surface. This emission of electrons is one of the factors that determines whether or not a gas discharge is to be self-sustaining, and is, therefore, of considerable interest in the study of conduction of electricity through gases. In studying the various experiments on the electron emission taking place under ideal conditions, one must, however, take care in applying the results to gas discharges in which the emission phenomena are involved, for the electron emission is greatly influenced by the gas condition of the emitting surface. Usually, experiments on emission involve a careful preliminary outgassing of the surfaces in order to obtain results for gas-free surfaces.

In connection with the transport of the charged particles from one place to another, there is the question of building regions of excess charge of one sort or the other, which in turn accelerate or decelerate a certain type of particles giving rise to an increase or decrease in heat transfer rates. Further, because of the metallic wall surface in a container having a discharge at different electrical potentials, these may cause several effects, which are the subjects of investigation in the present chapter.

8.1 Emission of Electrons and Ions

The electrons may be emitted at high temperatures by a process called *"thermo-ionic emission"*, by high electric fields, or by bombardment of some of the high speed charged particles accelerated in electromagnetic fields, or by chemical or photo-chemical effects. The thermoionic emission is effected by emission of bonded electrons in a crystalline structure of a solid body surface due to supply of a large quantity of heat externally to the surface. While examining the process, let us assume that the solid surface of the body and the electrons are at a temperature T. In case the separation of the electrons would have taken place at the absolute zero degree Kelvin, where a *"latent heat of evaporation of electrons"* at absolute zero, L_o, has to be supplied, then the value of the latent heat of evaporation at the temperature T, which is designated as L, can be obtained from the relation

$$L = L_o - \int_0^T C_1 \mathrm{d}T + \int_0^T C_2 \mathrm{d}T + \int_0^T C_3 \mathrm{d}T \qquad (8.1)$$

where C_1 is the molar specific heat of the metal and electrons in the bounded state combined, C_2 is the molar specific heat of the metal alone, and $C_3 = (5/2)R^*$ is the molar specific heat of the free electrons. Thus this mechanism pertains to the one in which the metal and the bonded electrons are cooled to the absolute zero degree temperature, the bonded electrons are freed, and both the free electrons and the metal are brought back to the temperature T. Assuming average values of C_1, C_2 and C_3, and further assuming that $C_1 = C_2$, we may write

$$L = L_o + \frac{5}{2}R^*T \ , \ \mathrm{J.kmole}^{-1} \ . \qquad (8.2)$$

Now from the *Clapeyron's equation* of vapor pressure, as obtained from *chemical thermodynamics*

$$L = R^*T^2 \frac{\partial(\ln p_e)}{\partial T} \ , \ \mathrm{J.kmole}^{-1} \qquad (8.3)$$

where p_e is the partial vapor pressure of the free electrons. Combining (8.2) and (8.3), and integrating with respect to temperature, we get

$$\ln p_e = -\frac{L_o}{R^*T} + \frac{5}{2}\ln T + \ln C_1 \qquad (8.4)$$

and thus

$$p_e = C_1 T^{5/2} \exp^{-L_o/(R^*T)} \ . \qquad (8.5)$$

Herein C_1 can be computed from the known partial pressure of the electrons at a particular temperature.

Now the equation of state for the free electrons is

$$p_e = n_e k_B T \ . \qquad (8.6)$$

These come out of the surface with a mean kinetic speed

$$v_e = \left(\frac{8k_B T}{\pi M_e} \right)^{1/2} \ . \qquad (8.7)$$

The particle flux of the electrons is given by the relation

$$\dot{n}_e = \frac{1}{4} n_e v_e \tag{8.8}$$

where the factor $(1/4)$ is the product of two $(1/2)$s, one of which accounts for one-half of the sphere (a hemisphere) for emission that can be put on the emitting surface, and the other for the integral of cosines of the angle of the direction of emission from the normal to the surface. Since the electron current density is $j_e = e\dot{n}_e$, one gets from Eqs. (8.6–8.8) the relation

$$j_e = \frac{e}{4} n_e v_e = \frac{e}{4} \left(\frac{p_e}{k_B T} \right) \left(\frac{8 k_B T}{\pi M_e} \right)^{1/2} . \tag{8.9}$$

Combining the above relation further with Eq. (8.5), we get

$$j_e = C T^2 \exp^{-e\phi_o/(k_B T)} , \; \mathrm{Am}^{-2} \tag{8.10}$$

where

$$C = eC_1/\sqrt{2\pi M_e k_B} , \; \mathrm{A.m}^{-2}\mathrm{K}^{-1} \tag{8.11}$$

and $\phi_o = $ *work function* in volts.

The work function is obtained in the following manner. If L_o is the work required to remove one kmole of electrons and W is the work required to remove an electron, then the number of electrons per kmole is $N_A = L_o/W = R^*/k_B$, where R^* is the *universal gas constant* and k_B is the *Boltzmann constant*. Noting $L_o/R^* = e\phi_o/k_B$, (8.11) is obtained. Now from the tabulation of C and ϕ_o in Table 8.1 for various substances, it can be seen that for most metals, $C = 6.105 \; \mathrm{A(mK)}^{-2}$ and $\phi_o = 1.7$ to 4.7 volts. It is noted from (8.10) that the exponential term goes to zero as $T \to 0$ and it goes to one as

Table 8.1. Comparison of maximum current densities of several substances

Mater.	C [Am^{-2}K^{-2}]	ϕ_o [V]	T$_M$ [K]	$e\phi_o/k_B T_M$	j_{max} [Am^{-2}]	k W/(mK)$^{-1}$
C	5.94e4	4.82	4,000	14.0	0.79	–
Cu	7.6e3	4.30	1,356	36.8	1.4e-11	392.5
Hg	–	4.53	234	225	–	10.38
K	–	2.00	336	69	–	94.2
Li	–	2.55	453	64	–	64.8
Na	–	2.28	371	71.3	–	134.2
Pt	1.7e8	6.30	2,046	35.7	2.18e-7	69.7
Th	6.02e3	3.35	–	10.7[a]	143.3[a]	–
W	6.02e3	4.52	3,653	14.3	2.97	199

[a]For tungsten with Thorium added, let $T_M = 3{,}653$K

$T \to \infty$. Therefore the current density is a continuous increasing function of temperature. It also shows that for a given surface temperature a lower value of the work function gives a higher current density. It is, therefore, necessary to have a high current density if the material has a high operational temperature and low work function, and the maximum operational temperature has to be the melting or the sublimation temperature T_M. From the value of C and ϕ_o given by *Cobine* [7] (values given in Table 8.1, $e\phi_o/(k_B T_M)$ were calculated for several materials), which can be used to evaluate the suitability of a material for thermoionic emission of electrons. Further, the table includes heat conductivity coefficient of the material, k.

It may be noted from this comparison that since $e\phi_o/(k_B T_M)$ must be as small as possible among the metals and non metals given in the table, tungsten and carbon cathodes give reasonably high electron current density, and initially this can be further improved by adding one or two percent thorium to the tungsten. Since alkali metals and also mercury have low melting temperatures, for these the cathode can be in liquid state and the electron emission may be by mechanisms other than the thermo-ionic emission. Maximum current density in Table 8.1 at the maximum melting temperature is given, at which the value of the maximum current density, j_{max}, is calculated. For most of the metals the maximum current density is too small for a pure thrmoionic emission, except for the thoriated tungsten cathode. For the carbon electrode the electron emission mechanism seems partly connected with the sublimation of carbon.

While external heating is used for low power discharges, for high power discharges, the heat (to heat up the surface) is supplied by the discharge itself, necessitating a very high rate of heat flux in a very small region of order two to five millimeter diameter. Assuming that in this small heat region, a constant heat flux $\bar{q}$ (in Wm^{-2}) is applied, an analysis of the temperature distribution near the surface has been carried out by Mehta [83]. Let a semi-infinite radius circular plate having finite thickness h be heated by a uniform heat flux $\bar{q}$ over the area of radius r_o, which is investigated with the intention of finding the role of cooling at the opposite side of the plate.

The Laplace equation of a steady heat conduction in cylindrical coordinates and symmetrical to the axis is

$$\nabla^2 T(r, z) = 0 \tag{8.12}$$

with the boundary conditions that

$$kT_z(r, 0) = \bar{q} \text{ for } r \leq r_o;$$
$$= 0 \text{ for } r > r_o \tag{8.13}$$

where k is the heat conductivity coefficient of the metallic surface, T is the excess temperature with respect to that at h. Now the infinite *Hankel transformation* (Mehta [83]) is given by

$$\bar{T} = \int_0^\infty r J_o(pr) T(r, z) \mathrm{d}r \tag{8.14}$$

$$T = \int_0^\infty p\bar{T}(p, z) J_o(pr) \mathrm{d}r \tag{8.15}$$

where J_o is the *Bessel function* of the zeroth order. Applying *Hankel transformation*, (8.14), to (8.12), and inverting with (8.15), the general solution is

$$T(r, z) = \int_0^\infty pC \exp^{-pz} \exp^{-p(z-2h)} J_o(pr) \mathrm{d}r \tag{8.16}$$

and the boundary conditions become

$$\int_0^\infty f(p) J_o(rp) \mathrm{d}p = \bar{q}/k \quad \text{for} \quad r \le r_o \quad \text{and} \ = 0 \text{ for} \quad r > r_o \tag{8.17}$$

where

$$f(p) = Cp^2 \left(1 + \exp^{-2hp}\right) . \tag{8.18}$$

From the boundary conditions the constant C and the function f(p) are determined by using a *Watson expression* (*Mehta* [83]), and one gets the solution of the *Laplace equation* as

$$T(r, z) = \frac{\bar{q}r_o}{k} \int_0^\infty \frac{1}{p} \left(1 + \exp^{-2hp}\right) J_1(r_o p) J_o(rp) \left[\exp^{-pz} - \exp^{-p(z-2h)}\right] \mathrm{d}p . \tag{8.19}$$

This equation is evaluated conveniently by expressing it in terms of series expansion of *hyperbolic functions* and further in terms of *Watson expressions*. Using the proper boundary conditions it is found that at $z = 0$, the maximum temperature at the center line ($r = 0$) is $\bar{q}r_o/k$, and the mean temperature over the area $r < r_o$ is $8\bar{q}r_o/(3\pi)$. For the determination of the temperature distribution one can terminate the series expansion after two terms because of the very rapid convergence of the series, and one gets an expression for the temperature distribution in terms of *Riemann's zeta function*. Since the mathematical procedure is somewhat involved, only the results are discussed.

It can be seen from Fig. 8.1 that the maximum temperature on the center line of the plate, if its thickness is $h \to \infty$, is $(T - T_h) = \bar{q}r_o/k$, where $T_h = 0$ is the uniform excess temperature of the other surface at $z = h$. As a result of the finite thickness of the plate, the surface temperature at the heating side comes down, and the maximum temperature for the plate thickness $h = 1$ cm at $r = 0$, $z = 0$ is no more than $0.7\bar{q}r_o/k$. It shows further that beyond $h > 4$ cm the effect of cooling on the other surface is negligible.

An estimate of the anode heat load, for a given current, cathode material and for thermoionic emission can now be made. From the cathode material

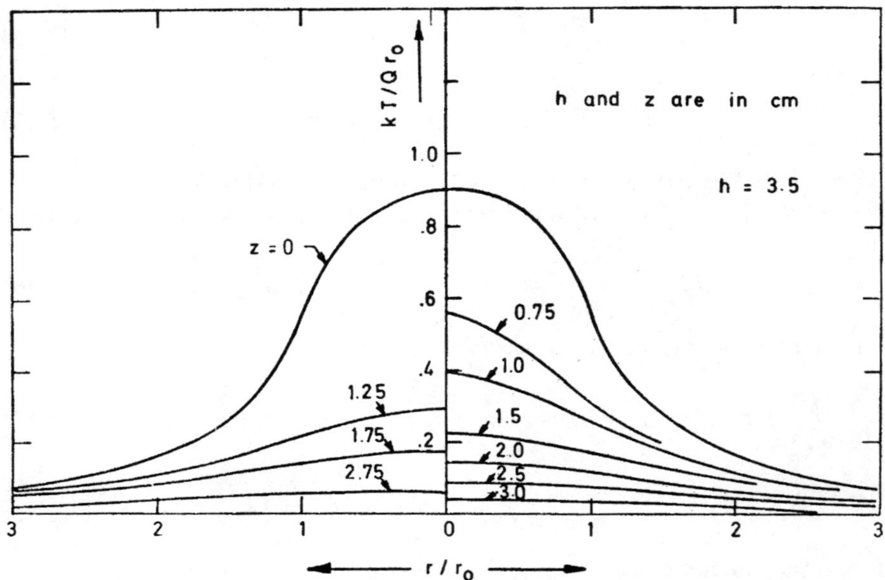

Fig. 8.1. Temperature profile in a circular plate

data given in Table 8.1 and for a given temperature distribution, for example,
the parabolic temperature distribution

$$\frac{(T - T_h)}{(T_M - T_h)} = 1 - r^{*2} \tag{8.20}$$

where $r^* = r/r_o$ and T_M = melting (maximum) temperature of the cathode,
one can easily find the distribution of the electric current density, j. From
(8.12), one can thus write for the total current

$$I = 2\pi r_o^2 \int_0^1 jr^* \mathrm{d}r^* = C_1 r_o^2 \tag{8.21}$$

where $C_1 = 1.14$ A.mm^{-2} for a pure tungsten cathode and $C_1 = 41$ A.mm^{-2}
for a thoriated tungsten cathode. Thus it is seen that in the electric current
range between 60 to 500 amps, the thoriated tungsten cathode, under the
thermionic emission gives a reasonable cathode spot radius $r_o = \sqrt{I/C_1}$,
and is used for further calculation. For the cathode, the average heat flux is

$$\bar{q} = k(T_M - T_h)/r_o = k(T_M - T_h)\sqrt{C_1/I} \tag{8.22}$$

and the total heat load to the cathode is

$$Q = \pi r_o^2 \bar{q} = \pi k(T_M - T_h)\sqrt{I/C_1} \ . \tag{8.23}$$

This is quite a large value and leads to the conclusion that immediately adjacent to the cathode there must be a very steep temperature gradient to facilitate the large heating and electron emission. Dividing Q in (8.23) by the total current I, one gets $U_1 = (Q/I)$, and by subtracting the work-function ϕ_o to it, one gets the actual heat load U_2 on the cathode. For an electric current of 200 amps, the electrons emitted from a thoriated tungsten cathode, one may thus calculate the following parameters: the cathode spot radius $r_o = 2.21$ mm, average heat flux $\bar{q} = 302$ W.mm^{-2}, heat load $Q = 4.63$ kW, $U_1 = 23.15$ volts and $U_2 = 19.8$ volts. It may be noted that U_1 and U_2 are not directly related to the cathode fall. The analysis leads to quite high values of U_1 and U_2. However, certain conclusions regarding the cathode spot radius, the minimum cathode radius and length (plate thickness) appear reasonable, but the temperature at the cathode spot may be with a higher order parabolic distribution than two.

In case the temperature of the solid metal surface is not too high and is not sufficient to have a pure thermoionic emission of the electrons, it is necessary to have a strong electric field to *pull out* the electrons so that it may overcome the force acting due to image field on the surface $E_i = e/(4x^2\epsilon_o)$. The factor $(1/4)$ is due to the force acting on one side of the wall, as well as due to the integration of the cosine of the angle with the normal to the surface, and ϵ_o is the dielectric constant. An external field E exerts a force eE on the free electron, and at a distance $x = x_o$ from the cathode, these forces are equal. Thus, an electron, reaching the distance x_o from the cathode, may escape from the influence of the surface. Since at $x = x_o$, $E_i = E$, it follows that

$$x_o = \sqrt{e/(4\epsilon_o|E|)} \ . \tag{8.24}$$

Since the applied electric field exerts a force on the escaping electron, it changes the effective work function of the surface to a value ϕ'_o, which is determined by calculating the effective energy that must be spent to release the electron from $x = 0$ to $x = x_o$

$$
\begin{aligned}
e\phi'_o &= e \int_0^{x_o} |E_i - E|\mathrm{d}x \\
&= e \left[\int_0^{\infty} |E_i|\mathrm{d}x - \int_{x_o}^{\infty} |E_i|\mathrm{d}x - \int_0^{x_o} |E_i|\mathrm{d}x \right] \\
&= e \left[\phi_o - \int_{x_o}^{\infty} \frac{e}{4\pi x^2 \epsilon_o}\mathrm{d}x - \int_0^{x_o} |E_i|\mathrm{d}x \right] \ .
\end{aligned}
\tag{8.25}
$$

Note that

$$\int_{x_o}^{\infty} \frac{e}{4\pi x^2 \epsilon_o}\mathrm{d}x = \frac{e}{4 x_o \epsilon_o} = \frac{1}{2}\sqrt{\frac{e|E|}{\epsilon}} \ . \tag{8.26}$$

Further assuming that there exists a constant external electric field between $x = 0$ and $x = x_o$,

$$\int_0^{x_o} |E|\mathrm{d}x = |E|x_o = \frac{1}{2}\sqrt{\frac{e|E|}{\epsilon_o}} \; . \tag{8.27}$$

Thus the effective work-function is given by the relation

$$\phi_o' = \phi_o - \sqrt{e|E|/\epsilon_o} \tag{8.28}$$

which must be put substituted in (8.10) to get the modified equation for the current density of electrons by *thermoionic* plus *field emission*. Under the action of the electrostatic field alone the electrons may be pulled out if

$$|E| > \epsilon_o\phi_o^2/e = 5.51 \times 10^7 \phi_o^2 \; , \; \mathrm{Vm}^{-1} \tag{8.29}$$

if ϕ_o is taken in [volts]. While for an arc plasma this high electric field is never applied externally, fields of this order may develop between very fine pointed electrodes, separated by a distance of 1 cm at a potential difference of 1,000 volts only.

It may be noted that the emission of electrons by bombardment of the charged particles can be significant at high electric fields like in glow discharges and may also be significant for arc plasma, in general. If δ is the ratio of the number of secondary electrons per incident charged particle, it is evident that this is a function of both bombarded material as well as electric field near the surface. In an arc plasma, where the cathode potential is at a somewhat lower potential than the plasma, electron bombardment at the cathode is unlikely, and the secondary emission of electrons at the cathode by the positive ion bombardment may be quite within the realm of possibilities. In case of an insulator, there may be a positive charge build up, if the number of the emitted secondary electrons is larger than the number of the incident primary electrons, leading to the establishment of active hot spots.

A related topic of electron emission at the cathode and such absorption at the anode is what happens near the solid surface of electrodes by way of potential or electric field distribution when potential is applied externally with respect to the plasma. In the following section we would now discuss the sheath effects that occur near the surface which affects the energy transfer from the hot gas to the cold surface. This energy or heat transfer is now discussed.

8.2 One-Dimensional Sheath Effects

There are few studies modelling the physical phenomenon occurring near the electrodes and the insulated wall. For slightly negative electrostatic (*Langmuir*) probes the usual theory assumes a constant ion number density distribution near the wall, while the electron number density falls exponentially towards the wall. The main difficulty in such a model is that even for an anode current density in the practical range (10^7 to 10^8 Am^{-2}), the electric potential must fall continuously from the plasma toward the anode, and the anode should be negative with respect to the plasma. This is quite different from the experimentally observed phenomenon whereby, near an anode, the electric potential first increases towards the anode and then falls slightly, so that the net anode potential is still positive. There have, however, been efforts (*Hsu, Etamadi* and *Pfender* [69]; *Sanders* and *Pfender* [103]) to use for electrons in an adverse field the old Langmuir model, but the ion distribution is determined on the basis of conservation of ion flux. The present discussion follows a tentative model by this author (*Bose* [46]).

For this purpose we consider the plasma as consisting of electrons, atoms and singly-charged ions, and the region under consideration is bounded by an outside boundary, where the temperatures and pressure are prescribed, and at the wall (Fig. 8.2). The entire domain is divided into a collision-dominated or continuum region, for which the energy equations of the electrons and the heavy particles are solved, and an electron free-fall region, where the Poisson equation is solved along with the flux equations for the charged particles. In addition, some of the basic assumptions are: (1) the plasma is in a steady state, (2) the heat and electric conductivity coefficients are scalar quantities (small magnetic field), and (3) the radiation loss is neglected. The electric current density is now allowed to be either parallel to the wall, so that the electric field E parallel to the wall is uniform in the direction normal to the wall ($E = E_o = j/\sigma = j_o/\sigma_o$), or it is constant in the direction normal to the wall ($j = j_o$). Here j is the current density and σ is the electrical conductivity. On the outer boundary the given parameters are: the temperature $T_o = T_{ho} = T_{eo}$, the pressure p, and the magnitude and direction of the current density j_o, which is carried mainly by the electrons due to the externally applied electric field. The subscripts h and e refer to the heavy particles and the electrons, respectively, and the subscript o refers to the outer boundary.

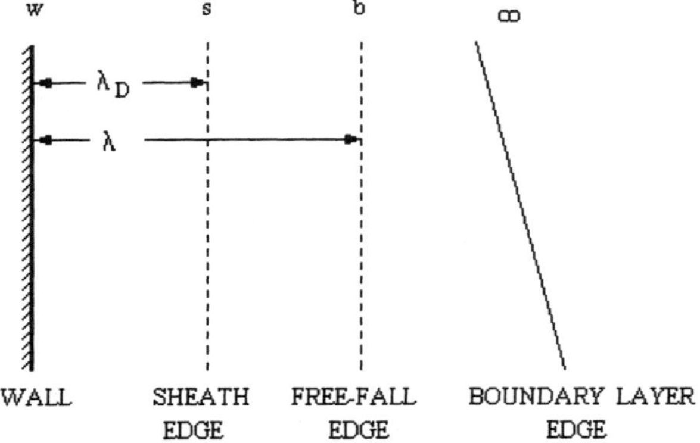

Fig. 8.2. Different regions near a wall

In addition, the following motions of the particles are considered. Due to ambipolar diffusion the electrons and ions move in pairs in the direction of lower mole fraction of the charged particles, and the neutrals move in the opposite direction, but in an electric field the electrons move from the region of lower potential to that of higher potential, and the ions move in the opposite direction. Since in an electric field the electrons move much faster than the ions, these are mainly responsible for carrying the current, for which the convention is that the direction of the electric current is opposite the direction of movement of the electrons. We now consider the situation in the continuum region, where quasi-neutrality is assumed ($n_i = n_e$), where these are the number density of the electrons and ions, respectively).

The extent of the various regions of interest in Fig. 8.2 is delineated by calculating the mean free paths of various specie, λ_j, and the Debye shielding distance λ_D. These have been computed for an argon plasma for two-different pressures and single-temperature equilibrium model, and they have been presented in Fig. 8.3. The results show that, in the pressure range under consideration, the electron and atom mean free paths (atom mean free paths

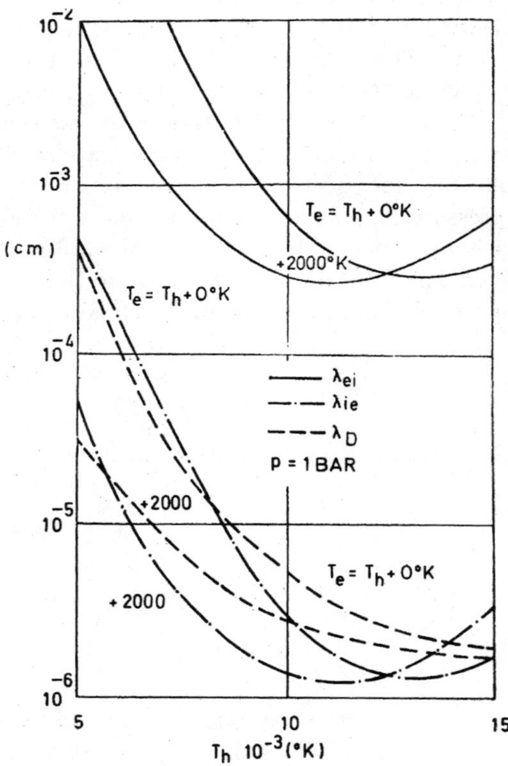

Fig. 8.3. Mean free paths and Debye shielding distance for argon plasma

not shown) are one to two orders of magnitude larger than the *Debye length*, but the mean free path of the ions is of the same order of magnitude as the Debye length. These order of magnitude estimates show that the electrons in the collisionless region are affected by the electric field distribution near the wall only.

Initially we consider the classical *Langmuir model*, which is valid strictly for a very low-density gas without a boundary layer. For this purpose, let us consider a one-dimensional model with the wall surface having a negative external potential (ϕ_w <0). It is now assumed that the charged particles fall freely without collision from the free-fall edge, beyond which there is quasi-neutrality, and no electric field exists in this region. Now the potential distribution near the body surface, *Poisson equation* in one-dimensional case, is

$$\frac{d^2\phi}{dx^2} = -\frac{e}{\epsilon_o}(n_i - n_e) \tag{8.30}$$

where e is the elementary charge, ϵ_o is the dielectric constant in vacuum, and n_i and n_e are the particle number density of the (singly-charged) ions and the electrons, respectively.

The above equation is made non-dimensional by introducing the following non-dimensional variables:

$$\phi^* = -e\phi/(k_B T_e); n_i^* = n_i/n_{eb}; n_e^* = n_e/n_{eb}; x^* = x/\lambda_D \tag{8.31}$$

where λ_D is a characteristic distance and is called the *Debye shielding distance* defined as follows:

$$\lambda_D = \sqrt{\frac{k_B T_e \epsilon_o}{e^2 n_{eb}}} . \tag{8.32}$$

Thus, the above differential equation (8.30) becomes

$$\frac{d^2\phi^*}{dx^{*2}} = (n_i^* - n_e^*) . \tag{8.33}$$

If the wall surface is positively charged, then the ion-number density will fall, and the accelerating electron number density will fall also. On the other hand, for strongly negatively charged wall surface the electron number density to wall may fall to zero, but the number density of the accelerating ions may fall slightly. The third case, which will now be considered, is that of the slightly negatively charged wall surface. For this case, $n_i^* = 1$, $n_e^* \approx \exp^{-\phi^*}$ and (8.33) is written as

$$\frac{d^2\phi^*}{dx^{*2}} = 1 - \exp^{-\phi^*} \approx \phi^* . \tag{8.34}$$

The general solution of Eq. (8.34) is

$$\phi^* = C_1 \exp^{-x^*} + C_2 \exp^{x^*} \tag{8.35}$$

with boundary conditions, $x^* = 0$: $\phi^* = 1$ and $x^* \to \infty$: $\phi^* = 0$. Thus one finds that $C_1 = 1$ and $C_2 = 0$, and the non-dimensional potential distribution is given by the relation

$$\phi^* = C_1 \exp^{-x^*} \; . \tag{8.36}$$

Substituting (8.36) back into (8.33), one can get the distribution of the electron number density for a slightly negatively charged surface as

$$n_e^* = 1 - \exp^{-x^*} \; . \tag{8.37}$$

Equation (8.37) indicates that at the wall surface ($x^* = 0$), the electron particle density $n_e^* \approx 0$ for all slightly negatively charged surface, but independent of the applied potential. This is, of course, not true, since the result can not be valid even for the limiting case of the zero negative potential of the surface, and hence this has been discussed further below. However, (8.36) gives approximately the potential distribution, and it is seen that the (negative) potential falls exponentially to about one percent of its value at about four times λ_D. Thus one gets an order of magnitude of the distance by which the potential applied on the surface drops to a very small value (*"shielding distance"*). Actual value of the surface potential can now be estimated for the special case of no electric current being collected at the surface (*floating potential*). For this case the ion-flux must be equal to the electron flux reaching the surface. While the ion-flux, according to the model, can be evaluated at the *sheath edge* (at a distance of approximately four times λ_D from the wall) and remains constant in the region, the collected electron flux undergoes a change from the sheath edge to the wall. Thus, by equating the two fluxes at the wall surface, we write

$$\frac{1}{4} e n_{eb} \left(\frac{8 k_B T_{hb}}{\pi M_h} \right)^{0.5} = \frac{1}{4} e n_{eb} \left(\frac{8 k_B T_{eb}}{\pi M_e} \right)^{0.5} \exp^{e\phi_w / (k_B T_{eb})} \tag{8.38}$$

in which the subscript "h" refers to the heavy particles (ions). It is now possible to get the wall potential from the above equation as

$$\phi_w = \frac{k_B T_{eb}}{2e} \ln \left(\frac{T_{hb}}{T_{eb}} \frac{M_e}{M_h} \right) = \frac{T_{eb}}{23200} \ln \left(\frac{T_{hb}}{T_{eb}} \frac{M_e}{M_h} \right) \; . \tag{8.39}$$

While the current density at the floating potential is equal to zero, at the *plasma* or *space potential* ($\phi_w = 0$) the maximum current density (positive since maximum current is carried by the electrons moving towards the wall) is given by the relation

$$j_{\max} = e n_{eb} (c_{eb} - c_{hb}) / 4 \tag{8.40}$$

where c_{eb} and c_{eb} are the mean kinetic speed for the electrons and the ions at the free-fall edge "b", respectively. In the intermediate current density region ($0 < j/j_{\max} < 1$), the current density is given by the relation

$$j = en_{eb} \left(c_{eb}\, \exp^{e\phi_w/(k_B T_{eb})} - c_{hb} \right)/4 \qquad (8.41)$$

and the corresponding wall potential can be written as

$$\phi_w = \frac{k_B T_{eb}}{e} \ln\left[\left(\frac{T_{hb}}{T_{eb}}\frac{M_e}{M_h}\right)^{1/2} + \left\{ 1 - \left(\frac{T_{hb}}{T_{eb}}\frac{M_e}{M_h}\right)^{1/2} \right\} \frac{j}{j_{\max}} \right] . \qquad (8.42)$$

As a sample case, for equal temperature of the ions and the electrons, the wall floating potential for a particular gas can now be evaluated easily. For argon at $T_{eb} = 10{,}000$ K, for example, the result is $U_o = -4.83$ volts. However, the above estimate is only tentative even for the special case of the *floating potential*, since in the continuum region of the plasma the electron temperature at the sheath edge may be quite different from the one at a large distance from the surface (*"free-stream"*). Hence a direct numerical simulation has been attempted (*Bose* [46]). The physical model being considered is to assume in the continuum region the two-temperature model (electron temperature different from the heavy particles temperature) without convective terms in the differential equations. This insures that there is sufficient number of free electrons in the wall region; the local mole fraction of various specie are computed with the help of two-temperature equilibrium model discussed in Chap. 6. The equations of energy for the electrons and the heavy particles are

$$(k_{te}T_{ex})_x - 3(m_e/m_h)(k_B\varGamma_{eh})(T_e - T_h) + (j^2/\sigma) = 0 \qquad (8.43)$$

and and the heavy particles are

$$(k_{th}T_{hx})_x + 3(m_e/m_h)(k_B\varGamma_{eh})(T_e - T_h) = 0 . \qquad (8.44)$$

In above equations the subscript "t" in the heat conductivity coefficient k refers to the total value, that is, it includes *pure* heat conduction and *reactive heat conduction* due to *ambi-polar diffusion* of the j-th species (j = h,e for heavy particles and electrons, respectively). Further, m_j is the mole mass of the j-th species, T_j is the temperature, $\varGamma_{eh}$ is the collision frequency for collisions between the electrons and heavy particles, σ is the electrical conductivity, and j is the electric current density. In addition the subscript "x" refers to the derivative with respect to this coordinate and k_B is the *Boltzmann constant*.

Equations (8.43) and (8.44) are second order coupled differential equations and can be solved for the temperatures prescribed at the two-ends of the continuum region. Although in the free-stream region the temperature for the electrons and the heavy particles are given quantities, these are obtained at the *"free-fall edge"* by matching the analysis in the sheath region.

Now, at the interface between the continuum and *free-fall* regions (shown with "b" in Fig. 8.2 and referred henceforth with subscript "b"), the electron energy flux at the *free-fall edge* is due to pure conduction, $-k_{ce,b}(T_{ex})_b$, and diffusion, $(5/2)k_B T_{eb}(n_e V'_e)_b$, where V'_{eb} is the *diffusive speed*. In the case where we consider a positive potential at the wall, the electron flux at this interface reaches the wall entirely and the electrons can acquire, in addition to the average thermal energy $(2k_B T_{eb})$, an additional energy flux $e\phi_w$. On the other hand, for a negative potential of the wall only a fraction of the electron-flux reaches the wall. Similar is the case of the ions, and we can consider both the ions and the electron flux in the sheath region, which is assumed to be entirely within the *free-fall* or *collisionless region*. The individual species particles are accelerated or decelerated depending on the sign of the potential of the wall with respect to the potential at the *"free-fall edge"*.

Let the mean kinetic speed of the j-th species with mass M_j and temperature T_j at the free-fall edge be c_j, and we write

$$\frac{1}{2}M_j c_j^2 = \frac{3}{2}k_B T_{jb} = e\widehat{\phi}_{jb} \tag{8.45}$$

where $\widehat{\phi}_{jb}$ is the potential equivalent of the kinetic energy for the j-th species at the boundary "b". Thus, for the accelerating case (increasing the speed of the particles in the direction of motion), we may write

$$\frac{1}{2}M_j(c_j^2 - c_{jb}^2) = \frac{3}{2}k_B(T_j - T_{jb}) = e(\widehat{\phi}_{jb} + \phi) \ . \tag{8.46}$$

In addition, in the collisionless zone there is no collisional reaction, and $d(n_j c_j) = 0$ especially for the accelerating case.

Thus, for the accelerating case, we get the two relations in differential form as

$$\frac{dT_j}{dx} = \pm\frac{2}{3}\frac{e}{k_B}\frac{d\phi}{dx} \tag{8.47}$$

$$\frac{dn_j}{dx} = \pm\frac{1}{3}\frac{en_j}{k_B T_j}\frac{d\phi}{dx} \ . \tag{8.48}$$

The signs on the right-hand side of the two-equations are such that the temperature increases and the number density decreases in the direction of motion of the accelerating charged particles. For the retardation case, it is usually assumed that the particles at a given reference temperature T_{jb} must undergo a reduction in the number density, since the slower particles can not overcome the potential barrier. Hence the usual relation for the number density n_j in the retardation case with respect to the reference value at the free-fall edge n_{jb} is given by the relation

$$n_j/n_{jb} = \exp^{-e|\phi - \phi_b|/(k_B T_{jb})} \ . \tag{8.49}$$

Thus, in the differential form, the two equations, valid for the retardation case, are

$$\frac{dT_j}{dx} = 0 \tag{8.50}$$

$$\frac{dn_j}{dx} = \pm\frac{en_j}{k_B T_j}\frac{d\phi}{dx} . \tag{8.51}$$

Further, the potential distribution in the sheath region is given by the *Poisson equation* (8.30).

Computations in the above study (*Bose* [46]) were done for argon for different values of the (one-dimensional) electric current density, separately for the continuum and collisionless regions and matching at the interface. In the collisionless region three different options were used: (a) the electron number density calculated from (8.47, 8.48) or (8.50, 8.51) but $n_i = n_{eb}$; (b) n_i and n_e are calculated according to these equations without any restrictions; and (c) n_i and n_e are calculated according to these equations but subject to the restrictions that $n_i < n_e$ and $n_i = n_e$. It is evident that the option (a) is equivalent of the Langmuir model, except that the results are obtained from the solution of the differential equations. From extensive numerical experimentation it is concluded that the option (a) gives reasonable result for the cathode or insulated wall (at floating potential) except that it gives more realistic wall potential in comparison to the Langmuir model, option (b) gives totally unrealistic wall potential for the wall as the anode, and option (c) is reasonable for the anode. For the wall as cathode with thermoionic emission the option (a) gives somewhat reasonable wall potential in comparison to the Langmuir model. These numerical experimentation throws, therefore, important challenges in bringing out new theories for studying the sheath region.

Finally we discuss the case of no current flowing through the wall, that is when the wall is at the (negative) floating potential, which develops because of the higher mobility of the electrons. While the ion-flux, according to the Langmuir theory, remains constant across the collisionless region, only a fraction of the electron flux reaches the wall. At the interface of the two regions, but in the collision-dominated region side, we can now write the energy balance equation as

$$-k_{ce,b}(\nabla T_e)_b + \frac{5}{2}k_B T_{eb}n_{eb}\mathbf{V}'_{eb} = (2k_B T_{eb} + |e\phi_w|)(n_{eb}\mathbf{V}'_{eb}) . \tag{8.52}$$

The above equation does not consider the contribution of the electric current density due to the externally applied electric field. In addition, from consideration of the continuity of electrons at the free-fall edge, the ambi-polar diffusion term is now taken as

$$-n_{ce,b}\mathbf{V}'_{eb} = -n_b D_{amb}(\nabla x_e)_b = -\frac{1}{4}n_{eb}c_{eb}\exp^{-e|\phi_w|/(k_B T_{eb})} \tag{8.53}$$

where $|\phi_w|$ is the floating potential of the wall. Equation (8.52) can now be written as

$$(\nabla T_e)_b = -K(\nabla x_e)_b \qquad (8.54)$$

where

$$K = \frac{5}{2}\frac{k_B T_{eb}(nD_{amb})_b}{k_{ce,b}}\left[0.2 - 0.4\frac{|e\phi_w|}{k_B T_{eb}}\right] . \qquad (8.55)$$

In the above expression since the pressure dependency of the number density and ambi-polar diffusion coefficient tend to cancel each other, the value of K is somewhat independent of pressure. Further, the gradient of the electron temperature at the free-fall edge of the reacting (chemically and thermally at non-equilibrium) continuum region is approximately equal to zero since, near the wall $x_e \to 0$ and also the gradient of $x_e \to 0$. On the other hand, for the gradient of x_e in the chemically equilibrium but thermally non-equilibrium case, we write

$$(\nabla T_e)_b = -\left[\frac{K(\partial x_e/\partial T_h)_b}{1 + K(\partial x_e/\partial T_e)_b}\right](\nabla T_h)_b . \qquad (8.56)$$

In a practical calculation the grid point right next to the wall is so chosen that it would be well within the continuum region, and thus all the quantities in the right hand side of Eq. (8.56) can be evaluated. From sample calculations with argon plasma it was found, that the quantities under [ ] is a very small quantity since the numerator is much smaller than the denominator (of the order of 10^{-4}), and, therefore, for all practical purposes it is possible to state that the gradient of electron temperature in the region of a wall at floating potential is equal to zero.

8.3 Heat Transfer

The validity of the continuum theory is assumed in the main flow field to a location b, which is at one mean free path of the collected particles from the surface. This mean free path is, as discussed in the previous section, much longer than the distance across which the applied electric potential on the surface is shielded (sheath region). At moderate pressures ($p > 1$ atm) it is found that the electron mean free path λ_e is much longer than the Debye shielding distance λ_D, whereas the ion mean free path $\lambda_i \approx \lambda_D$. Now, the heat flux at any surface depends on a conduction term $-k_c \nabla T_j$, and a term due to the transport of enthalpy by diffusion $\sum \rho_j \mathbf{V}'_j h_j$. For charged particles in an electromagnetic field, the diffusion velocity consists of effects due to *pure diffusion* due to the concentration gradient, which is characterized by a common *ambi-polar diffusion coefficient* D_{amb}, and the *electric field diffusion velocity* $\mathbf{V}'_{fj}$. This second effect can be written as follows:

$$\sum \rho_j \mathbf{V}'_{fj} h_j = \frac{5}{2}k_B \sum j_j T_j/q_j \approx \frac{5}{2}\frac{k_B}{e}j_e T e \qquad (8.57)$$

where q_j is the charge of the particle and e is the elementary charge. If in the following $\mathbf{q}$ is designated as heat flux, then the convective heat flux at the free-fall edge b (Fig. 8.2) is given by the relation

$$q_b = -k_h \nabla T_h - k_e \nabla T_e - \rho D_{amb} E_{mi} \nabla Y_j - \frac{5}{2} \frac{R^*}{m_e} D_{amb} T_e \nabla Y_e + \frac{5}{2} \frac{k_B}{e} j T_e \quad (8.58)$$

where $Y_j = \rho_j/\rho$, T_h = heavy particles (neutrals, ions) translational temperature, T_e = electron translational temperature, R^* = universal gas constant, E_{mi} = ionization potential per unit mass ions, and m_e = mole mass of the electrons. It may be noted that the gradients and the properties in this equation are evaluated at the free-fall edge b. In addition, k_h and k_e are not the heat conductivity of the pure gas, but they specify only the contributions (due to mole or mass fraction) of the heavy particles and electrons, respectively. In Chap. 7, it has been shown already that $\tilde{k}'_{ce}$ is of the order of $(x_e k'_{ce})$ and that $\tilde{k}'_{ch}$ is of the order of $(x_a k'_{ca} + x_i k'_{ci})$, where the mole ratio is $x_j = n_j/n$ (j = a, i, e for atom, ion and electron, respectively) and k'_{cj} denotes the heat conductivity coefficient by translation of *pure gases*. In many cases the mole fraction of the electrons x_e at the wall is very small and the effect due to the second term in the earlier equation may be neglected.

The heat flux at the wall is now the heat flux at b, $\mathbf{q}_b$, plus the energy gained (or lost) in the free-fall region $\delta\dot{e}$, and the radiative heat flux $\mathbf{q}^R$ to the surface, and thus one may write

$$\mathbf{q}_w = \mathbf{q}_b - \Delta\dot{e} + \mathbf{q}^R \ , \ \mathrm{Wm}^{-2} \ . \quad (8.59)$$

Next we examine the energy gain or loss in the sheath.

(a) *Species in a decelerating field*: To compute the energy gain (or loss) in the sheath $\Delta\dot{e}$, a potential U_w is applied to the body surface to retard the j-th species and to attract the k-th species. The former reaches the wall in free-fall provided that the random velocity at "b" is $v_j > (2e|\phi_w|/M_j)^{0.5}$, where M_j is the mass of a single particle of the j-th species. The flux of the j-th species reaching the wall and their associated energy flux at "b" is, therefore, found by integrating v_j from $(2e|\phi_w|/M_j)^{0.5}$ to infinity and not from zero to infinity. Thus

$$\dot{\mathbf{n}}_{jw} = -\frac{1}{4} n_{jb} v_{jb} \exp^{-e|\phi_w|/(k_B T_{jb})} \ , \ \mathrm{m}^{-2}\mathrm{s}^{-1} \quad (8.60)$$

$$q_{jb} = -\frac{1}{4} n_{jb} v_{jb} [e|\phi_w| + 2k_B T_{jb}] \exp^{-e|\phi_w|/(k_B T_{jb})} \ , \ \mathrm{Wm}^{-2} \quad (8.61)$$

where the *mean kinetic speed* is

$$v_j = \left(\frac{8k_B T_j}{\pi M_j}\right)^{0.5} \ . \quad (8.62)$$

The loss of energy in the retarding field is $\Delta\dot{e}_j = -\dot{n}_{jw}e|\phi_w|$ [Wm^{-2}] and the net energy flux from the j-th species to the wall is

$$\mathbf{q}_{jw} = 2\dot{n}_{jw}k_B T_j \ , \ \text{Wm}^{-2} \ . \tag{8.63}$$

(b) *Species in an accelerating field*: As in the previous case, for the accelerating k-th species, the flux of particles at the wall as well as the associated energy fluxes at "b" and "w", and the gain are:

$$\dot{\mathbf{n}}_{kb} = \dot{\mathbf{n}}_{kw} = -\frac{1}{4}n_{kb}v_{kb} \ , \ \text{m}^{-2}\text{s}^{-1} \tag{8.64}$$

$$\mathbf{q}_{kb} = -\frac{1}{4}n_{kb}v_{kb}(2k_B T_{kb}) = 2k_B T_{kb}\dot{\mathbf{n}}_{kw} \ , \ \text{Wm}^{-2} \tag{8.65}$$

$$\mathbf{q}_{kw} = \dot{\mathbf{n}}_{kw}(2k_B T_{kb} + e|\phi_w|) \ , \ \text{Wm}^{-2} \tag{8.66}$$

$$\Delta\dot{e}_k = \dot{\mathbf{n}}_{kw}e|\phi_w| \ , \ \text{Wm}^{-2} \ . \tag{8.67}$$

(c) *Species emitted from the wall*: In the case of a mass flux of the j-th species being emitted from the wall, namely $\dot{\mathbf{n}}_{jw}$ [m^{-2}s^{-1}], there is an associated energy flux at the wall

$$\mathbf{q}'_{jw} = \dot{\mathbf{n}}_{kw}(2k_B T_{kb} + e|\phi_w|) \ , \ \text{Wm}^{-2} \tag{8.68}$$

where E_{jw} is the average kinetic energy of the particles just emitted, and ϕ_w is the *work-function* of the material. The free electrons, on the other hand, *condensing* on the wall, give rise to a heat flux that is expressed as

$$\mathbf{q}_{ew} = \dot{\mathbf{n}}_{kw}e|\phi_w| = -\mathbf{j}_e|\phi_w| \ , \ \text{Wm}^{-2} \ . \tag{8.69}$$

(d) *Recombination at the wall*: Furthermore, there is also a possibility that a flux exists of the j-th species, which recombine with their counterparts at the wall and release an energy flux

$$\mathbf{q}_{jw} = \dot{\mathbf{n}}''_{jw}(I_j - e|\phi_w|) \ , \ \text{Wm}^{-2} \tag{8.70}$$

where I_i is the *ionization potential*.

We investigate now several heat transfer cases.

At the anode with reference to the previous analysis, and since the current density $\mathbf{j} = en_{eb}v_{eb}/4$, the heat flux is

$$\mathbf{q}_{Anode} = \mathbf{q}_j + \mathbf{j}(\phi_A + \phi_w) + \mathbf{q}^R = -k_h(\nabla T_h)_b + \frac{5}{2}\frac{k_B}{e}\mathbf{j}T_{eb} + \mathbf{j}(\phi_A + \phi_w) + \mathbf{q}^R \tag{8.71}$$

where ϕ_A is the *anode fall* and ϕ_w is the *work-function* of the anode material.

Generally, the ion current attracted at the cathode is very small, and the current is maintained by the thermoionic and/or the field emission described earlier. A simple estimation of the different emission mechanisms shows clearly that the thermoionic emission is indeed the dominating mechanism in high intensity arcs. Neglecting the enthalpy of the emitted electrons, the total heat flux to the wall is

$$\mathbf{q}_{Cathode} = -k_h(\nabla T_h)_b + \mathbf{j}\phi_w + \mathbf{q}^R \ . \tag{8.72}$$

When a conducting wall is placed between the electrodes, but insulated from either of them, it is capable of circulating an electric current. Near the cathode, the wall is negative with respect to the plasma, and the ions may drift to the wall and recombine giving rise to an effective electric current $\mathbf{j}$ flowing through the wall. The heat flux to the wall is

$$\mathbf{q}_{Anode} = \mathbf{q}_b - \dot{n}_{ew}e\phi_w + \dot{n}_{iw}(I_i - e\phi_w) + \mathbf{q}^R = -k_h(\nabla T_h)_b + \frac{5}{2}\frac{k_B}{e}\mathbf{j}T_{eb} + \mathbf{j}\frac{I_i}{e} + \mathbf{q}^R \tag{8.73}$$

where I_i is the ionization energy of the ions in volts. Unfortunately it is difficult to estimate the current density distribution. For segmented or non-conducting walls, $\mathbf{j} = 0$, and the equation reduces to the usual heat transfer relation for a plasma without a current flow.

9 Production of High Temperature Gases

High temperature gases can be produced in different ways, for example, by an isentropic compression, by an adiabatic compression in shocks, by combustion, by electrical means or by thermo-nuclear reaction, etc. For calculating the temperature generated by combustion, in general, one has to proceed in two steps. First, one has to calculate, at given pressure for combustion, the specific enthalpy per unit mass of the reacting components before combustion, and also of the equilibrium composition after combustion as a function of temperature. In an adiabatic combustion, the specific enthalpy of the product must be the same (provided the specific enthalpy is absolute, that is it includes the heat of reaction as its part) as that of the initial reacting components, and this allows determination of the temperature after reaction. As an example, the stoichiometric reaction of carbon monoxide with oxygen is taken, for which, in Fig. 9.1, the equilibrium composition (mole fraction) and in Fig. 9.2 the enthalpy are plotted against the temperature for different pressures. Before combustion, the ideal gas mixture is not dependent on pressure, and it gives in the latter figure a single line. After the combustion, however, the equilibrium composition is dependent on pressure. To find the adiabatic flame temperature, first the enthalpy of the gas mixture before combustion is found, and then only one has to move to the right at constant enthalpy. Thus for the reaction of the stoichiometric mixture of the carbon monoxide and oxygen, initially at room temperature of 300 K, the *adiabatic flame temperature* after reaction at pressures of 10^{-2}, 1 and 10^2 bar are 2,780 K, 3,210 K and 3,240 K, respectively. Thus the effect of higher dissociation is to reduce the adiabatic flame temperature. At the same time it is noted that at high pressures, an increase in the pressure does not lower the adiabatic flame temperature proportionately.

We shall now examine the range of temperatures that can be reached by an isentropic compression. As an example, nitrogen is taken whose specific heat ratio at around the room temperature is $\gamma = 1.4$. In the range of temperatures, in which the value of γ remains essentially constant and equal to 1.4, the temperature ratio (T_2/T_1) is dependent only on the pressure ratio (p_2/p_1), both of which are linked by the relation

$$\frac{T_2}{T_1} = \left(\frac{p_2}{p_1}\right)^{(\gamma-1)/\gamma} . \tag{9.1}$$

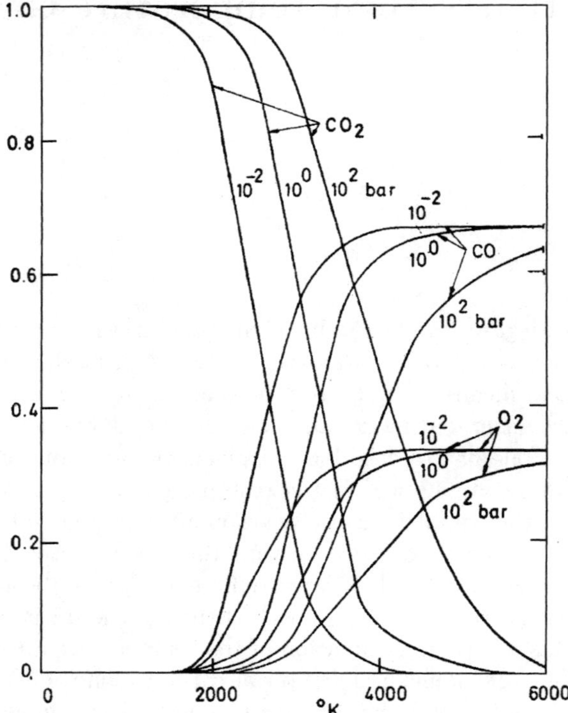

Fig. 9.1. Equilibrium mole fraction of Components of stoichiometric reaction of carbon monoxide with oxygen

Since, however, at higher temperatures the gases dissociate, it is necessary to consider the initial state $(p_1,\ T_1)$. Keeping $p_1 = 10^{-2}$ bar and for two different initial temperatures of 10^3 and 10^4 K, the *compression ratio* for nitrogen is plotted against the *temperature ratio*, and are shown in Fig. 9.3. It may be noted from the figure that for a *compression ratio* of 100, if $T_1 = 300$, 1000 and 10000 K, the respective temperature after compression of nitrogen with real gas properties (γ is not constant) are 1122, 2800 and 13600 K. Calculation of this and other processes can be studied best with the help of thermodynamic charts, as has been described for air plasma in the next section.

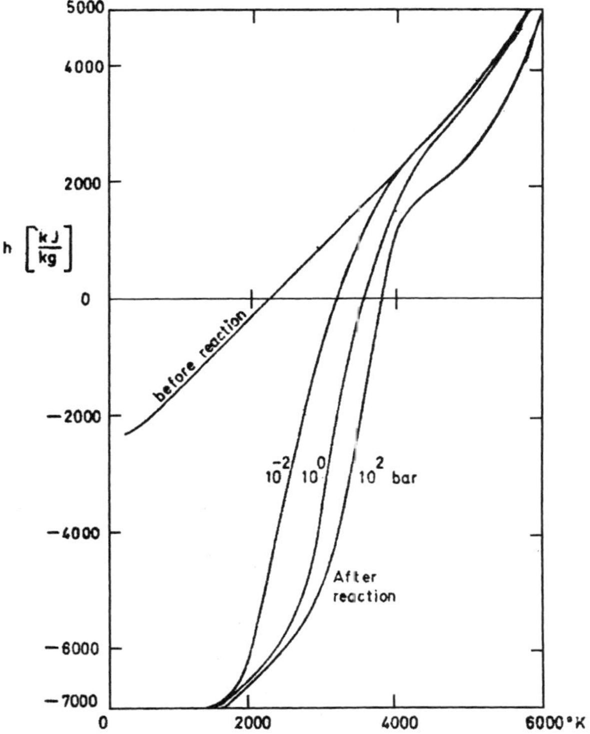

Fig. 9.2. Enthalpy-temperature plot before and after reaction for stoichiometric reaction of carbon monoxide with oxygen

9.1 Thermodynamic Charts for Air Plasma

For the purpose of studying various thermodynamic processes use of an enthalpy-entropy chart is well known. Such a chart for equilibrium air plasma is shown in Fig. 9.4, in which the constant temperature (isotherm) lines are shown between 2,000 K and 30,000 K, and the constant pressure (isobar) lines between 10^{-5} and 100 bars.

While an enthalpy-entropy chart is useful in studying an isentropic flow, for example in a (convergent or convergent-divergent) nozzle, one requires also the density information. Unfortunately, plotting of constant density lines in above chart makes the chart very difficult to read, especially when the plots are in one color. This is again difficult in studying the shock parameters with the help of such charts. A better method, shown for the first time independently by *Spalding* [106] and *Knoche* [113, 74], is by studying the processes in a new ($h,\log_{10}\rho$) chart, which has also been shown for air in Fig. 9.5. This chart contains not only the isotherm and isobar lines, but also the constant

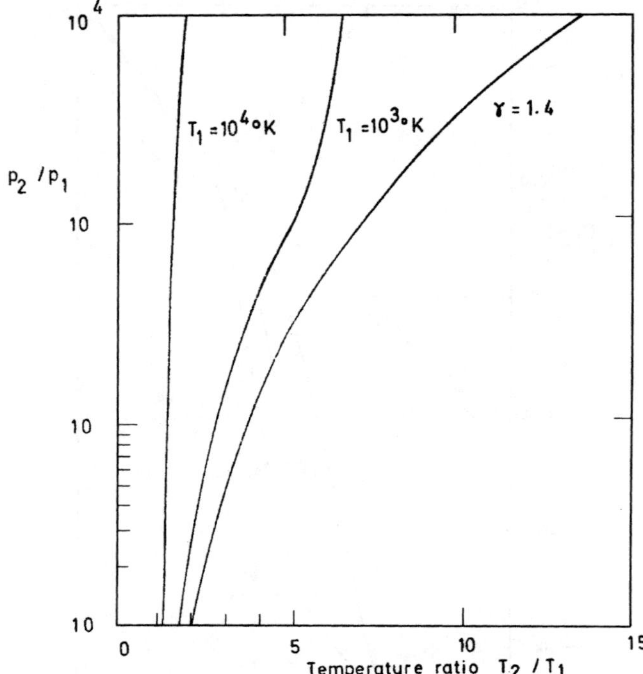

Fig. 9.3. Compression ratio vs. temperature Ratio for nitrogen for isentropic compression from $p_1 = 10^{-2}$ bar

entropy lines, which enables one to calculate the sonic speed of hot gas plasma very efficiently. On the basis of these charts we would now study, for a real gas (air) plasma some of the gas dynamic flow processes.

9.2 Isentropic Flow in a Nozzle

For the study of a real gas isentropic flow in a nozzle, the starting point is the knowledge of the stagnation state of pressure-temperature, (p^o, T^o), or the corresponding pressure-enthalpy, (p^o, h^o). One could then follow the constant entropy process in one of the *thermodynamic charts*, and for each value of enthalpy h, the corresponding density value ρ is noted and the *"one-dimensional"* gas speed

$$u = \sqrt{2(h^o - h)} \, , \, \text{ms}^{-1} \tag{9.2}$$

is calculated. This enables calculation of (ρu), which can be plotted against u. A positive slope denotes the subsonic region and a negative slope is for the supersonic region; the point of zero slope denotes the critical state at the

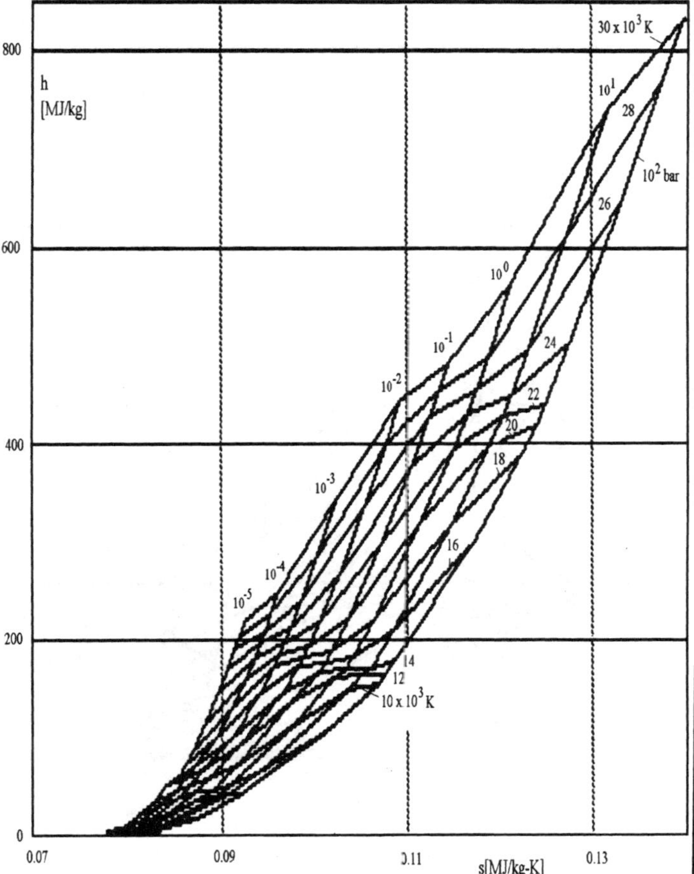

Fig. 9.4. Enthalpy-entropy chart for air plasma

nozzle throat, where the gas speed is equal to the sonic speed. One could now get the area ratio (A^*/A) by dividing the local (ρu) with the critical throat value $(\rho u)^*$.

In this connection, we note now the definition of the sonic speed as the square of the sonic speed being equal to the derivative of pressure with density at constant entropy change, or

$$a_s^2 = \left(\frac{\partial p}{\partial \rho}\right)_s \, . \tag{9.3}$$

Now from the *first law of thermodynamics*,

$$Tds = dh - \rho^{-1}dp \tag{9.4}$$

where h is the specific (per unit mass) enthalpy, s is the specific entropy and ρ is the density.

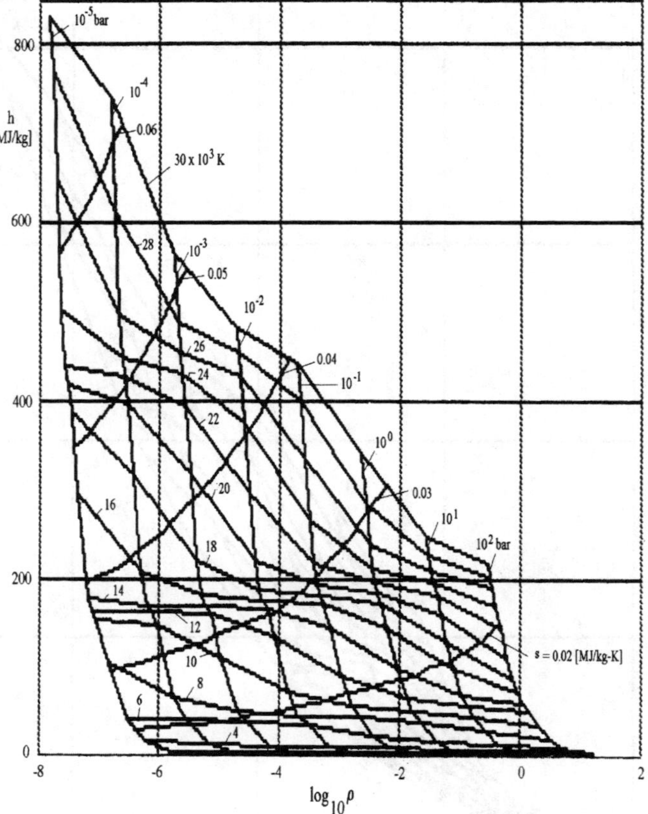

Fig. 9.5. Enthalpy-log(density) chart for air plasma

For a constant entropy case ($\mathrm{d}s = 0$), the left hand side of the above equation is made equal to zero, and hence $\mathrm{d}p = \rho\,\mathrm{d}h$. Therefore

$$a_s^2 = \left(\frac{\partial p}{\partial \rho}\right)_s = \left(\frac{\partial h}{\partial(\ln \rho)}\right)_s = 0.438\left(\frac{\partial h}{\partial(\log_{10}\rho)}\right)_s \tag{9.5}$$

and from the slope of the constant entropy line in (h, $\log_{10}\rho$) chart, one can easily get the local sonic speed. In fact, the method is particularly useful for high temperature dissociated and ionized gases where one does not have an exact value of the specific heat ratio or mole mass to determine the sonic speed. It may be noted again, that in Fig. 9.5 the constant entropy lines are plotted, the slope of which gives the square of the sonic speed.

9.3 Gas State After a Shock

The high temperature effect due to shock can best be demonstrated by plotting values of temperature T_2 and pressure p_2 behind the shock for different flow (supersonic) speeds ahead of the shock u_1, assuming the temperature and pressure before the shock as known. These results are calculated for air both as an ideal gas with constant specific heat, and also when the real gas properties are used, and they have been plotted in Fig. 9.6.

Noting that the typical orbit speed is about 7 km/s, at this speed the calculation of temperature and pressure behind the shock is of considerable interest to rocket engineers. It is seen that at such speeds, by considering air as an ideal gas and initial state of air is at sea level (p_1 = 100 bar, T_1 = 300 K, the temperature and the pressure are about 25,000 K and 600 bars, respectively. By considering dissociation and ionization, these are around 10,000 K and 700 bars, respectively. Thus the conclusion is reached that in a real gas, the temperature behind the shock is considerably smaller than that if it is computed for an ideal gas, but the pressure is somewhat

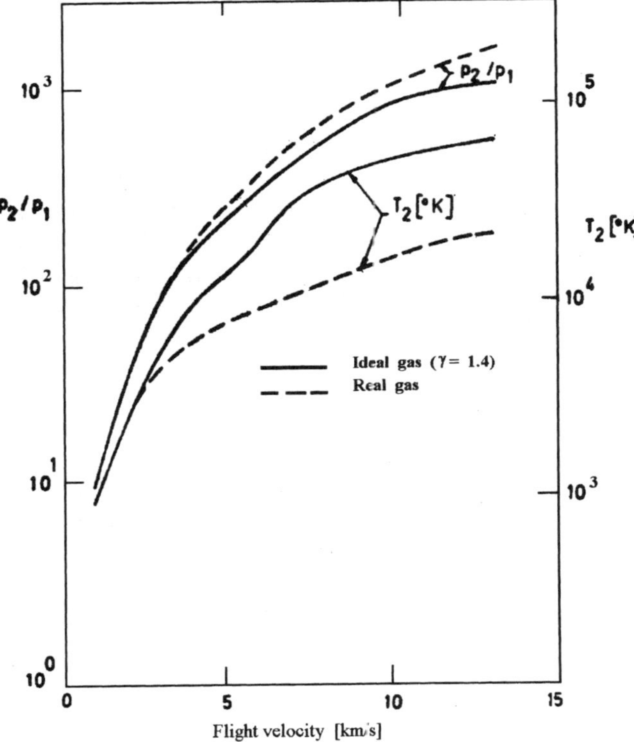

Fig. 9.6. Compression ratio across a normal shock and temperature behind the shock

larger. While in actual practice, the above high orbital speed is relevant for re-entry vehicles at high altitudes, and thus the absolute pressure behind the shock may be small, but while watching a re-entry vehicle one may always watch a long trail of glowing ionized air at temperatures estimated around 10,000 to 12,000 K.

For real gas properties, as mentioned already in the previous section, the subsonic and supersonic speeds can be determined by examining in the direction of larger static enthalpy the sign of $[\mathrm{d}(\rho u)/\mathrm{d}(h^o - h)]_s$. When it is positive it is a subsonic flow, but when it is negative it is a supersonic flow. In case the flow is supersonic, the state behind a normal shock can be calculated from expressions derived from the basic one-dimensional equations of mass, momentum and energy, and the equation of state. These basic equations are as follows:

Continuity:

$$\rho_1 u_1 = \rho_2 u_2 = \mu \tag{9.6}$$

Momentum:

$$p_1 + u_1^2 = p_2 + u_2^2 \tag{9.7}$$

Energy:

$$h^o = h_1 + u_1^2/2 = h_2 + u_2^2/2 \tag{9.8}$$

State:

$$\rho = \rho(p, T); h = h(p, T) \ . \tag{9.9}$$

In these equations μ is the mass flux rate $[\mathrm{kgm}^{-2}\mathrm{s}^{-1}]$, which can be computed exactly from the given state and speed values before the shock. Further, p is the pressure, ρ is the density, T is the temperature, and h is the specific enthalpy. The superscript "o" stands for the stagnation condition. In this, there are seven equations for conditions before and after the shock, subscripted 1 and 2, namely, the three conservation equations and four state equations. Also there are the following ten variables:

$$p_1, T_1, \rho_1 h_1, u_1, p_2, T_2, \rho_2, h_2, u_2 \ . \tag{9.10}$$

Thus one has to specify any three as independent variables – usually they are p_1, T_1 and u_1. At fairly moderate temperatures, where the specific enthalpy can be given as a simple relation with temperature, closed form solutions are available to obtain conditions behind the shock as a function of the earlier three independent variables. For non-ideal gases a more elaborate procedure is necessary.

Combining (9.6) and (9.8), one can write

$$h = h^o - \mu^2/(2\rho^2) \tag{9.11}$$

which is called the *Fanno equation*. Similarly combining (9.6) and (9.7), one can write

$$p - p_1 = \mu^2 \left(\frac{1}{\rho_1} - \frac{1}{\rho} \right) \tag{9.12}$$

which is called the *Rayleigh equation*. Unfortunately, the process of plotting of the results of the Fanno and Rayleigh equations in a thermodynamic chart is extremely tedious and time-consuming, and for every new value of μ and (p_1, ρ_1) new lines have to be drawn. This difficulty can be overcome by combining the three equations (9.6–9.8), and one obtains the *Rankine-Hugoniot equation*

$$h_1 - h = \frac{1}{2}(p - p_1) \left(\frac{1}{\rho_1} + \frac{1}{\rho} \right) \tag{9.13}$$

which does not contain the velocity, implicitly or explicitly, and as such can be used to study both the normal and oblique shocks. Thus a plot of this equation gives the locus of all points, which can be reached from the state (p_1, T_1), and is called the *Rankine-Hugoniot curve*. It has also the advantage, that it is of a fairly uniform shape both in enthalpy-entropy and enthalpy-log(density) charts. Thus these can be drawn from the beginning on these thermodynamic charts for a few initial states, and for other states one can easily draw interpolated curves. The state behind the shock is obtained from the Rankine-Hugoniot curve and satisfying the continuity equation (9.6). Sketch of these lines in an schematic enthalpy-entropy (h,s) chart are shown in Fig. 9.7 to illustrate the method.

A still better method, shown for the first time independently of each other by *Knoche* [74] and *Spalding* [106], using a (h, $\log_{10}\rho$) chart, in which the Rankine-Hugoniot lines are again drawn at a few places and interpolated in-between, is described now. For this purpose a regular Fanno-curve, made out of plexi-glass or similar material, can be used easily. This is explained in the following:

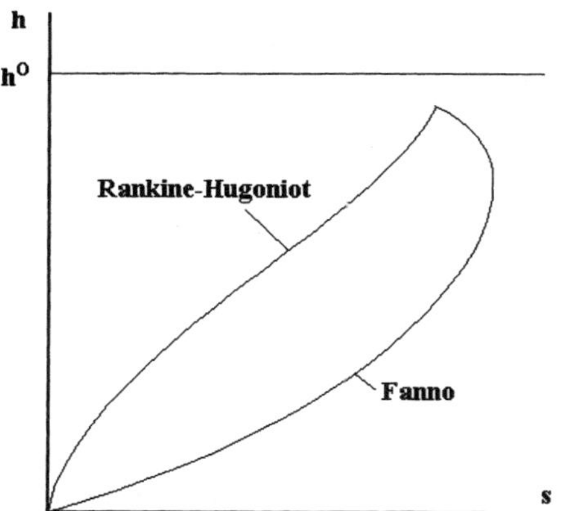

Fig. 9.7. Enthalpy-entropy chart with Rankine-Hugoniot lines

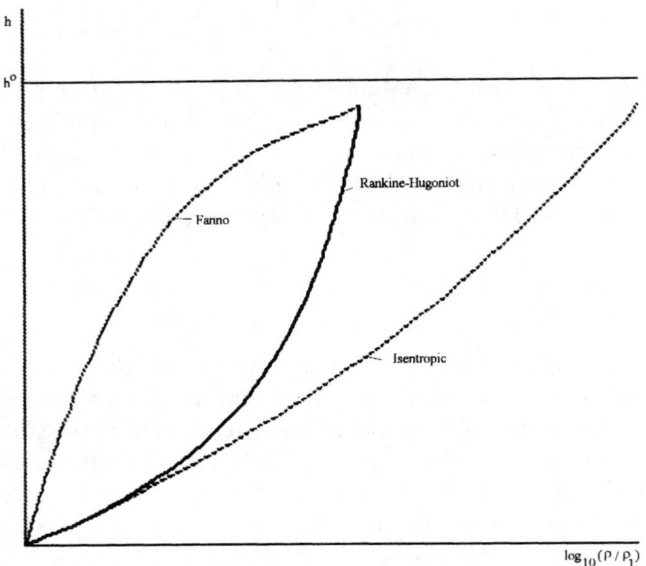

Fig. 9.8. Schematic sketch of Fanno lines to determine shock

From (9.6) and (9.11)

$$\frac{d[\ln(h^o - h)]}{d(\ln \rho)} = -2 \qquad (9.14)$$

and thus,

$$h^o - h = \exp^{-4.606 \log_{10} \rho} \ . \qquad (9.15)$$

Depending on the scales of the $(h, \log_{10} \rho)$ thermodynamic chart, one can use (9.15) to construct a Fanno-curve once for all, in which the abscissa is $\log_{10} \rho$ and the ordinate is $(h^o - h)$. By knowing the values of u_1, p_1 and T_1, one can easily calculate h^o, and then one can place the Fanno-curve on the thermodynamic chart with its origin at h^o and touching the state (p_1, T_1). The second intersection of this Fanno-curve with the plotted or interpolated Rankine-Hugoniot curve gives the state (p_2, T_2), as it has been shown schematically in Fig. 9.8.

9.4 Vibrational Relaxation Effects in Gas Dynamics

Vibrational relaxation effects of considerable interest are found in connection with hypersonic flights, non-equilibrium electric discharge reactors and gas lasers. In each of these applications, non-equilibrium of the molecular vibrational modes can often be a major influence, especially when the translational and rotational molecular energy modes are maintained at equilibrium at a relatively low temperature. For this purpose expressions for specific enthalpy and entropy (per unit mass) for a diatomic gas in translational-rotational mode on the one side and the vibrational mode on the other are now reproduced from Chap. 3 as follows:

(a) *Translation* and *rotation*:
enthalpy:

$$h = \frac{7}{2}RT + h_o \;,$$
(9.16)

where h_o is specific enthalpy at 0 K.
entropy:

$$s = R\left[-0.15548 + 3.5\ln T - \ln p + 1.5\ln m - \ln\Theta_r\right] \;,$$
(9.17)

where p is in bar.

(b) *Vibration*:
enthalpy:

$$h_v = \frac{1}{2}R\Theta_v\left[\coth\left(\frac{\Theta_v}{2T_v}\right) - 1\right] \;.$$
(9.18)

entropy:

$$s_v = R\left[\frac{\Theta_v}{2T_v}\coth\left(\frac{\Theta_v}{2T_v}\right) + \ln\left\{\frac{1}{2\sinh\left(\frac{\Theta_v}{2T_v}\right)}\right\}\right] \;.$$
(9.19)

In above Θ_r and Θ_v are characteristic temperature of rotation and vibration, values of which are given in Table 2.1 for several diatomic gases. Further, T and T_v are the translational-rotational temperature and the vibrational temperature, respectively. Now the specific entropy of state is given by the relation

$$s = s(p, T, T_v)$$
(9.20)

and if s is a total differential, then

$$\mathrm{d}s = \left(\frac{\partial s}{\partial p}\right)_{T,T_v}\mathrm{d}p + \left(\frac{\partial s}{\partial T}\right)_{p,T_v}\mathrm{d}T + \left(\frac{\partial s}{\partial T_v}\right)_{p,T}\mathrm{d}T_v \;.$$
(9.21)

For the adiabatic change of state, $\mathrm{d}s = 0$, and we get the relation

$$\left(\frac{\partial \ln p}{\partial \ln T}\right)_s = -\frac{T}{p}\left[\left(\frac{\partial s}{\partial T}\right)_{p,T_v} + \left(\frac{\partial s}{\partial T_v}\right)_{p,T}\left(\frac{\partial T_v}{\partial T}\right)_s\right]\bigg/\left(\frac{\partial s}{\partial p}\right)_{T,T_v} \;.$$
(9.22)

For an ideal diatomic gas the derivatives are

(a) *Translation* and *rotation*:

$$\left(\frac{\partial s}{\partial T}\right)_p = \frac{7}{2}\frac{R}{T}; \left(\frac{\partial s}{\partial p}\right)_T = -\frac{R}{p} \;.$$
(9.23)

(b) Vibration:

$$\left(\frac{\partial s}{\partial T_v}\right) = \frac{R\Theta_v^2}{4T_v^3 \sinh^2\left(\frac{\Theta_v}{2T_v}\right)} \; ; \; \left(\frac{\partial s}{\partial p}\right) = 0 \; . \tag{9.24}$$

Hence under adiabatic condition,

$$\left(\frac{\partial \ln p}{\partial \ln T}\right)_s = \frac{7}{2} + \frac{T}{T_v}\left[\left(\frac{\Theta_v}{2T_v}\right)\frac{1}{\sinh\left(\frac{\Theta_v}{2T_v}\right)}\right]^2 \frac{\partial T_v}{\partial T} \; . \tag{9.25}$$

For $T = T_v$, the above equation becomes

$$\left(\frac{\partial \ln p}{\partial \ln T}\right)_s = \frac{7}{2} + \left[\left(\frac{\Theta_v}{2T_v}\right)\frac{1}{\sinh\left(\frac{\Theta_v}{2T_v}\right)}\right]^2 \; . \tag{9.26}$$

From thermodynamics, the left hand side of Eq. (9.26) is equal to $[\gamma/(\gamma - 1)]$, where γ is the specific heat ratio, and the numerical results of calculation are given in Table 9.1. The results show a considerable reduction in the value of γ for diatomic gases (without considering dissociation) at higher temperatures under equi-temperature ($T = T_v$) case.

Table 9.1. Effect of equilibrium temperature ($T = T_v$ on γ)

T/Θ_v	0.0	0.1	0.2	0.3	0.4	0.5	0.6	0.7	0.8	0.9	1.0
γ	1.4	1.399	1.374	1.342	1.322	1.310	1.303	1.299	1.296	1.294	1.292

For a very fast expansion in a convergent-divergent nozzle, the vibrational temperature may not change (T_v = constant) and the second term in the right hand side of (9.25) may be dropped . Since the state of the gas (pressure, temperature, gas speed) depends on γ, a knowledge of vibrational relaxation is thus of importance. While an exact calculation of a nozzle flow requires knowledge of the shape of the nozzle, the problem becomes simpler for the two limiting cases: (1) the vibrational temperature is not changing, and (2) the local vibrational temperature is equal to the local translational-rotational temperature.

For the above two cases, the starting equation is the one-dimensional energy equation

$$h^o = c_p T^o = h + \frac{u^2}{2} = \frac{7}{2}RT^o + \frac{1}{2}R\Theta_v\left[\coth\left(\frac{\Theta_v}{2T^o}\right) - 1\right]$$
$$= \frac{7}{2}RT + \frac{1}{2}R\Theta_v\left[\coth\left(\frac{\Theta_v}{2T^o}\right) - 1\right] + \frac{u^2}{2} \; . \tag{9.27}$$

For the case 1 ($T_v = T_o$, the initial stagnation temperature), the following equations can be derived easily:

gas speed:

$$u = \left[7RT^o \left(1 - \frac{T}{T^o} \right) \right]^{0.5} \tag{9.28}$$

pressure-temperature relation:

$$\left(\frac{\partial p}{\partial T} \right)_s = \frac{p}{T} \frac{\gamma}{\gamma - 1} \tag{9.29}$$

density:

$$\rho = p/(RT) \tag{9.30}$$

nozzle cross-section area:

$$A = \dot{m}/(\rho u) \tag{9.31}$$

where $\dot{m}$ = *mass flow rate*.

For case 2, $T_v = T$, and the corresponding equations are as follows:

gas speed:

$$u = \left[7RT^o \left\{ 1 - \frac{T}{T^o} + 0.4 \left(\frac{\Theta_v}{2T^o} \right) \left(\coth \left(\frac{\Theta_v}{2T^o} \right) - \coth \left(\frac{\Theta_v}{2T^o} \frac{T^o}{T} \right) \right) \right\} \right]^{0.5} \tag{9.32}$$

pressure-temperature relation:

$$\left(\frac{\partial p}{\partial T} \right)_s = \frac{p}{T} \left[\frac{7}{2} + \left[\left(\frac{\Theta_v}{2T^o} \frac{T^o}{T} \right) \frac{1}{\sinh \left(\frac{\Theta_v}{2T^o} \frac{T^o}{T} \right)} \right]^2 \right] \tag{9.33}$$

density: same as (9.30)

nozzle cross-section area: same as (9.31)

The solution for both the cases can be obtained in discrete steps from the initial state (T^o, p^o, $u = 0$), by solving, for pressure as a dependent variable, as a function of T with the help of any standard numerical procedure for initial value problems, for example the Runge-Kutta procedure. Results show for oxygen and stagnation state $T^o = 80$ K, $p^o = 1$ bar, that in comparison to case 1, for case 2 the mass flow rate increases by 0.639 per cent, the exit pressure (for a given throat area to exit area ratio) increases by 3.02 per cent and the temperature increases by 10 per cent.

Subsequently, computations are done for oxygen for a normal shock with initial condition of $T_1 = 300$ K and $p_1 = 1$ bar, for the following two cases: (1) $T_{v2} = T_1$ (vibrational non-equilibrium), and (2) $T_{v2} = T_2$ (vibrational equilibrium), where the subscripts 1 and 2 refer to the condition before and after the normal shock. Computation procedure using the *Rankine-Hugoniot* and *Fanno* or *Rayleigh* relations, together with the equation of state in an iterative manner, and the results of calculation after the shock for the two cases as a function of initial Mach number, M_1, are shown in Table 9.2.

Table 9.2. Computed states after shock

M_1	Vibrational Non-equilibrium			Vibrational Equilibrium		
	T_2 [K]	u_2 ms^{-1}	ρ_2 kgm^{-3}	T_2 [K]	u_2 ms^{-1}	ρ_2 kgm^{-3}
2.0	505.9	247.5	3.42	499.1	241.4	3.51
2.5	655.9	256.9	4.12	635.4	244.2	4.34
3.0	845.2	277.0	4.59	800.0	254.7	4.99

9.5 Electrical Breakdown in Gases

If in the laboratory, an electric potential difference is applied between two electrodes, a small electric current of the order of a fraction of a microampere may start flowing between them. This is due to the electron emission caused by irradiation by a few ionized particles of cosmic origin. With increasing potential difference the electric current in the gap may be first saturated, and only with higher potential difference there may be an increase in the current flow due to further secondary ionization by collision. Discharges of this type are called the Townsend discharges. These take place because of an external ion source, and thus they are not self-sustaining, if the ions of external origin are removed in some manner. However, in the presence of normal external ions and with increasing potential difference to reach a critical voltage U_s, which is about 3,500 volts in one cm gap, the current increases very rapidly and a spark results in one self-sustaining discharges like the glow or arc. The glow discharges are, for comparatively large potential difference between the electrodes, completely covered with the glow. If, however, the potential difference is increased further, an arc, sustained by emitted electrons from the cathode by thermoionic emission, burns stable in an electric field of about one volt/cm and a current of more than 10 amps. In summary the voltage-current characteristic of all the earlier mentioned discharges are given in Fig. 9.9, and the possibility of having a stable discharge in one or other regions depends on the characteristic of the external electric power source, the characteristic of the discharge, and the value of the external impedance. This may be examined for a simple electric circuit given in Fig. 9.10a, consisting of a power source of infinite capacity to keep the potential difference U_o constant for any current, but no internal resistance, an external resistance R, and an electric discharge between the two electrodes having a voltage drop in the discharge U. For a falling voltage-current characteristic in the discharge (Fig. 9.10b) at the point A, a small increase in the value of U has to be compensated by a further increase in the potential drop across the resistor by increasing the current. Since at point A, this is not possible, the discharge is unstable. However, by similar reasoning one can show that at point B, the discharge is stable. The

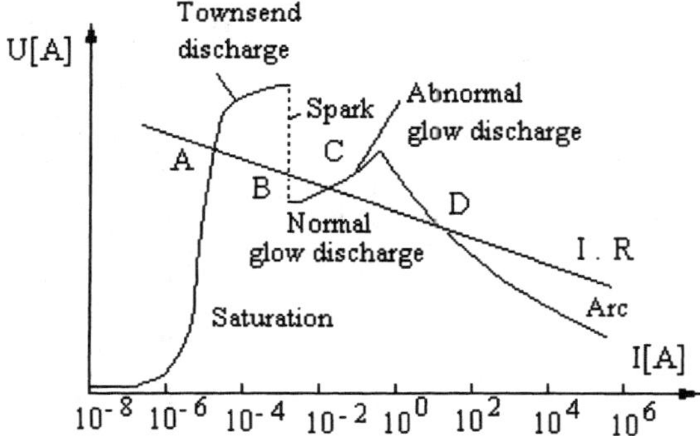

Fig. 9.9. Schematic voltage-current characteristic of a discharge in various regimes of operation

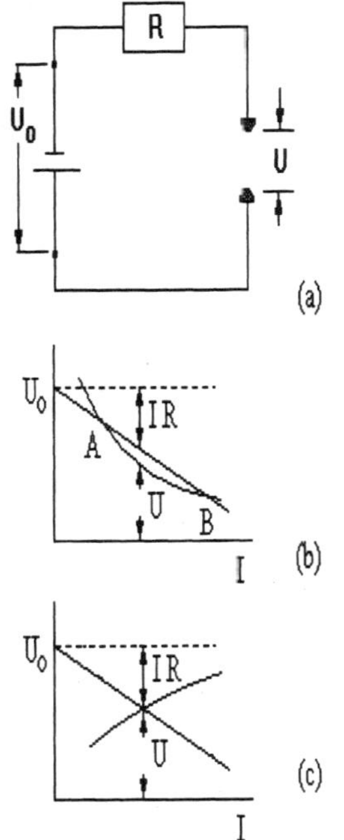

Fig. 9.10. On stability of a discharge. (**a**) A simple circuit; (**b**) a discharge with a falling U-I characteristic, and (**c**) a discharge with a rising characteristic

stability criteria may be stated mathematically as $(\mathrm{d}U/\mathrm{d}I) + R > 0$. This discharge characteristic given in Fig. 9.10c is always stable.

While the previous investigation is done for a simple circuit consisting of only an external resistor, a discharge itself may have some inductivity L and capacitance C. Thus a simple experimental setup consisting of a power source, an external resistor and a pair of electrodes may have an equivalent

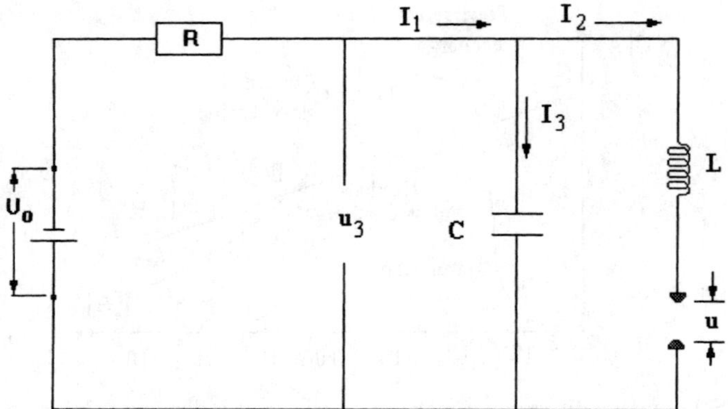

Fig. 9.11. Schematic equivalent circuit of an electric discharge

circuit in Fig. 9.11. Generally, for this circuit $I_1 = I_2 + I_3$. However, under steady state, $I_3 = 0$ and $I_1 = I_2 = I$, and there is a steady potential drop u. Thus, $U_o = u + IR$. If the fluctuation in current is denoted by a prime, under unsteady state,

$$U_o = (I + I_1')R + L\frac{\mathrm{d}}{\mathrm{d}t}(I + I_2') + \bar{u} + u_I I_2' \tag{9.34}$$

where

$$u_I = \left(\frac{\partial u}{\partial I}\right) . \tag{9.35}$$

Subtracting the steady state potential balance equation from the unsteady state potential balance equation, one gets

$$I_1'R + L\frac{\mathrm{d}I_2}{\mathrm{d}t} + u_I I_2' = 0 . \tag{9.36}$$

Now

$$I_3 = C\frac{\mathrm{d}u_3}{\mathrm{d}t} \tag{9.37}$$

and

$$u_3 = (\bar{u} + u_I I_2') + L\frac{\mathrm{d}}{\mathrm{d}t}(\bar{I} + I_2') . \tag{9.38}$$

Thus,

$$I_3 = C\left[\frac{\mathrm{d}}{\mathrm{d}t}(\bar{u} + u_I I_2') + L\frac{\mathrm{d}^2}{\mathrm{d}r^2}(\bar{I} + I_2')\right] = C\left[u_I\frac{\mathrm{d}I_2'}{\mathrm{d}t} + L\frac{\mathrm{d}^2 I_2'}{\mathrm{d}t^2}\right] . \tag{9.39}$$

Noting that

$$I_1 = I = \bar{I} + I_1' = I_2 + I_3 = \bar{I} + I_3 + I_2' \tag{9.40}$$

one gets

$$I_1' = I_3 + I_2' = C\left[u_I\frac{\mathrm{d}I_2'}{\mathrm{d}t} + L\frac{\mathrm{d}^2 I_2'}{\mathrm{d}t^2}\right] + I_2' . \tag{9.41}$$

Substituting the above equation into (9.36), and after some manipulation, one gets the differential equation

$$\frac{\mathrm{d}^2 I_2'}{\mathrm{d}t^2} + \left(\frac{1}{RC} + \frac{u_I}{L}\right) \frac{\mathrm{d}I_2'}{\mathrm{d}t} + \frac{I_2'}{LC}\left(\frac{u_I}{R} + 1\right) = 0 \ . \tag{9.42}$$

With a trial function $I_2' = A\exp^{\lambda t}$, one gets a quadratic equation $\lambda^2 + a\lambda + b$, whose roots are

$$\lambda = -\frac{a}{2} \pm \sqrt{\frac{a^2}{4} - b} \tag{9.43}$$

where

$$a = \frac{u_I}{L} + \frac{1}{RC}, b = \frac{1}{LC}\left(\frac{u_I}{R} + 1\right) \ . \tag{9.44}$$

If λ is real, that $a > 2\sqrt{b}$, then I_2' is a pure exponential function, but if $a < 2\sqrt{b}$, then λ is complex and so there will be oscillation; however, the process is stable if the real part is positive.

From these considerations it is clear that in Fig. 9.9, the points A and C are stable, but B and D are unstable, and this explains why the spark is an unsteady process.

For steady discharges the voltage drop across the external resistor, though necessary for stability, is an undesirable loss in power. It is, therefore, desirable that the voltage-current characteristic of the external power source should match the falling characteristic of the discharge to keep the loss in the external resistor as minimum. This is actually done in welding transformers, regarding which details are available in any book on arc welding. However, it may be mentioned here that there are arc configurations (with tubular electrodes, for example) which have neutral or slightly rising characteristic for which an external resistor is hardly necessary.

For a discharge of the *Townsend* type when voltage is first applied, the electric current (electron current) from the cathode to the anode increases slowly proportional to applied voltage at constant electron number density and an average velocity proportional to the applied voltage. The electron number density for this type of discharge at steady state, designated as n_e, is for a given external radiation source strength, the rate of production of electron number density $[\mathrm{m}^{-3}\mathrm{s}^{-1}]$ is postulated to be given by the relation of the type $n_e = n_e \exp^{-kt}$, where k is a constant in some kind of rate equation. If the electrons are drawn sufficiently slowly due to the applied voltage, then the two conditions for the solution of the rate equation for the production of the electrons are: $t = 0$, $n_e = 0$ and $t \to \infty$, $n_e = n_e^*$. Thus from the beginning of the constant irradiation by the external source, the electron number density is given by the relation

$$n_e = n_e^* \left(1 - \exp^{-kt}\right) \ . \tag{9.45}$$

However, if at a comparatively large applied voltage between the electrodes, the electrons are drawn towards the anode as soon as they are produced, and under the assumption that no electrons are produced by the effects of the applied voltage, the steady state number density will be smaller than n_e^*. Under such a condition, the electron current, of the order of a few microamperes, is saturated.

With increasing applied voltage between the electrodes, the primary electrons, gaining kinetic energy from the electric field, generate new electrons by collision.

If α = (number of ionizing collision)/(path distance), m^{-1}, and, in addition, if $\dot{n}_e$ and $(\dot{n}_e+d\dot{n}_e)$ are electron flux $[m^{-2}s^{-1}]$ entering and leaving the two control surfaces, respectively, perpendicular to the electron flow direction, then

$$d\dot{n}_e = d(n_e V_e) = \alpha' n_e dx = \alpha V_e n_e dx \tag{9.46}$$

where α' = number of ionizing collisions per unit time = αV_e. s^{-1}

Assuming a constant drift velocity of the electrons in a constant electric field, one gets the relation = $\alpha n_e \, dx$ which upon integration and after multiplication of both sides by eV_e gives the relation

$$j_e = j_{eo} \exp^{-x} \; . \tag{9.47}$$

At $x = 0$, the electron current is originated due to irradiation of the cathode due to the external source, and thus j_{eo} is the saturation current.

When the initial source of electrons is not due to irradiation of the cathode but due to ionizing radiation throughout the volume of the gas, then one can write the equation

$$d\dot{n}_e = \alpha' n_e dx = q dx \tag{9.48}$$

where q = number of electron-ion pairs produced per unit volume and time $[m^{-3}s^{-1}]$. An integration of the equation gives the relation

$$\alpha n_e = A \exp^{\alpha x} - q/V_e \; . \tag{9.49}$$

Implicit in the model is that at $x = 0$, $n_e = 0$. Consequently, $A = q/V_e$, and one can write

$$n_e = \frac{q}{\alpha V_e} (\exp^{\alpha x} -1) \tag{9.50}$$

From this expression, one can get the number density of electrons at the anode at $x = L$. From experimental results, it can be seen that the value of α depends on both the electric field and the pressure. Although the ionization by collision may be present for all electric field strengths, it does not set in sufficiently at one atmospheric pressure in air at an electric field of less than 30,000 volts/m. Thus, from the practical point of view, the situation $\alpha L \ll 1$ is quite justified. By expanding the exponential (αL) in series form and terminating the series after the second term, one can write for the saturation current, $j_{eo} = qeL$. Thus, one can write

$$j_e = e n_e V_e = \frac{j_{eo}}{\alpha L} (\exp^{\alpha x} -1) \; . \tag{9.51}$$

It is noted that since then the mean free path is inversely proportional to pressure, the electron ionization coefficient may be expressed in the form

$$\frac{\alpha}{p} = f(E/p),$$ (9.52)

where p is the pressure and $\mathbf{E}$ is the electric field. For air, argon and nitrogen, these values, taken from Cobine [7], are shown in Fig. 9.12. It is found that there is an optimum pressure at which the electric current becomes maximum. The condition for this can be found by differentiating Eq. (9.52), and one gets

$$\frac{d\alpha}{dp} = f\left(\frac{E}{p}\right) + p'\left(-\frac{E}{p^2}\right) = 0 .$$ (9.53)

This gives the optimum

$$\left(\frac{E}{p}\right)_{opt} = \frac{f(E/p)}{f'(E/p)} .$$ (9.54)

Now the exponential relation, valid at small voltages, is no longer valid at higher voltages. It is evident that some other method is playing a role in this

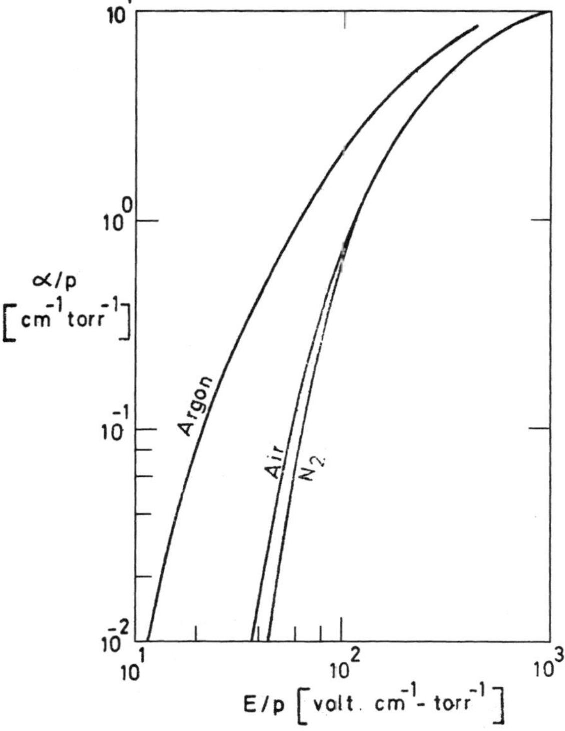

Fig. 9.12. Coefficient for field-intensified ionization by electrons

region. Townsend assumed that in this region the positive ions formed by the electron collision begin to gain sufficient energy from the field to ionize the gas by collision. This leads to a positive ion coefficient β, and to the equation for steady electric current density

$$j_e = j_{eo}\frac{(\alpha - \beta)\exp^{(\alpha-\beta)x}}{\alpha - \beta\exp^{(\alpha-\beta)x}}\,. \tag{9.55}$$

For small values of β, which occur for small value of (E/p), (9.55) leads to (9.51). However, in the case where β becomes sufficiently large for a large value of (E/p), the denominator becomes zero, and the current density becomes infinite. Under this condition a spark occurs. However, there could be more than one probable mechanism at high electric fields, which gives expressions for the electric current density of the same form as (9.55), and an occurrence of a spark is in no way a proof of the practical validity of this model.

According to *Paschen's law*, the sparking potential is dependent on the product of pressure and the electrode gap. For spark breakdown in air, the breakdown voltage is given in Fig. 9.13. However a spark breakdown depends on the frequency of applied voltage and is reduced by above to one-third if the frequency is changed from 0 to 500 cycles/sec.

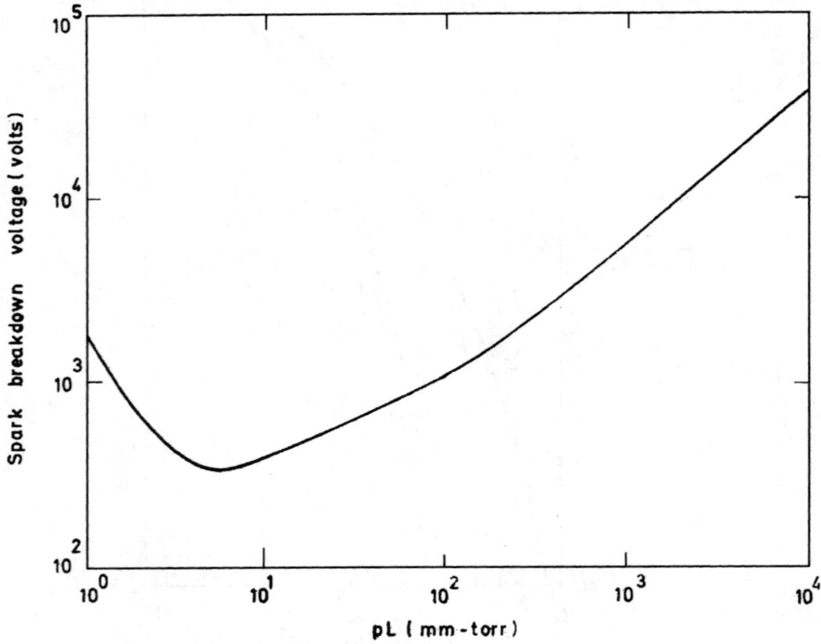

Fig. 9.13. Spark breakdown voltage for parallel plane path in air (temperature = 20°C)

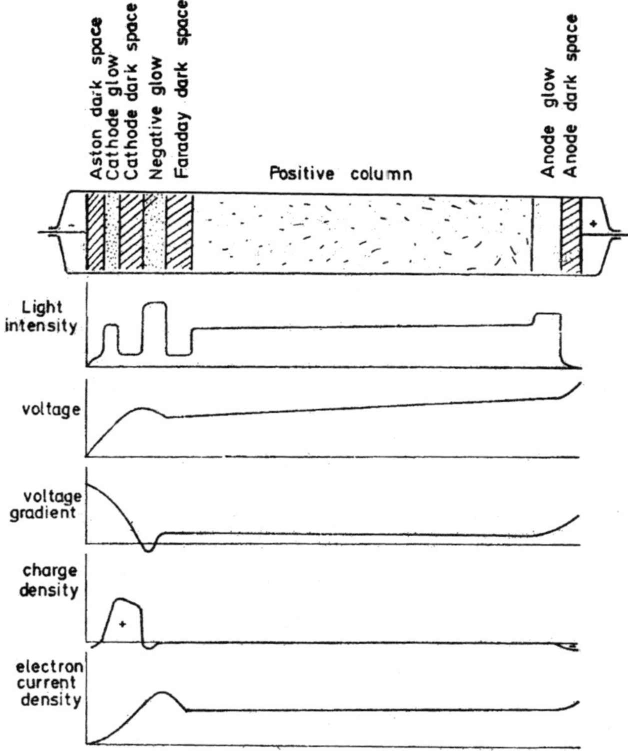

Fig. 9.14. Different regions of a glow discharge

One of the self-sustaining discharges is a glow discharge operating in the range between a cold Townsend discharge, and a high current arc. At low pressures (less than a few torr) the glow discharge is seen to consist of alternate dark and light regions. These different regions with their names and the characteristics of such a discharge in the region are given in Fig. 9.14. At the cathode there is a net negative charge produced by the emitted electrons. Since their initial velocity is low, the current is carried entirely by positive ions arriving at the cathode from the cathode dark space, which is a region of high positive ion density accounting for the high cathode drop and causing the electrons to accelerate through this region. For the region between the cathode and the end of the cathode dark space, which are apart by a distance l, the electric field is found experimentally to fall linearly, $E = C(l\text{-}x)$, where C is a constant. Since $E = \mathrm{d}U/\mathrm{d}x$, where U is the potential,

$$U = \int_0^x E\,\mathrm{d}x = C \int_0^x (l - x)\mathrm{d}x \ . \tag{9.56}$$

For $x = l$, $U = U_c$, the cathode potential, and thus, $C = 2U_c/l^2$, and

$$U = 2U_c \left[\frac{x}{l} - \frac{1}{2} \left(\frac{x}{l} \right)^2 \right] .$$

(9.57)

Further, from *Poisson equation*, applied for the one-dimensional case,

$$\frac{\mathrm{d}^2 U}{\mathrm{d}x^2} = -\frac{4\pi n_c}{\epsilon_o} = -\frac{2U_c}{l^2}$$

(9.58)

where $n_c = e(n_e - n_i)$ is the charge density, and thus, $n_c = U_c\epsilon_o/(2\pi l^2)$. Although the exact limits of the cathode drop region are somewhat uncertain, the product K, of l and gas pressure p, is found to be somewhat constant, and it is less dependent on the electrode material and more on the gas. The value of K for hydrogen is around 0.84 cm-torr, for helium 1.35, for argon 0.31, for nitrogen 0.36, for oxygen 0.27, and for air 0.33 cm-torr. These correspond to mean free paths between 50 and 100 cm. The normal cathode fall depends on the combination of the cathode material and the gas; in general, the alkali metals as cathode have lower cathode drops. For a typical gas like argon, the cathode drop for potassium as the cathode material is 64 volts whereas with copper as the cathode material it is 130 volts (*Cobine* [7]).

With low external resistance in the d.c. circuit, an arc, having a stable voltage-current characteristic may be established. An investigation of the potential distribution in an arc shows that there are relatively large potential drops near the cathode (cathode drop), and a region of a fairly uniform voltage gradient in between these regions called the *positive column*. An estimate of the electric field in the positive column without any convection is possible by considering the equation for current flow by integrating the Ohm's law equation over the cross-section (*Elenbaas-Heller model*, discussed in detail in Chap. 12),

$$I = 2\pi E \int_0^R \sigma r \mathrm{d}r$$

(9.59)

and the *equation of heat balance* for a fully-developed cylindrical arc is

$$\frac{1}{r} \frac{\partial}{\partial r} \left(r k \frac{\partial T}{\partial r} \right) = \sigma E^2$$

(9.60)

where $\sigma = $ *electrical conductivity* $[\mathrm{A(Vm)}^{-1}]$, $k = $ *coefficient of thermal conductivity* $[\mathrm{W(mK)}^{-1}]$, E is the electric field in the axial direction and r is the radial coordinate. While the solution of the two equations simultaneously for a given electric current is somewhat tedious, a quick estimate of the electric field is possible by considering the first equation alone. Assuming an average electrical conductivity of $10^4 \mathrm{A(Vm)}^{-1}$, current $I = 100$ to 500 amps and arc radius $R = 5$ to 10 mm, the electric field $E = 0.3$ to 6 V.cm^{-1}, and for a quick estimate of the electric field in the positive column a value 1 Vcm^{-1} can be considered.

Values of cathode and anode voltage drops in arcs have been given by *Cobine* [7] for a few electrodes and gases and for different current ranges. With air as the working medium, and copper, carbon or iron as electrodes in the electric current range one to 300 amps, the typical cathode drops are in the range 8 to 10 volts, and anode drops are in the range 2 to 12 volts. However, experiments conducted by *Bose* and *Pfender* [37] with copper anode and argon as gas at pressures 1 to 80 torr, and currents from 60 to 100 amps show a considerably less anode drop of the order of magnitude of 0 to 2 volts, which appears to be typical of high current arcs. In addition, there is increasing evidence of a small negative potential gradient in the immediate vicinity of an anode in such arcs, for which the model, developed by this author, was already discussed in the previous chapter.

9.6 High Frequency Discharges

High frequency discharges in the frequency range of radio frequency or *micro wave* are used often as a starter discharge for electric arcs. In a fluctuating electric (or electromagnetic) field, such as these, one could make an estimate of the energy contained in such fields and dissipation of such energy. For the guidance of the energy hollow metallic tubes of circular or rectangular cross-section (*wave guide*) are used, the minimum dimension being of the order of the wave length of the electromagnetic energy. Hence for a wave guide of reasonable dimension very high frequency energy is generated.

Starting equations for our analysis are Maxwell equations, (7.18–7.23), leading to the wave equation, which has been discussed later in Sect. 11.3. As a result, we consider the possibility of generating high frequency discharges. For such discharges we may write the continuity equation of the current

$$\frac{\partial n_c}{\partial t} = -\nabla \cdot \mathbf{j} \tag{9.61}$$

to get

$$\epsilon \nabla \cdot \left(\frac{\partial \mathbf{E}}{\partial t} \right) = \frac{\partial n_c}{\partial t} = -\nabla \cdot \mathbf{j} \ . \tag{9.62}$$

Now assuming

$$\mathbf{E} = \mathbf{E}^o \exp^{i2\pi\nu t} \quad \text{and} \quad \mathbf{j} = \mathbf{j}^o \exp^{i2\pi\nu t} \tag{9.63}$$

where "i" denotes the imaginary number, ν is the discharge frequency and the superscript "o" denotes the amplitude. (9.62) is written as

$$\nabla \cdot \mathbf{j}^o = -i\epsilon 2\pi\nu \nabla \cdot \mathbf{E}^o \ . \tag{9.64}$$

The only way both sides of the above equation can be equated if

$$\mathbf{j}^o = -i\epsilon 2\pi\nu \mathbf{E}^o \tag{9.65}$$

and the real part of the proportionality constant between the two (by taking the imaginary part of ϵ) can be considered as the *electrical conductivity coefficient* in a fluctuating electric field. The latter is determined from the momentum equation for electrons

$$M_e \frac{d}{dt}(n_e V_e) = -en_e \mathbf{E} - \Gamma_{eh} n_e M_e \mathbf{V}_e \qquad (9.66)$$

where Γ_{eh} is the collision frequency between the electrons and the heavy particles. The above equation is the same as (11.15) discussed in Chap. 11, except that we have neglected the velocity of the heavy particles in comparison to the one for the electrons and also the pressure gradient term. In addition, we consider the relation between the current density and the velocity as

$$\mathbf{j} = -en_e \mathbf{V}_e \ . \qquad (9.67)$$

With the definition of the *plasma frequency*

$$\nu_p = \frac{1}{2\pi} \sqrt{\frac{e^2 n_e}{\epsilon_o M_e}} \qquad (9.68)$$

and after some manipulation we get from (9.66, 9.68)

$$\mathbf{j}^o(\Gamma_{eh} + i2\pi\nu) = (2\pi\nu_p)^2 \epsilon_o \mathbf{E}^o \qquad (9.69)$$

and finally we get the expression for the electrical conductivity for a fluctuating electric field as

$$\sigma_f = \frac{4\pi^2 \epsilon \nu_p^2 \Gamma_{eh}}{\Gamma_{eh}^2 + 4\pi^2\nu^2} = \frac{e^2 n_e \Gamma_{eh}}{M_e(\Gamma_{eh}^2 + 4\pi^2\nu^2)} \ . \qquad (9.70)$$

It is shown that the effective electrical resistivity in comparison to the one in a stationary field can be quite large. In addition the energy dissipation will be dependent on the square of the amplitude of the electric field and hence large amount of electro-magnetic energy can be put to use in an high frequency discharge.

10 Diagnostic Techniques

Side by side with the production of high temperature gases one has to think about the measurement of gas thermodynamic state and flow parameters. Among the parameters those which need be measured are the total and static enthalpy, the gas velocity, the total and static pressure, temperature at different modes (translation, rotation, vibration, etc..) for different specie (electron temperature, heavy particle translational temperature, heavy particle excitation temperature, etc.). Out of the various flow variables mentioned above, measurement of the static and total pressure may pose little difficulty with small static pressure holes on the side wall of the channel or the body and water-cooled probes for the measurement of the total pressure, although at low pressures special pressure-measuring equipment (thermocouple probe, ion probe, etc.) may be necessary.

For the measurement of the temperature, one may choose between the probe and optical methods as follows:

(1) Probe method:
(a) Thermocouple and resistance thermometer.
(b) Total enthalpy-total pressure probe.
(c) Electrostatic probe.

(2) Optical method:
(a) Optical pyrometry.
(b) Line reversal technique.
(c) Spectral emission.
(d) Interferometric methods-optical interferometry, micro-wave technique, etc..

Regarding the use of the *thermocouple* and the *resistance thermometer*, there are many excellent treatises available and these need not be discussed here in detail. They are found to be extremely accurate, in fact much more accurate than other methods given in the above list. However they have three basic limitations. Firstly, they can work in a limited range with the maximum temperature around 3,000 K. Thus, they cannot be used in most of the combustion chamber for rockets and ramjets, and for gas plasmas. Secondly, they are incapable of measuring the temperature of gas mixtures, when at least one component is at a different temperature than the others. As an example, we may consider the temperature of common tube lights in which the electrons, strongly accelerated in the electric field in one particular direction, may have the kinetic energy equivalent to the electron temperature of a few hundred-thousand degree Kelvin. Although, since the high velocity is in one direction, it is difficult to define the temperature; the heavy particle in tube lights are known to have quite low random particle velocities corresponding to temperatures of the order of several hundred degree Kelvin only. Thirdly, these cannot be used easily in corrosive atmospheres.

The total enthalpy-total pressure probe, suggested by Professor Jerry Grey of the Princeton University, measures the total enthalpy fairly accurately (less than five percent inaccuracies), and appears at present to be the only method operating at pressures around one atm and temperatures between 3,000 and 10,000 K. However deduction of the temperature from the measured total enthalpy assumes the knowledge of the velocity (or at least the information that the gas kinetic energy is negligible), and also the knowledge of the enthalpy-temperature relationship. This obviously means that the method is restricted to gas mixtures in which there is equilibrium (the temperature of all specie are same) and at least some information about the state of the chemical equilibrium are available. Further, because of the need to cool these probes properly to prevent disintegration while operating at high temperatures, they have to be sufficiently large in size so that the complete local measurement is not possible. In addition, when these probe are used in corrosive atmospheres, special precautions have to be taken in design to save the vacuum pump and the pressure-measuring instruments which are accessories to such probes.

The electrostatic probes, in general, are very simple devices, that is, both simple to manufacture and to measure the translational temperature of charge particles. However, they can be operated only if the range in which there is sufficient ionization. Further, the equipment is reliable only at fairly low pressure of operation. The probe methods, described earlier, are highly accurate and reliable in the range of operation for which they are designed. However, they may distort the flow field and also an elaborate cooling arrangement is necessary for continuous measurements. These defects are absent in the case of optical probes. However, in general, the optical method are less accurate even if within the limited range of operation.

Among the optical probes use of bolometers or any other total radiation probes presume that the radiation characteristic of the investigating gas is known. Since gas radiation is dependent on both the state of the gas as well as the optical thickness, mere knowledge of the emissivity coefficient as a function of the gas is not enough. On the other hand, and in general, gases do not radiate even approximately like a black body and as such these probes are generally not used for high temperature gas research.

While the total radiation probes are completely insufficient for gases, an optical pyrometer is useful for some gases especially with carbon particles which radiate approximately like black body at least in the limited visible range of the spectrum. The equipment for this is relatively simple and inexpensive, and works in comparison with the luminosity of an electrically heated and calibrated wire; thus the maximum temperature for this is limited to about 3,000 K. Similar to this method of the *optical pyrometer* the line reversal technique needs comparison with the intensity of background light source, and if for such a source a tungsten strip lamp is used, then the maximum temperature is also about 3,000 K whereas with anode crater of a carbon arc being the background source the maximum temperature is 4,000 K. The method is, however, not dependent on the emissivity of the radiating gas and all that is needed is an inexpensive spectroscope and background source. However, introduction of commonly used sodium salt locally gives rise to problems, as well as spatial resolution of the temperature field.

For temperatures above 10,000 K for nitrogen and argon, and at different pressures the spectral line emission method have been used successfully for determination of both average and spatial distribution of temperature. The method, however, lacks accuracy (at temperatures mentioned earlier the accuracy is about ±500 K). At low temperatures *infrared spectroscopy* of rotational and rotational-vibrational bands are used. For these purposes it is necessary to make judicious choice of the spectral equipment needed and these are discussed in detail later. The temperature measurement methods with a short discussion on interferometry and the microwave technique, which have been used in a limited way for plasmas at around one atmosphere pressure are described later.

In case the temperature is known by one of the earlier described methods and also the total enthalpy probe is used, then at least in principle, the velocity can be determined. However, the method is highly inaccurate and has not been used. The methods of spark or streak photography have been used successfully in plasma research, especially for low velocity carbon arcs. Another promising method is the use of laser velocity meter, which has also been described later. For research on high temperature gases, one may also need suitable probes to measure electromagnetic fields, although these have not been discussed here.

10.1 Temperature Measurement-Probe Method

(a) *Total enthalpy-total pressure probe*: A schematic sketch of a total enthalpy-total temperature probe, placed in a supersonic stream of a high temperature gas, has been given in Fig. 10.1. The probe is amply cooled with water and the temperature rise of the water is accurately measured. A sample gas is sucked through the probe with the help of a vacuum pump whose design should be such so as to cause as little disturbance in the outside flow as possible. By measuring the pressure before the sonic orifice and making sure

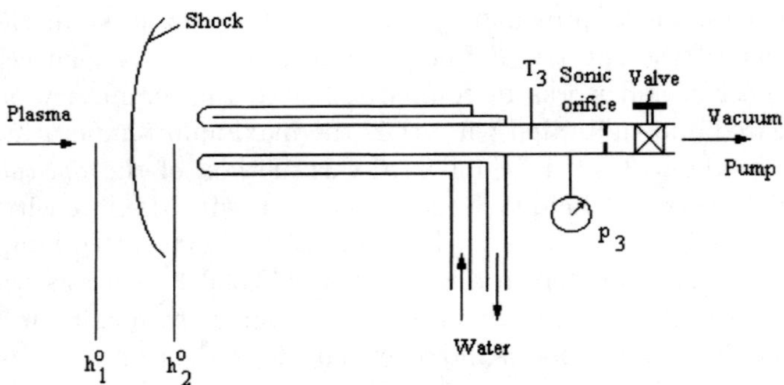

Fig. 10.1. Schematic design of a total enthalpy probe in a supersonic stream

that the pressure at the other end of the orifice is low enough to guarantee sonic condition at the orifice, the rate of the gas mass flow through the probe can be determined. Further the water flow rate is measured. The measurement starts by closing the valve and measuring the stagnation pressure p_{o2}. At that moment let the flow rate of water be $\dot{m}_w$ (kgs^{-1}) and $\Delta T'$ be the temperature rise of water. Now the valve is opened and the vacuum pump is started. The water mass flow rate is kept as before. Let the gas flow rate be $\dot{m}_g$ and the temperature rise of water now be $\Delta T''$. Further let h_3 be the enthalpy of the cooled gas. If $c_w = 4.187$ kJ(kgK)$^{-1}$ is the specific heat of water then, according to the energy balance,

$$\dot{m}_g(h_{o2} - h_3) = \dot{m}_w c_w(\Delta T'' - \Delta T') \tag{10.1}$$

and thus the total enthalpy of the gas is

$$h_{o2} = \frac{\dot{m}_w}{\dot{m}_g} c_w(\Delta T'' - \Delta T') + h_3 \ . \tag{10.2}$$

(b) *Electrostatic probes*: These probes which are very simple devices are generally called *Langmuir probes*. To understand their working let us consider the phenomenon near the wall surface, where an external potential ϕ_o is applied. It is assumed that the charged particles fall freely without collision from the free-fall edge beyond which there is the condition of quasi-neutrality and no electric field due to the applied potential. The theory for the potential distribution near the probe, which is slightly negative with respect to the plasma, has been described in Sect. 8.2. The analysis indicates that at the wall ($x^* = 0$) the electron particle density $n_e \approx 0$ for slightly negative probes but independent of the applied potential.

Now the number of electrons and ions falling freely per unit area and time without an applied potential on the probe are given by the relation

$$\dot{n}_{eb} = \frac{1}{4} n_{eb} v_{eb}; \dot{n}_{ib} = \frac{1}{4} n_{eb} v_{ib} \tag{10.3}$$

because of the quasi-neutrality condition at the sheath edge, $n_{eb} = n_{ib}$. In the above equations, the factor $(1/4)$ consists of two $(1/2)$s, in which one $(1/2)$ is because of the flux of electrons in one hemisphere and the other $(1/2)$ is due to the integration of cosines from all directions. Now, for the probe at the positive potential only the electrons may be collected and the electric current density is given by the relation

$$j = e\dot{n}_{eb} = \frac{1}{4}en_{eb}v_{eb} = \frac{1}{4}en_{eb}\sqrt{\frac{8k_BT_{eb}}{\pi M_e}}\ (\phi_o > 0)\ . \tag{10.4}$$

For slightly negative probes, a lesser number of electrons are collected, but the ion current density may still be small. Thus,

$$j = e\dot{n}_{eb}\exp^{e\phi_o/(k_BT_e)} = \frac{1}{4}en_{eb}v_{eb}\exp^{e\phi_o/(k_BT_e)}$$

$$= \frac{1}{4}en_{eb}\sqrt{\frac{8k_BT_{eb}}{\pi M_e}}\exp^{e\phi_o/(k_BT_e)}\ (\phi_o < 0)\ . \tag{10.5}$$

Further for strongly negative probes, the electrons are completely eliminated and the collected ion current is given by the relation

$$j_i = e\dot{n}_{ib} = \frac{1}{4}en_{eb}v_{ib} = \frac{1}{4}en_{eb}\sqrt{\frac{8k_BT_{ib}}{\pi M_i}}$$

$$(\phi_o \ll 0)\ . \tag{10.6}$$

Equations (10.4–10.6) have been shown schematically in Fig. 10.2. It is shown that these expressions in which the current densities are not dependent on the applied potential and thus they represent saturation electron and ion current

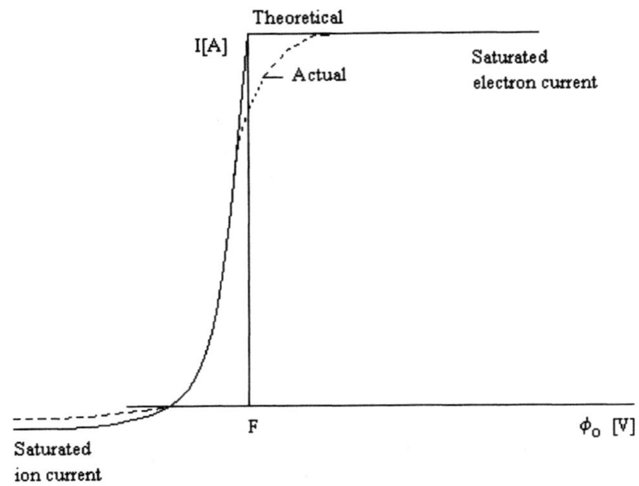

Fig. 10.2. Schematic current-voltage characteristic of an electrostatic probe

densities respectively. The ratio of the absolute value of current densities in such a case would be

$$\left|\frac{j_i}{j_e}\right| = \sqrt{\frac{T_{ib}}{T_{eb}}\frac{M_e}{M_i}}\ . \tag{10.7}$$

Now the electrons, which are moving through the thermal boundary layer, to be collected at the probe surface, may not lose their kinetic energy as fast as the ions to be collected in the same manner. Thus at moderate pressures around one bar, $T_{eb} \approx T_{e\infty} \approx T_\infty$, but $T_{ib} \approx T_w$. Thus

$$\left|\frac{j_i}{j_e}\right| = \sqrt{\frac{T_w}{T_\infty}\frac{M_e}{M_i}} \tag{10.8}$$

which is a very small quantity.

From (10.4) it is seen that the saturation electron current density is proportional to the number density of the electrons n_{eb}, and as may be expected, both should increase with pressure. However, it is found experimentally that the saturation electron current density decreases with increasing pressure. This can be explained by the fact that with increasing pressure the value of n_{eb} is reduced because of the combined effect of the smaller diffusion coefficient and higher reaction rate, so that near equilibrium condition is reached; their combined effect is to reduce n_{eb}. On the other hand it is found experimentally that at comparatively low pressures, the electron current density may be so large that because of large electron bombardment the probe may soon be corroded.

Since neither the probe collecting surface area for the charged particles nor the value of n_e can be determined accurately, (10.5) can be modified slightly to determine the electron temperature. From (10.5) one may write

$$\ln I = C + e\phi_o/(2.303 k_B T_{eb}) \tag{10.9}$$

where I is the electric current in amps. Thus the slope of a plot of current versus potential ϕ_o on a semi-logarithmic paper gives the value of $e/(2.303 k_B T_{eb}) = 5037/T_{eb}$. The method works fairly well at low pressures and has been used by this author (*Bose* and *Pfender* [37]) for argon plasma at pressures between 1 and 40 mm Hg and temperatures between 5,000 and 12,000 K. Its successful use for pressures around 1 bar for argon plasma for temperature around 10,000 K has also been reported in literatures.

In Fig. 10.2, the point F is the point at which the current densities of the falling electrons and ions are equal. By equating the absolute value of the current densities in (10.4–10.6) and after taking logarithms, one gets the expression to determine the required potential

$$\phi_{oF} = \frac{k_B T_{eb}}{2e}\ln\left(\frac{T_{ib}}{T_{eb}}\frac{M_e}{M_i}\right) \tag{10.10}$$

which is called the "*floating potential*". In one of the experiments with electrostatic probes, T_{eb} was determined from the exponential part of the current-voltage characteristics curve for argon plasma at pressures around 0.11 to

44.0 mm Hg, and the measured electron temperature was between 9,000 and 12,500 K. The plasma potential was also determined from the asymptotic extrapolation of the exponential part of the current-voltage characteristics estimated to be accurate within -0.2 volts. However, the determination of the electron temperature to heavy particle temperature ratio $\theta_b = T_{eb}/T_{ib}$ from these experimental data and (10.9) showed scattered results. Thus, it was to be concluded that the electrostatic probes, although very simple to use, must be used with caution as a diagnostic tool.

10.2 Temperature Measurement-Spectroscopical Methods

The basic choice of equipment necessary for spectral measurements is based on the following considerations: (i) Range of wavelength to be investigated: ultraviolet, visible, near and far-infrared, radio range, etc. (ii) Type of dispersion : prism or grating (iii) Method of recording : photograph (spectrograph), photosensitive tubes (spectral photometer or recording spectrograph), visual (spectroscope), etc.

At temperatures up to about 7,000 K for hetero-polar diatomic molecules and radicals, the useful spectra are those of rotational and rotation-vibrational bands in the near and far-infrared, but for plasmas the useful spectra are generally in the visible and ultraviolet range. While for the visible range one could use glass opticals, for operation in the ultraviolet it is necessary to work with quartz opticals. However, these cannot be used in the infrared region and a number of alkali-salt opticals are available. The main difficulties with these salt opticals are those regarding their proper setting up and also that most of these salt opticals are highly hygroscopic and have to be kept in specially heated containers. The advantage of such opticals of course, is the ease with which one can remove the scratches, etc. on the optical surface.

For the choice of a spectral measuring equipment, it is necessary to have one with a high resolution, that is, the possibility to recognize two spectral lines near each other, especially if it is necessary to study the molecular band structure. At large wavelengths the grating spectrograph are superior to the prism spectrographs from the view-point of resolution although the former are much more expensive. Thus, one has to be careful in the choice of the instrument and should chose on the basis of the spectral range intended to be studied.

For identification of some known spectra, or for comparison of intensities in a limited spectral range, as it is used under the method *"line reversal techniques"*, it is enough to have simple spectroscope in the visible range. However, for complicated measurements choice has to be made between *spectrograph* and *spectral photometer*.

The spectrograph with photographic films or plates, sensitive in the suitable wavelength range, can give in single exposure the entire intensity distribution. However, each film or plate is different from the other since difference in emulsion, developer temperature and age, and many other factors influence the density (darkness) on the plate. It is, therefore, necessary to take on the same plate of the spectra under investigation, additional spectra for a known gas for identification purposes, and also stepped continuous spectra of known source under possibly the same exposure time for calibration purposes. It is well-known fact that the shutter speed and intermittency of the light source affect the sensitivity of the photographic plate also. The calibrated stepping down in the intensity of the known source is done either with the help of a step filter or the so-called rotating sector. After developing the exposed plate or film the lines are identified by enlarging the entire spectrum on a projection screen followed by the measurement of the density of the spectra under investigation as well as the continuous band of the standard source to obtain the intensity of the spectra.

The spectral photometers are not restricted to the limited range of sensitivity of a photographic plate. However, they cannot be used for fluctuating light sources, and the spatial intensity distribution is possible only by measuring at different points at different times. On the other hand the photo-sensitive detectors have a fairly good and linear sensitivity over a large wavelength range and need not be calibrated each time.

At the outset it may be mentioned that use of the prism or optical grid based spectral equipment are not used for productive spectral analysis anymore, and intensified CCD (Closed Circuit Digital) cameras provide time resolved, high resolution spectral measurements. There are several other powerful spectroscopic techniques available, including dye Laser, Laser induced fluorescence, cavity-ring down spectroscopy, etc. which are normally discussed for advance level spectroscopical analysis but they are outside the scope of this book. However discussion further on optical measurements through prism-based equipment in this section is being done to introduce the reader to some preliminary techniques.

Among the spectral temperature measurements there are several methods as follows:

(a) *Temperature measurement by the line reversal technique*: The experimental set up for the line reversal technique is given in Fig. 10.3. On the matt glass screen of the spectrograph is recorded the continuous spectra of the background black body source, on which are superimposed twin (D-) lines of sodium at 5,890 and 5,896 Angstrom(Å). A part of the intensity of the background source is absorbed by the flame at the earlier-mentioned spectral lines. If a_D is the absorption coefficient at the wavelength of the D-lines and I_D is the intensity of the flame, then the recorded intensity is

$$I_{D(\text{Record})} = I_{D(\text{Flame})} + I_{D(\text{Black})} \left[1 - a_{D(\text{Flame})}\right] . \tag{10.11}$$

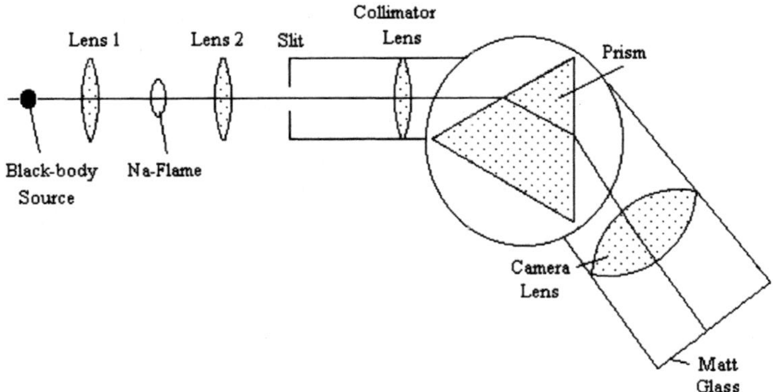

Fig. 10.3. External setup for line reversal technique

One has to change the intensity of the background source in such a manner that the sodium D-lines are neither lighter nor darker than the continuous spectra of the background source. Under this condition, *Kirchoff's law* is satisfied, which states that a body absorbs as much fraction of the incoming radiation as it emits, and noting

$$I_{D(\text{Record})} = I_{D(\text{Flame})} \tag{10.12}$$

results in the relation

$$a_{D(\text{Flame})} = I_{D(\text{Flame})}/I_{D(\text{Black})} \ . \tag{10.13}$$

This relation also gives the valid definition of the absorption coefficient. Thus if the recorded intensity is the same as the intensity of the black body, then as a direct consequence of the analysis the temperature of the flame is the same as the temperature of the background source, which is generally calibrated for the temperature as a function of the electric current to heat the source.

The method, although it appears to be very simple on the surface, may be quite difficult to carry out in case local measurement of the temperature is attempted since it is extremely difficult to keep the yellow sodium flame confined to a small region. On the other hand, if the sodium flame is allowed to spread over the entire flame, then one may measure only an average temperature.

(b) *Infrared measurement of rotational bands*: The intensity of the lines in the rotational and rotation-vibrational bands depends on the number of particles in the particular energy level, from which the spontaneous emission takes place, and the intensity is given by the relation

$$I_L = C(2J+1)\exp^{-J(J+1)\Theta_r/T} \tag{10.14}$$

where I_L is the intensity [$\text{Wm}^{-2}steradian^{-1}$], J is the rotational quantum number, Θ_r is the *characteristic rotational temperature* and C is a proportionality constant. It may be noted that the line intensity I_L is determined by integrating the intensity over the line width, and therefore,

$$I_L = \int_L I d\nu \ . \tag{10.15}$$

By taking the logarithm of Eq.(10.14), one gets

$$\log\left(\frac{I_L}{2J+1}\right) = \log C - \frac{J(J+1)\Theta_r}{2.3026T} \ . \tag{10.16}$$

Thus a plot of $I_L/(2J+1)$ on the logarithmic scale of a semi-logarithmic paper versus $J(J+1)\Theta_r/2.3026$ will have a slope inversely proportional to the temperature. Although the method is theoretically very simple the actual plotting of the points, on a semi-logarithmic paper does not show all the points to lie on a straight line, and there could be considerable scattering. However, an accuracy of ± 100K is estimated in this method.

(c) *Spontaneous line emission*: The intensity of a line due to the spontaneous transition of a bound electron in the ionization state i (for neutral, i $= 0$) and the energy level within this ionization, to the energy level n is considered so that the transition under consideration is (i,m) $\rightarrow$ (i,n). Now the intensity of such a line I_L is directly proportional to the number density of particles in (i,m), namely, $n_{i,m}$. The proportionality constant for the transition is A_{mn}, which is the *transition probability* for spontaneous emission. Thus,

$$I_L \approx A_{mn} n_{i,m} \ . \tag{10.17}$$

Now the ratio of the number density of particles in (i,m), $n_{i,m}$, to the total number of particles in the i-th ionization state is given for the Boltzmann statistic by the relation

$$n_{i,m} = \frac{g_{i,m} \exp^{-E_{i,m}/(k_B T)}}{\sum_m g_{i,m} \exp^{-E_{i,m}/(k_B T)}} \ . \tag{10.18}$$

Further from dimensional considerations (10.17) is multiplied by the factor $h\nu l/(4\pi)$ where l is the length (thickness) of the radiating gas column. Thus

$$I_L = \frac{1}{4\pi} A_{mn} n_i \frac{g_{i,m} \exp^{-E_{i,m}/(k_B T)}}{\sum_m g_{i,m} \exp^{-E_{i,m}/(k_B T)}} h\nu l \ , \ \text{Wm}^{-2}\text{sterad.}^{-1} \ . \tag{10.19}$$

In Equation (10.19) if i $= 0$, the number density of the neutrals (atoms) n_a decreases with temperature but the factor $\exp^{-E_{i,m}/(k_B T)}$ increases continuously with temperature. Further the partition function

$$Z = \sum_m g_{i,m} \exp^{-E_{i,m}/(k_B T)} \tag{10.20}$$

is not strongly dependent of temperature. The intensity, therefore, due to the product of two factors, one increasing and the other decreasing with temperature has a maximum with temperature. This fact is also true for single- and multiple-charged ions also, except that the number density of these initially increase with temperature, and then decrease.

Determination of the temperature by measuring the intensity of a line and using (10.19) is possible directly if the value of A_{mn} is known. This is called the "*absolute method*". In case the absolute value of A_{mn} is not known accurately, even a knowledge of the relative values can be used to determine the temperature in the following manner. From (10.19), one can denote the relative intensities of the lines, both from the same ionization state, as

$$\frac{I_{L1}}{I_{L2}} = \frac{A_{mn1}g_{i,m1}}{A_{mn2}g_{i,m2}} \exp^{(E_{i,m2}-E_{i,m1})/(k_BT)} \quad . \tag{10.21}$$

If, however, A_{mn} is not known, but the point (radius for an axi-symmetric plasma not on axis) at which there is maximum intensity, is known, one can use modified version of (10.19). By taking the logarithm of (10.19), one gets

$$\ln I_L = C - \frac{E_{i,m}}{k_BT} + \ln n_i \tag{10.22}$$

which has a maximum at a particular temperature. By comparing the intensities at all points and moving from the point of maximum intensity, one can determine the temperature distribution.

(d) *Continuum emission*: For the free-free and free-bound type of transitions *Cramer's theory* gives the spectral distribution of the emitted radiation. These have been discussed in Chap. 4. From this, for the continuum radiation, one gets a saw-tooth distribution of intensity, each tooth corresponding to a free-bound transition to the respective bound energy level. The slope of the saw-tooth, as shown in Fig. 10.4, is proportional to T^{-1}.

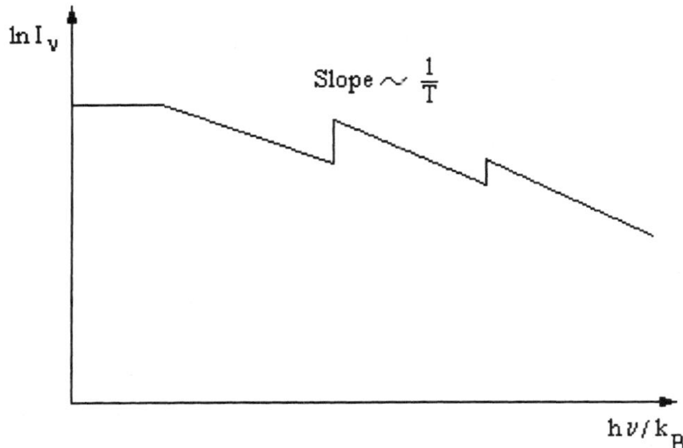

Fig. 10.4. Schematic Spectral Distribution of Intensity in Continuum

(e) *Broadening of spectral lines*: It has been known for a long time that the broadening of spectral line is a complicated function of the number density of the charge particles, and thus of pressure and temperature. Thus, the measurement of the profiles combined with the theory is an interesting method to determine temperature.

On the various mechanisms for *line broadening* those due to Doppler and Stark effects are significant practically. While the discussion here is kept very brief, for a detailed discussion the reader is referred to the book edited by *Huddlestone and Leonard* [12]. However, from some calculated values of the line width it can be seen that *Doppler broadening*, which does not depend on the electron number density, becomes important at low electron densities, whereas the Stark-broadening is important at high electron number densities ($n_e > 10^{16}$ cm^{-3}).

If the velocity component of the radiating particle parallel to the direction of observation is w, then the wavelength shift due to the Doppler effect is $\Delta\lambda/\lambda = w/c$, where c is the velocity of light. By assuming the random nature of the atomic motion one gets for the half-width (the wavelength width at which the line intensity is half of the maximum value) by the Doppler effect as

$$\Delta\lambda_{Doppler} = 7.16 \times 10^{-7}\lambda\sqrt{T/m} \ . \tag{10.23}$$

In this equation $\Delta\lambda_{Doppler}$ is obtained in Å, λ is in Å, T is in K and m is the mole mass. Thus (10.23) show that the Doppler-broadening is pronounced for lines of light element at high temperatures.

For the practical application of the Stark effect the tabulations of width and shift parameters are of primary interest since they furnish all the essential information for electron density determination. It is based on the theory that according to the linear Stark-broadening theory, the broadening $\Delta\lambda \propto$ electric field E, which is due to the microelectric field of the charge particles. Now the mean distance between the charge particles is $\propto n^{-1/3}$. Since the Coulomb force $\propto e^2/\text{distance}^2 \propto eE$, the electric field $E \propto e/\text{distance}^2 \propto en_e^{2/3}$. To determine the proportionality constant normalized *electric field* $E^* = E/E_o$ is introduced, where $E_o =2.61en_i^{2/3}$ volts.m^{-1} and $E^* = E^*(\Delta\lambda)$. The measured result of line width against electron density for a few element have been presented in the earlier mentioned book of Huddlestone and Leonard. However the method has not become practically useful for measurement of other elements.

(f) *Sideways temperature measurement of non-uniform axi-symmetric plasmas* (Fig. 10.5). Implicit in the method is that the absorption within the gas may be neglected. Let the intensity $I_L(x)$ [Wm^{-2}.sterad^{-1}] be measured sideways of different layer of gas. It is assumed that the radiation intensity of volumetric radiation is dependent on radius only that is, $i =i(r)$ [Wm^{-3}.sterad^{-1}]. Thus,

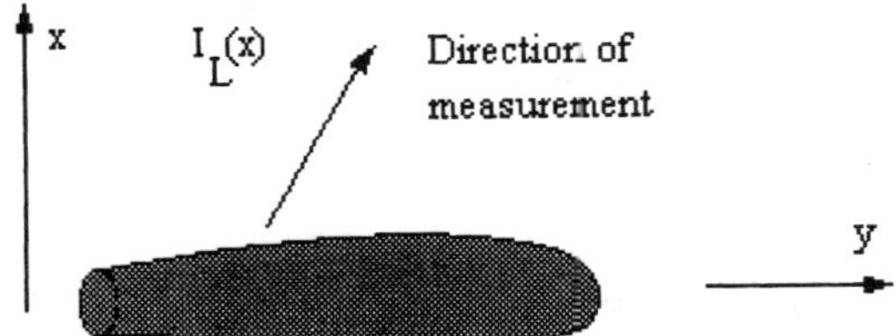

Fig. 10.5. Sideways radiation measurement for an axi-symmetric plasma column

$$I_L = 2 \int_0^y i(r)\mathrm{d}y = 2 \int_0^r \frac{i(r)r}{\sqrt{r^2 - x^2}}\mathrm{d}r \ . \tag{10.24}$$

Note that in the following I is equivalent of I_L.

For an analytically prescribed $I(x)$ it is possible to get $i(r)$ by *Abel's inversion method*. However, a numerical method, as given in the following, is quite convenient for evaluation of $i(r)$ from the measured $I(x)$. For this purpose, let the cylinder be divided into concentric circles, each concentric circular strip is assumed to radiate uniformly. Now the measured length of the radiating gas column δ, as given in Fig. 10.6, has two subscripts, the first giving the concentric circle number and the second giving the index number of position. Thus one notes the following set of equations.

$$\begin{aligned}
I_4 &= i_4\delta_{44} \\
I_3 &= i_4\delta_{43} + i_3\delta_{33} \\
I_2 &= i_4\delta_{42} + i_3\delta_{32} + i_2\delta_{22} \\
I_1 &= i_4\delta_{41} + i_3\delta_{31} + i_2\delta_{21} + i_1\delta_{11} \ .
\end{aligned} \tag{10.25}$$

From this set of equations the values of i_1 to i_4 are determined easily from the known measured value of I_1 to I_4 and δs. While the method has been

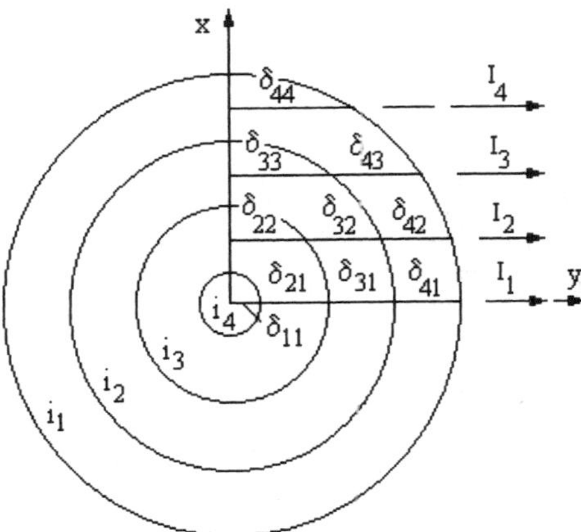

Fig. 10.6. Schematic diagram for sideview volumetric emission measurement

described for axi-symmetric plasmas the method has also been used success-fully for non-axi-symmetric plasma as well, by simultaneously measuring I from different directions.

10.3 Temperature Measurement-Interferometric Methods

The interferometric method works on the principle that when a ray of light, which is *coherent* in time and space is split, and the split rays are further allowed to past through two different media of common thickness L, the relative velocities of these two split rays are in the ratio of the refractive indices of the two media and consequently there is a phase shift between the two rays as they are allowed to recombine. In case the phase shift are exactly opposite, there are dark interference fringes and if they are in phase there are bright fringes. These alternate dark and bright interference fringes (Fig. 10.7) are obtained if

$$\int_0^L \Delta\mu dl \approx \bar{\Delta}\mu L = z\lambda \tag{10.26}$$

where $\Delta\mu$ is the difference between the refractive indices in both media, z is the number of fringes, and λ is the wavelength. For the characteristic length of the order of 1 cm and 5,000 Å in the visible range, the average difference in the refractive index $\Delta\mu$ is 0.0005 to get 10 fringes.

The *total refractivity* in either medium depends on the refractivity index of the electrons, ion and neutrals, μ_j, that is,

$$\mu = 1 + \sum(\mu_j - 1) \tag{10.27}$$

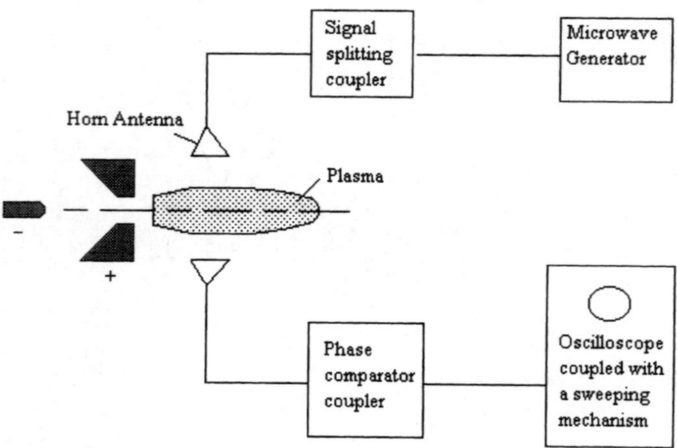

Fig. 10.7. Schematic diagram of a microwave generator

where j stands for electrons, ions, neutrals etc. As long as one is away from the spectral regions of strong absorption one can talk about the phase refractive index, which is given for neutrals and ions by the relation

$$\mu_j = 1 + \alpha_j n_j \tag{10.28}$$

where α_j is a coefficient and n_j is the particle number density $[\mathrm{m}^{-3}]$ of the j-th species. Typical values of α_j for components of argon plasma at around 5,463 Å wavelength, as taken from Huddleston and Leonard [12], are as follows: for neutral argon in ground state A_o, $\alpha_a = 1.036 \times 10^{-29}$ m^3 and for single charge argon A^+, $\alpha^+ = 7.15 \times 10^{-30}$ m^3.

For the electrons, however, the refractive index is given by the relation

$$\mu_e = 1 - \frac{1}{2}\left(\frac{\nu_p}{\nu}\right)^2 \tag{10.29}$$

where ν is the wave frequency of the radiated ray of light, and ν_p is the *electron plasma frequency* given by the relation

$$\nu_p^2 = \frac{1}{4\pi^2}\frac{e^2 n_e}{\epsilon_o M_e} \tag{10.30}$$

where n_e $[\mathrm{m}^{-3}]$ is the electron number density. Substituting this expression for ν_p in (10.29) and noting that the wavelength λ in Angstrom (Å) unit is related to frequency in sec^{-1} by relation $\lambda = 3 \times 10^{18}/\nu$, (10.29) now becomes

$$\mu_e - 1 = -4.477 \times 10^{-36}\lambda^2 n_e . \tag{10.31}$$

In (10.31) λ is in Angstrom units and n_e is in m^{-3}. For wavelength region around 5,463 Å, as observed for neutral argon, a coefficient for electrons, α_e, can also be defined as done earlier, whose value is $\alpha_e = 4.477 \times 10^{-25}$ m^3; in microwave region, the value is much larger. Now for typical plasma consisting of single charge ions, electrons and neutrals the degree of ionization α gives the ratio of electron, ion and neutral number density to the total number density by the relation

$$\frac{n_e}{n} = \frac{n_i}{n}\frac{\alpha}{1+\alpha}; \frac{n_a}{n} = \frac{1-\alpha}{1+\alpha} . \tag{10.32}$$

It can be shown that for strongly ionized plasma only the electron concentration can be determined from the fringes and without any knowledge of the non-electronic contribution.

Equation(10.29) is valid under the condition that the frequency of collision between electrons and heavy particles, ν_{coll} is much smaller than the wave frequency ν, and there is no magnetic field (the electron cyclotron frequency = 0). For no magnetic field, but ν_{coll}/ν not small, the expression to be used is

$$\mu^2 = \frac{1 - \left(\frac{\nu_p}{\nu}\right)^2 \left(1 + i\frac{\nu_{\text{coll}}}{\nu}\right)}{\left[1 + \left(\frac{\nu_{\text{coll}}}{\nu}\right)^2\right]} \tag{10.33}$$

where i is the imaginary number. Further for wave propagation parallel to the magnetic field and collision frequency much smaller than the wave frequency, the expression to be used is

$$\mu^2 = \frac{1 - \left(\frac{\nu_p}{\nu}\right)^2}{\left[1 - \frac{\nu_{\text{cycl}}}{\nu}\right]} \tag{10.34}$$

where the cyclotron frequency, ν_{cycl} for the charged particles (electrons, ions), may be computed with the help of relation given in Chap. 7.

An estimate in calculating the sensitivity, for example, to be able to measure 1/100th of fringe spacing, is given by the minimum electron number density acceptable. For this purpose, $(\mu_e - 1)_{\min} = 10^{-2}\lambda/L$, where L is the optical path length and λ is the wave length. Equating the above expression with (10.31), and for strongly ionized plasma we get

$$n_{e,\min} = 2.23 \times 10^{13}/(L\lambda) \tag{10.35}$$

where $n_{e,min}$ is the minimum electron density in m^{-3}. For a wavelength in optical range and $L = 1$ cm, $n_{e,\min} = 4.5 \times 10^{21}$ m^{-3}. However in the microwave region ($\nu = 10^{11}$ to 10^7 s^{-1}) corresponding to wave length ($\lambda = 3$ mm to 30 m), and for the same value of L, $n_{e,\min} \approx 10^{12}$ to 10^{16} m^{-3}.

10.4 Velocity Measurement by Laser-Doppler Velocimeter

The velocity is determined by measuring the Doppler shift in the frequency (or wavelength between the incident and the scattered light due to suspended particles, which is given by the formula

$$\Delta\nu = \frac{2u}{\lambda_o} \sin\theta \tag{10.36}$$

where u is the velocity of the scattered particles, λ_o is the wavelength of the incident light and θ is the semi-angle of the split rays. The equipment for the measurement of the velocity is given in Fig. 10.8. Measurement is done by mixing the scattered light beam with an undisturbed beam of the incident light into a photomultiplier and analyzing the spectrum for beam frequency given by (10.36) with the help of an oscilloscope or *spectrum analyzer*.

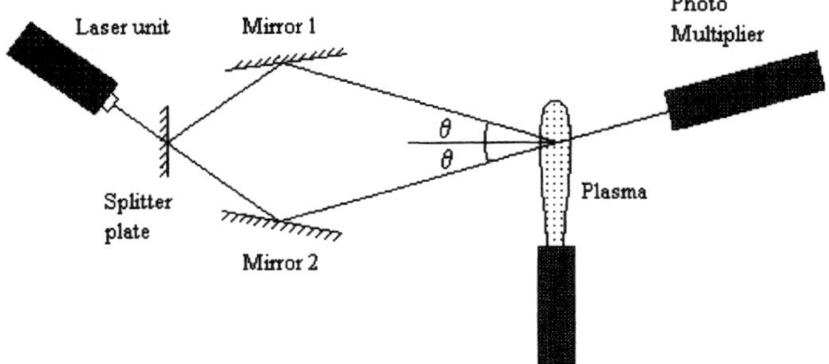

Fig. 10.8. Schematic diagram of a laser doppler velocimeter

10.5 Exercise

10.5.1 Compute the saturation current density of electrons and ions, and the floating point potential of a typical plasma (mole mass of heavies = 40) at pressure 0.01 bar, temperature 10,000K and electron mole fraction 0.2. (Ans: 6.379e7 Am^{-2}; 2.365e5 Am^{-2}; -4.8259 V)

10.5.2 For a Balmer series line of hydrogen atom at 4864 Å, calculate the relative intensity distribution with temperature. [Hint: required hydrogen atom number density can be computed with the help of *Saha equation* without considering molecular hydrogen. For population calculation at the emitting energy level Boltzmann distribution can be taken].

11 High Temperature Gas Dynamics

Finally, in this and next chapters, we come to discussion of some of the special gas dynamics problems in high temperature gases. Unfortunately, it is necessary to conduct our investigation under highly restricted conditions to be able to solve some of the simplest problems. While, in general, a gas mixture containing electrons, ions and neutrals (gas plasma) is considered for such an investigation, the general results are, however, also applicable to non-ionized or dissociated gases. For the purpose of our analysis, the following quantities are assumed to be known: (1) the temperatures T_e and T_h of the electron and the heavy particles, respectively; (2) the total kinetic pressure of the gas mixture p and the partial pressure p_j of the individual species; (3) the mass-averaged gas velocity $\mathbf{V}$; (4) the kind of gas; and (5) the magnitude and direction of any externally applied electromagnetic fields. Furthermore, the analysis is based on the following quite reasonable assumptions, that (1) the electric current is carried mainly by electrons; and (2) a quasi-neutrality condition exists in general.

Firstly, we present the basic equations of state and conservation of species mass, global mass, species momentum, global momentum and energy. The momentum equations are used to derive the generalized *Ohm's law*. These and *Maxwell's equations* are investigated with respect to their similarity parameters and special effects under the assumption of infinite electrical conductivity are discussed. Further discussion on the application of these equations has been deferred to the next chapter, however.

11.1 Basic Equations

The equation of state for a mixture of gases at different translational temperatures T_j is

$$p = k_B \sum T_j n_j + p_R \qquad (11.1)$$

where n_j is the number density of the j-th species, k_B is the Boltzmann constant and p_R is the radiation pressure, which is significant only for a black-body radiation at high temperatures. For most of the cases of normal application in the laboratory the assumption $p_R = 0$ can safely be made.

While formulating the gas dynamic conservation equations, either a balance of the conserving properties in a volume element can be made, or the more sophisticated Maxwell-Boltzmann integro-differential equation can be used; for the purpose of clarity it has been thought useful to employ here the first method. It is, however, necessary to describe some of the velocities and their interrelations as follows:

$\mathbf{w}_j$ = velocity of particles of the j-th species with respect to the laboratory coordinate;

$\mathbf{V}_j$ = mass-average velocity of the j-th species;

$\mathbf{v}_j$ = kinetic velocity of particles of the j-th species with respect to the mass-average velocity of the j-th species = $\mathbf{w}_j - \mathbf{V}_j$;

$\mathbf{V}$ = global mass-average velocity;

$\mathbf{V}^*$ = global molar-averaged velocity;

$\mathbf{V}'_j$ = relative velocity of the j-th species with respect to the global mass-average velocity = $\mathbf{V}_j - \mathbf{V}$; and

$\mathbf{V}^{*\prime}_j$ = relative velocity of the j-th species with respect to the molar-averaged velocity = $\mathbf{V}_j - \mathbf{V}^*$.

The average velocity of a particular species in a gas mixture depends on three factors: 1. All the specie move globally in certain direction depending on the (gas dynamic) pressure gradient, which is, very often, due to external boundary conditions imposed for a particular flow; 2. the relative speeds of various specie (diffusion) due to gradient in number density of these specie; and 3. the motion of the charged particles in electro-magnetic fields. In a high temperature gas inside a confined space bounded by colder wall, calculation of the local equilibrium composition will show more dissociated and ionized particles in the interior region of the gas and more undissociated (and unionized) particles near the wall. Hence the dissociated (and ionized) specie will move from interior regions to the wall and recombine in the colder region, and there will be movement of undissociated (and unionized) specie towards the hotter region. For ionized specie, in particular the electrons and higher order ions, they move together from hotter to colder regions, and there can be flux of the electrons and lower order ions (or neutrals) moving in the opposite direction; the phenomenon is called the *"ambi-polar diffusion"*. Further to this if charged specie are put in an electric field then the (negative) electrons move from the regions of lower electric potential (cathode) towards those of higher electric potential (anode), and the (positive) ions move in the opposite direction. One can, therefore, define a sort of global-averaged velocity of all the specie together, in which all the three above part movements of various specie are taken into account. Such an averaging can be done either based on the mass-density fraction or on the molar fraction. Further discussion on this is done on the basis of the discussion by this author elsewhere (*Bose* [45]).

While a simplified approach using the rigorous kinetic theory to calculate the viscosity and the heat conductivity coefficients can be taken in a fairly straight forward manner, as discussed in Chap. 7 of this book, it is quite difficult to take exactly the same approach for the calculation of the diffusion coefficients and the related properties. This is, because of the additional complication of the relative velocities between the individual species, which are needed for calculating the diffusion coefficients. Hence the ultra-simplified approach (*Hirschfelder*, et al. [10]) of Chap. 7 is taken. It is evident that the mass-averaged velocity and the molar-averaged velocity are not the same, just like the respective relative velocities. We would, however, discuss the mass-averaged velocity first, followed by the molar-averaged velocity.

The relative mass-averaged velocity of the j-th species, $\mathbf{V}'_j$, is due to two different effects, and has thus two parts. The first part is due to diffusion by the mass-density gradient, $\mathbf{V}'_{dj}$, and the second part is due to the drift of the charged particles in the gas mixture due to electromagnetic fields $\mathbf{V}'_{fj}$, the so-called *"field velocity"*. Thus, the mass-average velocity of the j-th specie is the sum of the global mass-averaged velocity plus the relative mass-averaged velocity of the j-th specie

$$\mathbf{V}_j = \mathbf{V} + \mathbf{V}'_j = \mathbf{V} + \mathbf{V}'_{dj} + \mathbf{V}'_{fj} \ . \tag{11.2}$$

From the definition of the mass-averaged velocity

$$\mathbf{V} = \frac{1}{\rho} \sum \rho_j \mathbf{V}_j \tag{11.3}$$

where $\rho = \sum \rho_j$ is the total mass density and ρ_j is the mass density of the species. It follows that

$$\sum \rho_j \mathbf{V}'_j = \sum \rho_j \mathbf{V}'_{dj} + \sum \rho_j \mathbf{V}'_{fj} = 0 \ . \tag{11.4}$$

While in the above expression the sum of the diffusive flux due to concentration gradient and electric field must be equal to zero, a simpler model is the one in which the sum of the the the above two flux individually are equal to zero. These two are considered separately.

First we examine the validity of the expression for the diffusion part. The mass diffusion velocity due to the gradient of the mass density of a species is given by the relation

$$\mathbf{V}'_{dj} = - \left(\frac{\rho}{\rho_j} \right) D_j \nabla \left(\frac{\rho_j}{\rho} \right) \tag{11.5}$$

where D_j is the *mass-diffusion coefficient* of the j-th species and for ionized gases specially, $D_j = D_{amb}$, the ambi-polar diffusion coefficient. Thus, specially for the ionized gas

$$\sum \rho_j \mathbf{V}'_{dj} = -\rho D_{amb} \nabla \left(\sum \frac{\rho_j}{\rho} \right) = 0 \ . \tag{11.6}$$

Since, $\sum \rho_j V'_{dj} = 0$, it follows that the condition

$$\sum \rho_j \mathbf{V}'_{fj} = 0 \tag{11.7}$$

is to be satisfied. An estimate of the field velocities for a typical case of an argon plasma at 1 bar in an electric field of 100 V/m shows for the ion that there is a field velocity of about 1 m/s, whereas for the electrons it is about 600 m/s; these field velocities are many orders of magnitude larger than the mass diffusion velocity due to the concentration gradient for charged particles (*"ambipolar diffusion"*).

Now one may define the two different electric currents, namely, the convection current of the j-th species

$$\mathbf{j}_{cj} = n_j q_j \mathbf{V}_j \tag{11.8}$$

which is the total charge carried by the j-th species per unit area and time, and the field current

$$\mathbf{j}_j = n_j q_j \mathbf{V}'_{fj} \tag{11.9}$$

which is diffusive in nature and is part of the convection current. Since for a *quasi-neutral plasma*,

$$\sum q_j n_j = 0 \tag{11.10}$$

the *total current density* is

$$\mathbf{j} = \sum \mathbf{j}_{cj} = \sum q_j n_j \mathbf{V}_j = \sum q_j n_j \mathbf{V}'_{fj} = \sum \mathbf{j}_j \ . \tag{11.11}$$

Considering the particle flux entering and leaving a volume element, the species continuity equation for the j-th species can be written as

$$\frac{\partial n_j}{\partial t} + \nabla\cdot(n_j \mathbf{V}_j) = \dot{n}_{j,\text{reaction}} \ , \ \text{m}^{-3}\text{s}^{-1} \tag{11.12}$$

where the sign of the right hand term gives the production or the removal of the particles of the j-th species by reaction. Equation (11.12) is multiplied with M_j, mass of a single particle, to get

$$\frac{\partial \rho_j}{\partial t} + \nabla\cdot(\rho_j \mathbf{V}_j) = \dot{m}_{Rj} \ , \ \text{kgm}^{-3}\text{s}^{-1} \tag{11.13}$$

where $\dot{m}_{Rj}$ is the mass rate of production of the j-th species by reaction. Equation (11.13) is added for all specie, and since $\sum \dot{m}_{Rj} = 0$ and and (11.4) are valid, we get the global continuity equation

$$\frac{\partial \rho}{\partial t} + \nabla\cdot(\rho \mathbf{V}) = 0 \ , \ \text{kgm}^{-3}\text{s}^{-1} \ . \tag{11.14}$$

Now for the j-th species, the species momentum equation is

$$\frac{\partial}{\partial t}(\rho_j \mathbf{V}_j) + \nabla\cdot(\rho_j \mathbf{V}_j \mathbf{V}_j) = -\nabla p_j + \nabla \cdot \tau_j + \mathbf{F}_{jf} + \mathbf{F}_{j,\text{coll}} \tag{11.15}$$

where

$$\tau_j^{rs} = -\frac{2}{3}\mu_j(\nabla \cdot \mathbf{V}_j)\delta^{rs} + \mu_j\omega_j^{rs} \qquad (11.16)$$

$$\delta^{rs} = \text{Kronecker delta}: \delta^{rr} = 1, \delta^{rr} = 0 \qquad (11.17)$$

$$\omega_j^{rs} = \frac{\partial V_j^r}{\partial x^s} + \frac{\partial V_j^s}{\partial x^r} \qquad (11.18)$$

the *electromagnetic volume force*

$$\mathbf{F}_{jf} = n_j q_j[\mathbf{E} + (\mathbf{V}_j \times \mathbf{B})] \qquad (11.19)$$

and the *collisional volume force*

$$\mathbf{F}_{j,\text{coll.}} = \mathbf{F}_{j,\text{elast.}} + \mathbf{F}_{j,\text{inelast.}} \qquad (11.20)$$

in which elastic and inelastic (reactive) components of the collisional volumetric force are given by the relations

$$\mathbf{F}_{j,\text{elast.}} = \frac{2M_j M_k}{(M_j + M_k)}(\mathbf{V}_k - \mathbf{V}_j)\Gamma'_{jk} \qquad (11.21)$$

$$\mathbf{F}_{j,\text{inelast.}} = \mathbf{V}_j \dot{m}_{Rj} \approx \mathbf{V}\dot{m}_{Rj} . \qquad (11.22)$$

Further in some instances it is convenient to define an alternate definition of the shear stress term by including pressure as

$$\tau_j^{*rs} = -p\delta^{rs} + \tau_j^{rs} \qquad (11.23)$$

so that the pressure term is hidden to shorten the equations.

Through (11.16, 11.17) the shear stress on the j-th species has been introduced in a formal manner in analogy to the case of the single species gases, the underlying assumption being that the gradient of the relative mass-average velocities of the different specie and their effect on the shear stress for the j-th species is negligible. The formal formulation in this manner has the obvious advantage of clarity with which the ultimate result not being much different from that obtained through more rigid mathematical approach. Further it may be noted that the *dynamic viscosity coefficient* μ_j in the equation for stress is not that of a pure gas, but it has already the effect of the mole fraction of the j-th species in it, as already explained in Chap. 7.

Now the left hand side of (11.15), taking into account (11.13), can be expanded as follows:

$$\frac{\partial}{\partial t}(\rho_j\mathbf{V}_j) + \nabla\cdot(\rho_j\mathbf{V}_j V_j) = \rho_j\left[\frac{\partial}{\partial t}\mathbf{V}_j(V_j\nabla)\cdot\mathbf{V}_j\right] + \mathbf{V}_j\left[\frac{\partial\rho_j}{\partial t} + \nabla\cdot(\rho_j\mathbf{V}_j)\right]$$

$$= \rho_j\left[\frac{\partial}{\partial t}\mathbf{V}_j(V_j\nabla)\cdot\mathbf{V}_j\right] + \mathbf{V}_j\dot{m}_{Rj}$$

$$= \rho_j\frac{D}{Dt}\mathbf{V}_j + \mathbf{V}_j\dot{m}_{Rj}$$

where the *substantive differential quotient* is given by the relation

$$\frac{\mathrm{D}}{\mathrm{D}t} = \frac{\partial}{\partial t} + (\mathbf{V}\nabla) \ . \tag{11.24}$$

Thus, the *species momentum equation*, (11.15) becomes

$$\rho_j \frac{\mathrm{D}\mathbf{V}_j}{\mathrm{D}t} = -\nabla p_j + \nabla \cdot \tau_j + n_j q_j [\mathbf{E} + (\mathbf{V}_j \times \mathbf{B})] + \frac{2M_j M_k}{(M_j + M_k)}(\mathbf{V}_k - \mathbf{V}_j)\Gamma'_{jk} \ . \tag{11.25}$$

It may be noted further that neither in (11.15) nor in (11.25) any other type of volumetric forces, for example, the gravitational buoyancy force or centrifugal or coriolis force, have been considered at this stage. Now, the global momentum equation is derived from the species momentum equation by adding for all specie, for which the term by term added results are:

$$\sum \mathbf{F}_{j,\text{elast.}} = \sum \frac{2M_j M_k}{(M_j + M_k)}(\mathbf{V}_k - \mathbf{V}_j)\Gamma'_{jk} = 0 \ \text{(since } \Gamma'_{jk} = \Gamma'_{kj})$$

$$\sum \mathbf{F}_{j,\text{inelast.}} = -\sum \mathbf{V}\dot{m}_{Rj} = 0 \ , \ \sum \nabla p_j = \nabla p$$

$$\sum \frac{\partial}{\partial t}(\rho_j \mathbf{V}_j) = \frac{\partial}{\partial t}\sum(\rho_j \mathbf{V}_j) = \frac{\partial}{\partial t}(\rho\mathbf{V})$$

$$\sum \nabla\cdot(\rho_j \mathbf{V}_j V_j) = \nabla\cdot\left[(\rho\mathbf{V}V) + \sum(\rho_j \mathbf{V}'_j V'_j)\right] \approx \nabla\cdot[(\rho\mathbf{V}V)$$

$$\text{(some other terms drop off since } \sum \rho_j V'_j = 0)$$

$$\sum \nabla \cdot \tau_j = \nabla\cdot\sum\left[-\frac{2}{3}\mu_j(\nabla \cdot \mathbf{V}_j)\delta^{rs} + \mu_j^{rs}\right]$$

$$= \nabla\cdot\sum\left[-\frac{2}{3}\mu_j(\nabla \cdot \mathbf{V})\delta^{rs} + \mu_j^{rs}\right] = \nabla \cdot \tau$$

$$\text{(since } \mu = \sum \mu_j)$$

$$\sum \mathbf{F}_{jf} = \sum n_j q_j [\mathbf{E} + (\mathbf{V}_j \times \mathbf{B})] = \sum (n_j q_j \mathbf{V}_{fj})\times\mathbf{B} = \mathbf{j} \times \mathbf{B}$$

where $\mu_j = x_j \mu_j^{\text{pure}}$ and μ_j^{pure} is the viscosity coefficient of the pure species. It is multiplied with its mole fraction to obtain its contribution for the species momentum equation. Thus the *global momentum equation* becomes

$$\rho\frac{\mathrm{D}\mathbf{V}_j}{\mathrm{D}t} = -\nabla p + \nabla \cdot \tau + (\mathbf{j} \times \mathbf{B}) \tag{11.26}$$

where for the shear stress term we could use a similar expressions as (11.16–11.23) with the species index j removed. We can also derive further the left hand side of (11.26), as we did for the species momentum equation (11.15 to get

$$\frac{\partial}{\partial t}(\rho\mathbf{V}) + \nabla\cdot(\rho\mathbf{V}V) = \rho\left[\frac{\partial}{\partial t}\mathbf{V} + (V\nabla)\cdot\mathbf{V}\right] + \mathbf{V}\left[\frac{\partial\rho}{\partial t} + \nabla\cdot(\rho\mathbf{V})\right] = \rho\frac{\mathrm{D}\mathbf{V}}{\mathrm{D}t} \tag{11.27}$$

and an alternative formulation of the global momentum equation is

$$\rho\frac{\mathbf{DV}}{\mathbf{D}t} = -\nabla p + \nabla \cdot \tau + \mathbf{j} \times \mathbf{B} = \nabla \cdot \tau^* + \mathbf{j} \times \mathbf{B} \ . \tag{11.28}$$

As we have already done in the case of the continuity and momentum equations, the balance of the energy flux entering and leaving a volume element, first for a species, is considered and the following *species energy equation* can be written:

$$\frac{\partial}{\partial t}(\rho_j E_j^o) + \nabla \cdot (\rho_j \mathbf{V}_j E_j^o) = \nabla \cdot (k_j \nabla T_j) + \nabla \cdot (\mathbf{V}_j \tau_j^*) + \dot{Q}_{jf} + \dot{Q}_{j,\text{coll}} - \epsilon_{Rj} \tag{11.29}$$

where for the j-th species,

$$E_j^o = \text{total specific energy} = \frac{3}{2}\frac{R^*}{m_j}T_j - E_{mj} + \frac{1}{2}V_j^2, \ \text{Jkg}^{-1} \tag{11.30}$$

and

R^* = universal gas constant ; m_j = mole mass of j-th species
E_{mj} = ionization energy per unit mass of j-th species
$\dot{Q}_{jf}$ = energy production due to product of electric current density and external electric field = $\mathbf{V}'_{fj} \cdot [n_j q_j (\mathbf{E}' + \mathbf{V}_j \times \mathbf{B})] = \mathbf{j}_j \cdot \mathbf{E}'$
$\dot{Q}_{j,\text{coll}}$ = energy production due to collision, and
ϵ_{Rj} = radiative energy loss.
In addition for definition of τ_j^*, see (11.23).

The ionization energy, by convention, is assumed to be zero for the atoms and free electrons. Further, the collisional energy transfer may be due to elastic and non-elastic (reacting!) collisions; the former due to the difference of mean kinetic speed (translational temperature!), and the latter due to loss in energy to the species due the reaction, $\dot{m}_{Rj} E_J^o$. Since, according to the present model, it is assumed that all specie share the total energy of the gas mixture in proportion due to their relative mass fraction, this loss in energy to the species due to reaction is already included in the balance of the energy flux and need not be repeated. Thus for the collisional energy production, we may write

$$\dot{Q}_{j,\text{coll}} = \frac{3M_j M_k}{(M_j + M_k)^2} k_B (T_k - T_j)\Gamma'_{jk} \ . \tag{11.31}$$

Further in (11.29), the heat conductivity coefficient k_j, similar to the *dynamic viscosity coefficient* μ_j, is not for a pure gas, and it has already the effect of the mole fraction of the j-th species.

Now we perform term by term addition of (11.29) for all specie to get the *global energy equation* as follows:

$$\sum \epsilon_{Rj} = \epsilon_R = \text{total radiative energy} \ , \ \sum \dot{Q}_{j,\text{coll}} = 0.$$

$$\sum \rho_j E_j^o = \rho E^o \text{(by definition of total energy)}$$

$$\sum \mathbf{j}_j \cdot \mathbf{E}' = \mathbf{j} \cdot \mathbf{E}'$$

$$\sum \rho_j \mathbf{V}_j E_j^o = \sum \rho_j (\mathbf{V} + \mathbf{V}_j') E_j^o$$

$$= \rho \mathbf{E}^o + \sum \rho \mathbf{V}_j' E_j + \frac{1}{2} V^2 \sum \rho_j \mathbf{V}_j' = \rho \mathbf{V} E^o + \sum \rho_j \mathbf{V}_j' E_j$$

$$\sum (\mathbf{V}_j \tau_j^*) = \mathbf{V} \tau^* - \sum p_j \mathbf{V}_j'$$

Thus, after term by term addition of (11.29), we get the global energy equation

$$\frac{\partial}{\partial t}(\rho E^o) + \nabla \cdot (\rho \mathbf{V} E^o) = \sum \nabla \cdot (k_j \nabla T_j) - \sum \nabla \cdot (\rho_j \mathbf{V}_j' h_j)$$
$$+ \nabla \cdot (\mathbf{V} \tau^*) + \dot{Q}_f - \epsilon_R . \tag{11.32}$$

It is to be noted that the second term $\sum \rho_j \mathbf{V}_j' h_j$ consists of sum of two terms, that is $\sum \rho_j \mathbf{V}_j' E_j$ and $\mathbf{V}_j' p_j$, which together is the energy flux due to diffusion. For the special case of a plasma at thermal and chemical equilibrium ($T_j = T$ and ρ_j, h_j are given as a function of T, p), the first two terms in the right hand side of (11.32) become

$$\sum \nabla \cdot (k_j \nabla T_j) - \sum \nabla \cdot (\rho_j \mathbf{V}_j' h_j)$$

$$= \nabla \cdot (k \nabla T) + \rho \sum D_j h_j \nabla \left(\frac{\rho_j}{\rho} \right) = \nabla \cdot (k_{c,d} \nabla T) \tag{11.33}$$

where the *total conductivity coefficient* due to *pure conduction* and *diffusion* is

$$k_{c,d} = k + \rho \sum D_j h_j \frac{\partial}{\partial} \left(\frac{\rho_j}{\rho} \right) . \tag{11.34}$$

As explained in Chap. 7, the *diffusive heat conduction* term can be evaluated for chemical equilibrium in general, but particularly for ionized plasma, the *diffusion coefficient* $D_j = D_{amb}$, and for this special case of a mixture of neutrals, electrons and singly-charged ions with arbitrary electron and heavy particles temperature and electric current flow, we have to consider not only the energy transport due to conventional diffusion, but also the diffusive nature of motion of charged particles in electro-magnetic fields. Thus for an ionized plasma with two temperatures, the first two terms in the right hand side of (11.32) become

$$\sum \nabla \cdot (k_j \nabla T_j) - \sum \nabla \cdot (\rho_j V_j' h_j = \nabla \cdot (k_h \nabla T_h) + \nabla \cdot (k_e \nabla T_e)$$

$$+ \nabla \cdot (\rho D_{amb} E_{mi} \nabla Y_i) + \frac{5}{2} \frac{R^*}{m_e} \nabla \cdot (\rho D_{amb} \nabla Y_e) - \frac{5}{2} \frac{k_B}{e} \nabla \cdot (\mathbf{j} T_e) \tag{11.35}$$

where the *mass fraction* $Y_j = \rho_j / \rho$. Thus the energy equation for a plasma containing the electrons, neutrals and the singly-charged ions becomes

$$\rho \frac{DE^o}{Dt} = \nabla\cdot(k_h\nabla T_h) + \nabla\cdot(k_e\nabla T_e) + \nabla\cdot(\rho D_{amb}E_{mi}\nabla Y_i) + \nabla\cdot(\mathbf{V}\tau)$$

$$+ \mathbf{j}\cdot\mathbf{E}' - \epsilon_R + \frac{5}{2}\frac{R^*}{m_e}\nabla\cdot(\rho D_{amb}\nabla Y_e)$$

$$- \frac{5}{2}\frac{k_B}{e}\nabla\cdot)(\mathbf{j}T_e) + \nabla\cdot(\mathbf{V}\tau^*) + \mathbf{j}\cdot\mathbf{E}' - \epsilon_R \ . \tag{11.36}$$

While the above equation is written in terms of the total internal energy, it is quite usual in fluid or gas dynamics to write the energy equation also in terms of static internal energy, or even in terms of enthalpy (static or total). Thus we do the same thing. In order to write the energy equation in terms of the static internal energy, we have to subtract the kinetic energy equation from the energy equation written in terms of the total internal energy. For this purpose, the global momentum equation, (11.28), is multiplied by the velocity component in the respective coordinate directions, and these are added for all the three dirctions. Thus we get the equation for the gas dynamic kinetic energy as

$$\rho\frac{D}{Dt}\left(\frac{V^2}{2}\right) = (\mathbf{V}\nabla)\cdot\tau^* + \mathbf{V}\cdot(\mathbf{j}\times\mathbf{B}) \tag{11.37}$$

where $\tau^* = \tau - p\delta^{rs}$. Thus subtracting (11.37) from (11.36), we get the energy equation in terms of the static enthalpy as

$$\rho\frac{DE}{Dt} + p\nabla\cdot\mathbf{V} = \sum\nabla\cdot(k_j\nabla T_j) - \sum\nabla\cdot(\rho_j\mathbf{V}'_j h_j) + (\mathbf{V}\nabla)\tau^* + \phi + \frac{j^2}{\sigma} - \epsilon_R$$
$$\tag{11.38}$$

since $\mathbf{j}\cdot\mathbf{E}' - \mathbf{V}\cdot(\mathbf{j}\times\mathbf{B}) = \mathbf{j}\cdot(\mathbf{E}' + \mathbf{V}\times\mathbf{B}) = j^2/\sigma$. Further $\phi = (\tau\nabla)\cdot\mathbf{V}$ is the *dissipation function*. Noting further that

$$E = h - \frac{p}{\rho} \quad \text{and} \quad E^o = h^o - \frac{p}{\rho} \tag{11.39}$$

we write the energy equation in terms of the static and total enthalpy as follows:

$$\rho\frac{Dh}{Dt} - \frac{Dp}{Dt} = \sum\nabla\cdot(k_j\nabla T_j) - \sum\nabla\cdot(\rho_j\mathbf{V}'_j h_j)$$
$$+ \phi + \frac{j^2}{\sigma} - \epsilon_R \tag{11.40}$$

and

$$\rho\frac{Dh^o}{Dt} - \frac{Dp}{Dt} - p\nabla\cdot\mathbf{V} = \sum\nabla\cdot(k_j\nabla T_j) - \sum\nabla\cdot(\rho_j\mathbf{V}'_j h_j)$$
$$+ \nabla\cdot(\mathbf{V}\tau^*) + \mathbf{j}\cdot\mathbf{E}' - \epsilon_R \ . \tag{11.41}$$

We now derive from the momentum equation the *generalized Ohm's law* by writing first the species momentum equation separately for the electrons and the singly charged ions, as well as the global momentum equation in which the shear stress terms are neglected. Further, in the electron momentum equation, the inertial terms are neglected. Thus the three momentum equations are,

Electron momentum:

$$-\nabla p_e - en_e[\mathbf{E}' + \mathbf{V}_e\times\mathbf{B}] + 2M_e[(\mathbf{V}_a - \mathbf{V}_e)\Gamma'_{ea} + (\mathbf{V}_i - \mathbf{V}_e)\Gamma'_{ei}] = 0 \quad (11.42)$$

Ion momentum:

$$\rho_i\frac{DV_i}{Dt} \approx \rho_i\frac{DV_i}{Dt} = -\nabla p_i + en_e[\mathbf{E}' + \mathbf{V}_e\times\mathbf{B}]$$
$$-2M_e(\mathbf{V}_i - \mathbf{V}_e)\Gamma'_{ei} - \frac{2M_iM_a}{M_i + M_a}(\mathbf{V}_i - \mathbf{V}_a)\Gamma'_{ia} \quad (11.43)$$

Global momentum:
$$\rho\frac{D\mathbf{V}}{Dt} = -\nabla p + \mathbf{j} \times \mathbf{B} \quad (11.44)$$

Now a velocity difference be defined as

$$\mathbf{V}^* = \mathbf{V}_e - \mathbf{V}_i = \mathbf{V}'_e - \mathbf{V}'_i \quad (11.45)$$

and thus (11.11) becomes
$$\mathbf{j} = -en_e\mathbf{V}^* \ . \quad (11.46)$$

Further since,
$$\sum \rho_j\mathbf{V}'_j = \rho_e\mathbf{V}'_e + \rho_i\mathbf{V}'_i + \rho_a\mathbf{V}'_a = 0 \quad (11.47)$$

thus,

$$\mathbf{V}'_a = -\left[\frac{n_eM_e}{n_aM_a}\mathbf{V}^* + \frac{n_e}{n_a}\mathbf{V}'_i\right] \ . \quad (11.48)$$

Now we introduce the following terms:

Electron radian cyclotron frequency: $\omega_e = eB/M_e$
Ion radian cyclotron frequency: $\omega_i = eB/M_i$
where B is the magnitude of the magnetic induction. Further,
Collision to radian cyclotron frequencies:

$$K_{ei} = \frac{\Gamma_{ei}}{\omega_e} = \frac{\Gamma'_{ei}}{\omega_e n_e} = \frac{\Gamma_{ei}M_e}{en_e B}$$

$$K_{ea} = \frac{\Gamma_{ea}}{\omega_e} = \frac{\Gamma'_{ea}}{\omega_e n_e} = \frac{\Gamma_{ea}M_e}{en_e B}$$

$$K_{ia} = \frac{\Gamma_{ia}}{\omega_i} = \frac{\Gamma'_{ia}}{\omega_i n_e} = \frac{\Gamma'_{ia}M_i}{en_e B} \ .$$

Atom fraction:
$$f = \frac{n_a}{n_a + n_e} \ . \quad (11.49)$$

Further, assuming *quasi-neutrality*, $\sum n_j q_j = 0$, the following intermediate terms can be derived:

$$-\mathbf{j}_e = e n_e(\mathbf{V}^* + \mathbf{V}_i') = \mathbf{j}_i - \mathbf{j}; \quad -\mathbf{j}_i = e n_e \mathbf{V}_i'$$

$$-M_e \mathbf{V}^* = \frac{M_e}{e n_e}\mathbf{j} = \frac{\mathbf{j}B}{n_e \omega_e}; \quad -M_e \mathbf{V}^* \Gamma_{ea}' = \mathbf{j}B K_{ea} \quad -M_e \mathbf{V}^* \Gamma_{ei}' = \mathbf{j}B K_{ei}$$

$$M_e(\mathbf{V}_e - \mathbf{V}_i)\Gamma_{ei}' = M_e \mathbf{V}^* \Gamma_{ei}' = -\mathbf{j}B K_{ei}$$

$$\mathbf{V}_a - \mathbf{V}_i = \mathbf{V}_a' - \mathbf{V}_i' = -\frac{n_e M_e}{n_a M_a}\mathbf{V}^* - \left(1 + \frac{n_e}{n_a}\right)\mathbf{V}_i'$$

With the help of these expressions the electron momentum equation, (11.42), becomes

$$\nabla p_e + e n_e[\mathbf{E}' + \mathbf{V}_e \times \mathbf{B}]$$
$$= 2M_e\left[\left\{\left(1 + \frac{n_e M_e}{n_a M_a}\right)\mathbf{V}^* + \left(1 + \frac{n_e}{n_a}\right)\mathbf{V}_i'\right\}\Gamma_{ea}' + \mathbf{V}^* \Gamma_{ie}'\right]$$

which may be further expanded into

$$\nabla p_e + e n_e[\mathbf{E}' + \mathbf{V}_e \times \mathbf{B}] - (\mathbf{j} - \mathbf{j}_i) \times \mathbf{B}$$
$$= 2\left[\mathbf{j}B K_{ei} + \mathbf{j}B K_{ea}\left(1 + \frac{n_e M_e}{n_a M_a}\right) - \mathbf{j}_i B K_{ea}\left(1 + \frac{n_e}{n_a}\right)\right] \ .$$

It is estimated that if n_a is large, the term $(n_e M_e)/(n_a M_a)$ is small, and if n_a is small K_{ea} is small. Thus let $(n_e M_e)/(n_a M_a) \ll 1$, and the electron momentum equation (11.42) becomes

$$\nabla p_e + e n_e[\mathbf{E}' + \mathbf{V}_e \times \mathbf{B}] - (\mathbf{j} - \mathbf{j}_i) \times \mathbf{B} = \left[\mathbf{j}B(K_{ei} + K_{ea}) - \frac{\mathbf{j}_i B K_{ea}}{f}\right] \ . \quad (11.50)$$

Similarly the ion momentum equation, (11.43), becomes

$$\rho_i \frac{D\mathbf{V}}{Dt} = -\nabla p_i + e n_e[\mathbf{E}' + \mathbf{V}_i \times \mathbf{B}] + \mathbf{j}_i \times \mathbf{B} + \mathbf{j}B K_{ei}\left[\frac{1 - f}{f}\frac{\Gamma_{ia}'}{\Gamma_{ie}'} - 2\right] - \mathbf{j}_i B \frac{K_{ia}}{f} \ . \quad (11.51)$$

Adding Equations (11.50, 11.51), we get

$$\rho_i \frac{D\mathbf{V}}{Dt} = -\nabla(p_i + p_e) + \mathbf{j} \times \mathbf{B} + \mathbf{j}B K_{ei}\left[K_{ei}\frac{1 - f}{f}\frac{\Gamma_{ia}'}{\Gamma_{ie}'} + 2K_{ea}\right]$$
$$- \mathbf{j}_i B \frac{K_{ia}}{f}(2K_{ea} + K_{ie}) \ . \quad (11.52)$$

Now further we write the global momentum equation. Since,

$$\frac{\rho_i}{\rho} = \frac{n_i M_i}{n_i M_i + n_a M_a} = \frac{n_i}{n_i + n_a} = 1 - f \quad (11.53)$$

the *global momentum equation*, Equation (11.44), becomes

$$\rho_i \frac{\mathrm{D}\mathbf{V}}{\mathrm{D}t} = (1-f)(-\nabla p + \mathbf{j} \times \mathbf{B}) \ . \tag{11.54}$$

Substituting (11.54) into (11.52) we get

$$(1-f)(-\nabla p + \mathbf{j} \times \mathbf{B}) = -\nabla(p_i + p_e) + \mathbf{j} \times \mathbf{B}$$
$$+\mathbf{j}B\left[K_{ei}\frac{1-f}{f}\frac{\Gamma'_{ia}}{\Gamma'_{ie}} + 2K_{ea}\right] - \mathbf{j}_i B \frac{K_{ia}}{f}(2K_{ea} + K_{ia}) \tag{11.55}$$

It is estimated again that in a similar manner as before, the term $K_{ei}(1-f)\Gamma'_{ia}/(f\Gamma'_{ia})$ may be neglected. Further since,

$$-[(p_i + p_e) - (1-f)p] = [-2fp_e + (1-f)p_a] = -fp_e \ . \tag{11.56}$$

Equation (11.55) becomes

$$-f\nabla p_e + f\mathbf{j} \times \mathbf{B} + 2\mathbf{j}BK_{ea} - \frac{\mathbf{j}_i}{f}B(2K_{ea} + K_{ia}) = 0 \tag{11.57}$$

from which we get an explicit expression for the ion current density

$$\mathbf{j}_i = \frac{f}{B(2K_{ea} + K_{ia})}\left[f(-\nabla p_e + \mathbf{j} \times \mathbf{B}) + 2\mathbf{j}BK_{ea}\right] \ . \tag{11.58}$$

Substituting the above equation into (11.50), we get further

$$0 = -\nabla p_e - en_e[\mathbf{E}' + \mathbf{V} \times \mathbf{B}] + \mathbf{j} \times \mathbf{B}$$
$$- \frac{f}{(2K_{ea} + K_{ia})}\left[f(-\nabla p_e + \mathbf{j} \times \mathbf{B}) + 2\mathbf{j}BK_{ea}\right] \times \frac{\mathbf{B}}{B}$$
$$+ 2\mathbf{j}B(K_{ea} + K_{ei})$$
$$- \frac{2K_{ea}}{(2K_{ea} + K_{ia})}\left[f(-\nabla p_e + \mathbf{j} \times \mathbf{B}) + 2\mathbf{j}BK_{ea}\right] \ . \tag{11.59}$$

Noting that $K_{ea} \ll K_{ia}$ and since,

$$B(K_{ea} + K_{ei}) = \frac{M_e}{en_e}(\Gamma'_{ei} + \Gamma'_{ea}) \tag{11.60}$$

we get after some rearrangement

$$0 = -\nabla p_e - en_e[\mathbf{E}' + \mathbf{V} \times \mathbf{B}] + \mathbf{j} \times \mathbf{B}$$
$$- \frac{f}{K_{ia}}[f(-\nabla p_e + \mathbf{j} \times \mathbf{B})] \times \frac{\mathbf{B}}{B} + \frac{2M_e}{en_e}(\Gamma'_{ei} + \Gamma'_{ea})\mathbf{j}$$
$$+ 2\mathbf{j}B(K_{ea} + K_{ei})$$
$$- \frac{2K_{ea}}{(2K_{ea} + K_{ia})}[f(-\nabla p_e + \mathbf{j} \times \mathbf{B}) + 2\mathbf{j}BK_{ea}] \ . \tag{11.61}$$

Further noting the definition of a *scalar electrical conductivity*

$$\sigma_o = \frac{e^2 n_e^2}{2M_e(\Gamma'_{ei} + \Gamma'_{ea})} \ . \tag{11.62}$$

Equation (11.61), after some rearrangement, becomes

$$\frac{\mathbf{j}}{\sigma_o} = (\mathbf{E}' + \mathbf{V} \times \mathbf{B}) + \frac{1}{en_e}\nabla p_e - \frac{1}{en_e}\mathbf{j} \times \mathbf{B} - \frac{f^2}{M_i\Gamma'_{ia}}\left[\nabla p_e \times \mathbf{B} + \mathbf{B} \times (\mathbf{j} \times \mathbf{B})\right] \ . \tag{11.63}$$

Equation (11.63) is known as the *generalized Ohm's law*. It has several terms in the right hand side. The first term gives the effect of the externally applied electric field, the second term takes care of the induced electric field $\mathbf{V} \times \mathbf{B}$, the third term is due to electron pressure gradient , the fourth term is the *Hall effect*, and the terms within [ ] account for the *"ion-slip"*; this last term is usually neglected since divided by very large Γ'_{ia}.

We would now discuss the implication of various terms in the above equation. For this we define first an effective electric field $\mathbf{E}$, consisting of an *external electric field* $\mathbf{E}'$ and an *induced electric field* $(\mathbf{V} \times \mathbf{B})$, and then neglect the electron pressure gradient term to get

$$\frac{\mathbf{j}}{\sigma_o} = \mathbf{E} - \frac{1}{en_e}\mathbf{j} \times \mathbf{B} - \frac{f^2}{M_i\Gamma'_{ia}}(\mathbf{j} \times \mathbf{B}) \ . \tag{11.64}$$

It can be seen from the above expression, that in the direction of $\mathbf{B}$, the second and third term in the right hand side of (11.64) are zero, and if there is an electric field $\mathbf{E}$ parallel to $\mathbf{B}$, then only the scalar value of electrical conductivity is effective in that direction.

First we multiply (11.64) by σ_o and get after some rearrangement

$$\mathbf{j} = \sigma_o\mathbf{E} - \beta_2\frac{\mathbf{j} \times \mathbf{B}}{B} - \beta_1\frac{\mathbf{B} \times (\mathbf{j} \times \mathbf{B})}{B^2} \tag{11.65}$$

where

$$\beta_1 = \frac{\sigma_o B^2 f^2}{M_i\Gamma'_{ia}} = \frac{f^2 n_e^2 \omega_e \omega_i}{2\Gamma'_e \Gamma'_{ia}} = \frac{f^2 \omega_e \omega_i}{2\Gamma_e \Gamma_{ia}} \ ,$$

$$\beta_2 = \frac{\sigma_o B}{en_e} = \frac{n_e \omega_e}{2\Gamma'_e} = \frac{\omega_e}{2\Gamma_e} \ . \tag{11.66}$$

It is seen that both βs' depend on the ratio or the square of the ratio of the cyclotron frequency to the collision frequency. Noting further that the term $\mathbf{B} \times (\mathbf{j} \times \mathbf{B})$ may be reduced to $\mathbf{j}B^2 - \mathbf{B}(\mathbf{B} \cdot \mathbf{j})$, (11.65) is reduced to

$$(1 + \beta_1)\mathbf{j} = \sigma_o\mathbf{E} + \beta_1\frac{\mathbf{B}}{B}\left(\frac{\mathbf{B}}{B} \cdot \mathbf{j}\right) - \beta_2\mathbf{j} \times \frac{\mathbf{B}}{B} \ . \tag{11.67}$$

Let us now assume that the electric field $\mathbf{E}$ and current density $\mathbf{j}$ have only components in x and y direction and the magnetic induction $\mathbf{B}$ has component only in the z-direction. Thus we consider $\mathbf{E} = \{E_x, E_y, 0\}$, $\mathbf{j} = \{j_x, j_y, 0\}$ and $\mathbf{B} = \{0, 0, B_z\}$. We write, therefore, the three equations

$$(1 + \beta_1)j_x + \beta_2 j_y = \sigma_o E_x \qquad (11.68)$$

$$(1 + \beta_1)j_y - \beta_2 j_x = \sigma_o E_y \qquad (11.69)$$

$$j_z = \sigma_o E_y \qquad (11.70)$$

We can now write the above three equations in a matrix equation form as

$$\mathbf{j} = \sigma \cdot \mathbf{E} \qquad (11.71)$$

where

$$\sigma = \sigma_o \begin{bmatrix} \dfrac{1+\beta_1}{(1+\beta_1)^2 + \beta^2} & \dfrac{\beta_2}{(1+\beta_1)^2 + \beta^2} & 0 \\[4mm] \dfrac{-\beta_2}{(1+\beta_1)^2 + \beta^2} & \dfrac{1+\beta_1}{(1+\beta_1)^2 + \beta^2} & 0 \\[4mm] 0 & 0 & 1 \end{bmatrix} . \qquad (11.72)$$

It is now worth noting, that by neglecting the *"ion-slip"* term, $\beta_1 = 0$, it leads to the expression we have derived in Chap. 7. In addition to the above Ohm's law, there are *Maxwell's equations*, which must be satisfied. The latter are rewritten again from Chap. 7 as follows:

$$\nabla \cdot \mathbf{D} = n_c \qquad (11.73)$$

$$\nabla \cdot \mathbf{B} = 0 \qquad (11.74)$$

$$\mathbf{D} = \epsilon \mathbf{E} \qquad (11.75)$$

$$\mathbf{B} = \mu \mathbf{H} \qquad (11.76)$$

$$\nabla \times \mathbf{E} = -\frac{\partial \mathbf{B}}{\partial t} \qquad (11.77)$$

$$\nabla \times \mathbf{H} = \mathbf{j} + \frac{\partial \mathbf{D}}{\partial t} \qquad (11.78)$$

where μ_o is the *magnetic permeability* in vacuum.

Taking divergence of (11.78) and taking help of (11.73), we write

$$\nabla \cdot (\nabla \times \mathbf{H}) = \nabla \cdot \mathbf{j} + \epsilon \frac{\partial}{\partial t}(\nabla \cdot \mathbf{E})$$

$$= \nabla \cdot \mathbf{j} + \frac{\partial n_c}{\partial t} = 0 . \qquad (11.79)$$

Since for a quasi-neutral plasma, the charge density $n_c = 0$, we conclude, therefore, that for a *"quasi-neutral plasma"* the divergence of electric current density must be equal to zero.

Further at an interface and from the relation at the interface, $\nabla \cdot \mathbf{D} = 0$ and $\nabla \cdot \mathbf{B} = 0$, and we may write that the normal component across interface for $\mathbf{D}$ and $\mathbf{B}$ must remain continuous, that is, $D_{n1} = D_{n2}$ and $B_{n1} = B_{n2}$. On the other hand, the tangential components of $\mathbf{E}$ and $\mathbf{H}$ remain continuous, that is $E_{t1} = E_{t2}$ and $H_{t1} = H_{t2}$.

At the outset, we should mention the application of the above derivations in a type of propulsive device, called *Hall Thruster"*. Such a device, consists of a co-axial plasma tube with anode at the closed end and magnetic field in the radial direction. In such a tube the ions, with cyclotron radius of the order of or larger than the axial length of the channel, are electrically accelerated in axial direction, but the electrons are trapped in the magnetic field, because of the very small cyclotron radius. Such devices were proposed in sixties (*Seikel* and *Reshotko* [104]), and since 1971 the russians have flown many thrusters based on the Hall Transfer Technology.

11.2 Magneto- and Electromagneto-gas-dynamic Approximations

The basic equations for ionized gases in electromagnetic fields are the Maxwell equations, (11.73–11.78), in which the displacement current density term, $\partial \mathbf{D}/\partial t$, can be neglected, if the ratio of the displacement current density to the current density due to the externally applied electro-magnetic field $\mathbf{E}$

$$\frac{\partial \mathbf{D}}{\partial t} \frac{1}{\sigma_o \mid E \mid} = \frac{\epsilon_o}{\sigma_o} \frac{\partial \mathbf{E}}{\partial t} \frac{1}{\mid E \mid} \ll 1 . \tag{11.80}$$

The condition, under which this is possible, is now discussed.

Let $\mathbf{E} = \mathbf{A} \sin (\omega t)$ and thus,

$$\frac{\epsilon_o}{\sigma_o} \left(\frac{\partial \mathbf{E}}{\partial t} \right)_{\max} \left(\frac{1}{\mid \mathbf{E} \mid} \right)_{\max} = \frac{\epsilon_o \omega}{\sigma_o} . \tag{11.81}$$

Assuming in the laboratory typical values of σ_o between 10^2 and 10^4 $\mathrm{AV}^{-1}\mathrm{m}^{-1}$, we find that

$$\frac{\epsilon_o \omega}{\sigma_o} = 10^{-12}\omega \text{ to } 10^{-14}\omega \tag{11.82}$$

in which the higher limit is for the smaller values of σ_o. Thus it is concluded that for all practical purposes for a laboratory plasma, even at microwave frequencies, the displacement current can be neglected.

The basic equations are now analyzed under the assumed conditions that (1) quasi-neutrality exists, (2) translational temperatures of all species are equal, (3) the effective heat conductivity $k_{c,d}$ used in the analysis consists of contribution due to *pure* conduction, as well as the recombination of the diffused particles, (4) the displacement current density is neglected, and (5) the species conservation equations are not considered.

Now the *Ohm's law*, with the ion-slip and electron pressure gradient neglected, is

$$\mathbf{j} = \sigma_o[\mathbf{E} + \mathbf{V} \times \mathbf{B} - \beta \mathbf{j} \times \mathbf{B}] \tag{11.83}$$

where $\beta = (en_e)^{-1}$.

From Equation (11.83),

$$\mathbf{E} = \frac{\mathbf{j}}{\sigma_o} - (\mathbf{V} \times \mathbf{B}) + \beta(\mathbf{j} \times \mathbf{B}) \ . \tag{11.84}$$

Combining this with (11.74),

$$\mathbf{E} = \frac{1}{\sigma_o \mu_o}(\nabla \times \mathbf{B}) - (\mathbf{V} \times \mathbf{B}) + \beta(\mathbf{j} \times \mathbf{B}) \ . \tag{11.85}$$

Taking a curl of this equation and substituting (11.73), we get the electro-magneto-gas-dynamic equation

$$\frac{\partial \mathbf{B}}{\partial t} + \frac{1}{\sigma_o \mu_o}\nabla \times (\nabla \times \mathbf{B}) - \nabla \times (\mathbf{V} \times \mathbf{B}) + \frac{\beta}{\mu_o}\nabla \times [(\nabla \times \mathbf{B}) \times \mathbf{B}] = 0 \ . \tag{11.86}$$

Equation (11.86), without the final Hall effect term, is called the induction equation, and it determines the change in the magnetic induction in space and time, as will be discussed.

It is sometime the practice to take the approximation that $\sigma_o \to \infty$, which also means that $\beta \to 0$. Under this assumption

$$\frac{\partial \mathbf{B}}{\partial t} = \nabla \times (\mathbf{V} \times \mathbf{B}) \ . \tag{11.87}$$

Equation (11.87) is identical with the hydrodynamic equation for the velocity (defined as the curl of the vector $\mathbf{V}$) of an ideal fluid (without friction), from which we learn that the vortices move together with the fluid. In an analogous manner it may be deduced that the magnetic induction should change as if its lines of force were rigidly tied to the plasma. Physically this means also that when the plasma moves in a magnetic field, the electric field $\mathbf{E} = \mathbf{j}/\sigma_o = \mathbf{V} \times \mathbf{B}$ is zero if $\sigma_o \to \infty$, and consequently the conducting fluid should not cross any magnetic lines of force; thus the magnetic induction is frozen into the plasma. On the other hand for stationary plasma ($\mathbf{V} = 0$) and neglecting the Hall current (due to the last term in (11.83), probably by segmenting the wall of the channel in which the plasma is flowing, (11.86) becomes

$$\frac{\partial \mathbf{B}}{\partial t} = \frac{1}{\mu_o \sigma_o}\nabla^2 \mathbf{B} \ . \tag{11.88}$$

This equation has the form of a diffusion equation and shows that the magnetic induction seeps through the plasma and is damped out in time of the order of $L^2\sigma_o\mu_o$. In general, under laboratory conditions, this time is extremely small.

While the expressions given in (11.86, 11.87) are of general nature, the condition for the pure magneto-gas-dynamic approximation is that the externally applied electric field $\mathbf{E}' = 0$, from which it follows that $\partial\mathbf{B}/\partial t = 0$. Thus from (11.83) and (11.87), and taking (11.76) into account, the magneto-gas-dynamic equation becomes

$$\frac{1}{\mu_o\sigma_o}\nabla\times\mathbf{B} = \mathbf{V}\times\mathbf{B} - \beta(\nabla\times\mathbf{B})\times\mathbf{B}$$

$$= \mathbf{V}\times\mathbf{B} - \beta[\nabla(\nabla\cdot\mathbf{B}) - \nabla^2\mathbf{B}] = \mathbf{V}\times\mathbf{B} + \beta\nabla^2\mathbf{B}\ . \quad (11.89)$$

Now we write down the following basic equations:

Conservation of mass:

$$\frac{\partial\rho}{\partial t} + \nabla\cdot(\rho\mathbf{V}) = 0 \quad (11.90)$$

Equation of motion, (11.28):

$$\rho\frac{D\mathbf{V}}{Dt} = -\nabla p + \nabla\cdot\tau + \mathbf{j}\times\mathbf{B} \quad (11.91)$$

Equation of Energy, (11.40):

$$\rho c_p\frac{DT}{Dt} - \frac{Dp}{Dt} - p\nabla\cdot\mathbf{V} = \nabla\cdot(k_{c,d}\nabla T) + \phi + \frac{j^2}{\sigma} - \epsilon_R \quad (11.92)$$

and *electromagneto-gas-dynamic equation*, (11.86) minus deleting the fourth term in the right hand side:

$$\frac{\partial\mathbf{B}}{\partial t} + \frac{1}{\sigma_o\mu_o}[\nabla\times(\nabla\times\mathbf{B})] - \nabla\times(\mathbf{V}\times\mathbf{B}) = 0\ . \quad (11.93)$$

We would discuss the second and the third term in the left hand side further. For this purpose, we use the two well-known expressions as follows:

$$\nabla\times(\nabla\times\mathbf{B}) = \nabla(\nabla\cdot\mathbf{B}) - \nabla^2\mathbf{B} \quad (11.94)$$

$$\nabla\times(\mathbf{V}\times\mathbf{B}) = (\mathbf{B}\cdot\nabla)\mathbf{V} - (\mathbf{V}\cdot\nabla)\mathbf{B} - \mathbf{B}(\nabla\cdot\mathbf{V}) + \mathbf{V}(\nabla\cdot\mathbf{B}) \quad (11.95)$$

In Cartesian coordinate system the above two expressions in x-direction can be written as follows:

$$[\nabla\times(\nabla\times\mathbf{B})]_x = \frac{\partial}{\partial x}\left(\frac{\partial B_x}{\partial x} + \frac{\partial B_y}{\partial y} + \frac{\partial B_z}{\partial z}\right) - \left(\frac{\partial^2 B_x}{\partial x^2} + \frac{\partial^2 B_x}{\partial y^2} + \frac{\partial^2 B_x}{\partial z^2}\right)$$

$$[\nabla\times(\mathbf{V}\times\mathbf{B})]_x = \left[B_x\frac{\partial u}{\partial x} + B_y\frac{\partial u}{\partial y} + B_z\frac{\partial u}{\partial z}\right] - \left[u\frac{\partial B_x}{\partial x} + v\frac{\partial B_x}{\partial y} + w\frac{\partial B_x}{\partial z}\right]$$

$$- B_x\left[\frac{\partial u}{\partial x} + \frac{\partial v}{\partial y} + \frac{\partial w}{\partial z}\right] + u\left(\frac{\partial B_x}{\partial x} + \frac{\partial B_y}{\partial y} + \frac{\partial B_z}{\partial z}\right)$$

According to Equation (11.76), $\nabla \cdot \mathbf{B} = 0$ and we can write (11.93) as

$$\frac{\partial \mathbf{B}}{\partial t} + (\mathbf{V} \cdot \nabla)\mathbf{B} + \mathbf{B}\nabla \cdot \mathbf{V} = \frac{1}{\mu_o \sigma_o} \nabla^2 \mathbf{B} + (\mathbf{B} \cdot \nabla)\mathbf{V} \qquad (11.96)$$

which can be written in cartesian coordinates for x coordinate as

$$\frac{\partial B_x}{\partial t} + \left[\frac{\partial(uB_x)}{\partial x} + \frac{\partial(vB_x)}{\partial y} + \frac{\partial(wB_x)}{\partial z} \right]$$
$$= \frac{1}{\mu_o \sigma_o} \left[\frac{\partial^2 B_x}{\partial x^2} + \frac{\partial^2 B_x}{\partial y^2} + \frac{\partial^2 B_x}{\partial z^2} \right] + \left[B_x \frac{\partial u}{\partial x} + B_y \frac{\partial u}{\partial y} + B_z \frac{\partial u}{\partial z} \right] . \quad (11.97)$$

For the additional condition of incompressible flow, $\nabla \cdot \mathbf{V} = 0$, (11.96) in x coordinate direction becomes

$$\frac{\partial \mathbf{B}}{\partial t} + (\mathbf{V} \cdot \nabla)\mathbf{B} = \frac{1}{\mu_o \sigma_o} \nabla^2 \mathbf{B} + (\mathbf{B} \cdot \nabla)\mathbf{V} \qquad (11.98)$$

which can be written in cartesian coordinates in x direction as

$$\frac{\partial B_x}{\partial t} + \left[u\frac{\partial B_x}{\partial x} + v\frac{\partial B_x}{\partial y} + w\frac{\partial B_x}{\partial z} \right]$$
$$= \frac{1}{\mu_o \sigma_o} \left[\frac{\partial^2 B_x}{\partial x^2} + \frac{\partial^2 B_x}{\partial y^2} + \frac{\partial^2 B_x}{\partial z^2} \right] + \left[B_x \frac{\partial u}{\partial x} + B_y \frac{\partial u}{\partial y} + B_z \frac{\partial u}{\partial z} \right] . \quad (11.99)$$

In order to understand the problem better, (11.90– 11.92, 11.98) are now non-dimensionalized by introducing non-dimensional variables (designated with superscript *):

$$x^* = x/L; t^* = Ut/L, \mathbf{V}^* = \mathbf{V}/U, \rho^* = \rho/\hat{\rho}, p^* = p/\hat{p}$$
$$T^* = T/\hat{T}, \mu^* = \mu/\hat{\mu}, k^* = k/\hat{k}, c_v^* = c_v/\hat{c}_v, \epsilon_R^* = \epsilon_R/\hat{\epsilon}_R$$
$$\sigma^* = \sigma_o/\hat{\sigma}_o, \beta^* = \beta/\hat{\beta}, \gamma^* = \gamma/\hat{\gamma}, \hat{p}/(\hat{\rho}\hat{c}_v\hat{T}) = (\hat{\gamma} - 1)$$
$$\mathbf{j}^* = \mu_o L\mathbf{j}/B_o, \chi^* = \tau/(\hat{\rho}U^2), \phi^* = \phi L/(\hat{p}U), \mathbf{B}^* = \mathbf{B}/B_o$$

where L, U and B_o are the characteristic length, the magnitude of the characteristic velocity and the magnitude of the characteristic magnetic induction, respectively. Further, the variables denoted by $(\hat{\,})$ are reference values at a known geometrical point, and $\gamma = $ *specific heat ratio*, $\mu_o = $ *magnetic permeability*, and $\mu = $ *viscosity coefficient*. From the non-dimensionalized equations the following non-dimensional numbers are obtained.

– *Reynolds number*: $\mathrm{Re} = \hat{\rho}UL/\hat{\mu}$

– *Mach number*: $\mathrm{M} = U/\sqrt{\hat{\gamma}(\hat{\gamma} - 1)\hat{c}_v T}$

– *Hall parameter*: $\mathrm{K} = \hat{\omega}/\hat{\Gamma} = \hat{\sigma}_o\hat{\beta}B_o$

- *Magnetic Reynolds number*: $R_\sigma = \mu_o \hat{\sigma}_o UL$

- *Prandtl number*: $\Pr = (\hat{\gamma}\hat{\mu}\hat{c}_v)/\hat{k}$

- *Radiation parameter*: $R_d = \hat{\epsilon}_R L/(\hat{\rho}\hat{c}_p \hat{T} U)$

- *Magnetic pressure parameter* (magntic force to dynamic pressure ratio): $R_H = B_o^2/(\mu_o \hat{\rho} U^2)$

- *Hartmann number*: $\mathrm{Rh} = \sqrt{\mathrm{Re}.R_\sigma.R_H} = B_o L\sqrt{\hat{\sigma}_o/\hat{\mu}}$

- *Joule heating parameter*: $\pi_1 = B_o^2/(\mu_o^2 L \hat{\sigma}_c \hat{\rho} U)$

- *Electric field parameter*: $\pi_2 = E/(UB)$

Accordingly, (11.90 – 11.93) in non-dimensional form ("asterisks" are dropped for clarity) become

Conservation of mass:

$$\frac{\partial \rho}{\partial t} + \nabla\cdot(\rho\mathbf{V}) = 0 \tag{11.100}$$

Equation of motion:

$$\rho\frac{D\mathbf{V}}{Dt} = -\frac{1}{\gamma M^2}\nabla p + \frac{1}{\mathrm{Re}}\nabla\cdot\chi + R_H\mathbf{j}\times\mathbf{B} \tag{11.101}$$

Equation of Energy:

$$\frac{\hat{\gamma}}{\hat{\gamma}-1}\gamma\rho c_v\frac{DT}{Dt} - \frac{Dp}{Dt} - p\nabla\cdot\mathbf{V}$$
$$= \frac{\hat{\gamma}}{\hat{\gamma}-1}\frac{1}{\mathrm{Re}.\mathrm{Pr}}\nabla\cdot(k_{c,d}\nabla T) + \phi + \pi_1\frac{j^2}{\sigma} - \frac{R_d}{\hat{\gamma}-1}\epsilon_R \tag{11.102}$$

and*Electromagnetic equation*:

$$\frac{\partial \mathbf{B}}{\partial t} + \frac{1}{R_\sigma}\nabla\times(\nabla\times\mathbf{B}) - \nabla\times(\mathbf{V}\times\mathbf{B}) = 0 \tag{11.103}$$

The above equations need to be solved under proper boundary conditions. Some of such applications of the basic equations have been discussed in the next chapter. However at this stage it is necessary to point out some of the special provisions one has to make while solving the equations numerically. For example during solution of (11.96) by the time-dependent method, the requirement of (11.76), that $\nabla\cdot\mathbf{B} = 0$ may be violated. Let us assume that a correction $\mathbf{B}'$ has to be applied to $\mathbf{B}$ to make $\nabla\cdot\mathbf{B} + \nabla\cdot\mathbf{B}' = 0$. Now let us put $\mathbf{B}' = -\nabla\varphi_M$, where φ_M is the *magnetic potential*. Therefore, we get the *Poisson equation* $\nabla^2\varphi_M = \nabla\cdot\mathbf{B}$, solution of which under appropriate boundary condition gives the correction $\mathbf{B}'$. Now the second example is solving Ohm's law, (11.83), in which $\mathbf{E}$ is the externally imposed electric field. By putting $\mathbf{E} = -\nabla\varphi$, the gradient of the externally applied potential, and taking divergence of both sides obviously, $\nabla^2\varphi = 0$, which is to be solved under proper boundary condition to get φ. However we must add $\mathbf{E}'$ to (11.83), which should be called an *induced electric field*, and write

$$\mathbf{j} = \sigma_o[\mathbf{E} + \mathbf{E}' + \mathbf{V} \times \mathbf{B} - \beta\mathbf{j} \times \mathbf{B}] \ . \tag{11.104}$$

Taking divergence of both sides, and noting that for quasi-neutral plasma $\nabla \cdot \mathbf{j} = 0$, we get

$$0 = \left[(\mathbf{E} + \mathbf{E}' + \mathbf{V} \times \mathbf{B} - \frac{1}{en_e}\mathbf{j} \times \mathbf{B})\nabla\right] \cdot \sigma_o + \sigma_o\nabla\cdot(\mathbf{E}' + \mathbf{V} \times \mathbf{B} - \beta\mathbf{j} \times \mathbf{B}) \ . \tag{11.105}$$

The first term in the right hand side will drop off in case of uniform electrical conductivity. Further by noting $\mathbf{E}' = -\nabla\varphi'$, we can get a Poisson equation of φ', which can be solved.

The boundary condition for a conducting wall is that either the potential is prescribed as constant for the particular wall, or is equal to zero, if the walls are short-circuited. On the other hand for a non-conducting wall, the electrical current density or gradient of electric potential normal to the wall are zero.

11.3 Wave Propagation

So far in this chapter we have discussed the plasma flow where the flowing medium is conducting. We would now discuss the situation for a dielectric medium, for which the electrical conductivity $\sigma = 0$ and consequently $\mathbf{j} = 0$. Under these conditions the Maxwell equations become

$$\mu\frac{\partial H}{\partial t} = -\nabla \times \mathbf{E} \tag{11.106}$$

$$\epsilon\frac{\partial E}{\partial t} = -\nabla \times \mathbf{H} \tag{11.107}$$

$$\nabla \cdot \mathbf{E} = 0 \tag{11.108}$$

$$\nabla \cdot \mathbf{H} = 0 \tag{11.109}$$

In cartesian coordinates these can be written as

$$\mu\frac{\partial}{\partial t}\begin{Bmatrix} H_x \\ H_y \\ H_z \end{Bmatrix} = \begin{Bmatrix} \frac{\partial E_y}{\partial z} - \frac{\partial E_z}{\partial y} \\ \frac{\partial E_z}{\partial x} - \frac{\partial E_x}{\partial z} \\ \frac{\partial E_x}{\partial y} - \frac{\partial E_y}{\partial x} \end{Bmatrix} \tag{11.110}$$

$$\epsilon \frac{\partial}{\partial t} \left\{ \begin{array}{c} E_x \\ E_y \\ E_z \end{array} \right\} = \left\{ \begin{array}{c} \frac{\partial H_y}{\partial z} - \frac{\partial H_z}{\partial y} \\ \frac{\partial H_z}{\partial x} - \frac{\partial H_x}{\partial z} \\ \frac{\partial H_x}{\partial y} - \frac{\partial H_y}{\partial x} \end{array} \right\} \tag{11.111}$$

$$\frac{\partial H_x}{\partial x} + \frac{\partial H_y}{\partial y} + \frac{\partial H_z}{\partial z} = 0 \tag{11.112}$$

$$\frac{\partial E_x}{\partial x} + \frac{\partial E_y}{\partial y} + \frac{\partial E_z}{\partial z} = 0 \tag{11.113}$$

Now taking the curl of (11.106) and taking (11.108) into account,

$$\mu \frac{\partial}{\partial t} (\nabla \times \mathbf{H}) = -\nabla \times (\nabla \times \mathbf{E}) = \nabla^2 \mathbf{E} - \nabla(\nabla \cdot \mathbf{E}) = \nabla^2 \mathbf{E} . \tag{11.114}$$

Considering (11.107) we get further

$$\mu\epsilon \frac{\partial^2 \mathbf{E}}{\partial t^2} = \nabla^2 \mathbf{E} . \tag{11.115}$$

Let us now put $c_o^2 = (\mu\epsilon)^{-1}$, which had the unit of $\mathrm{m^2 s^{-2}}$, and call c_o as the speed of propagation of a wave in a dielectric medium. Therefore the above equation is written as

$$\frac{1}{c_o^2} \frac{\partial^2 \mathbf{E}}{\partial t^2} = \nabla^2 \mathbf{E} . \tag{11.116}$$

Equation (11.116) is known as the *wave equation* for propagation of the electric field wave.

In a very similar manner we derive the wave equation of the magnetic field and write

$$\frac{1}{c_o^2} \frac{\partial^2 \mathbf{H}}{\partial t^2} = \nabla^2 \mathbf{H} . \tag{11.117}$$

Equations (11.116) and (11.117) can be written in cartesian coordinates as

$$\frac{1}{c_o^2} \frac{\partial^2 \mathbf{E}}{\partial t^2} = \frac{\partial^2 \mathbf{E}}{\partial x^2} + \frac{\partial^2 \mathbf{E}}{\partial y^2} + \frac{\partial^2 \mathbf{E}}{\partial z^2} \quad \text{and} \tag{11.118}$$

$$\frac{1}{c_o^2} \frac{\partial^2 \mathbf{H}}{\partial t^2} = \frac{\partial^2 \mathbf{H}}{\partial x^2} + \frac{\partial^2 \mathbf{H}}{\partial y^2} + \frac{\partial^2 \mathbf{H}}{\partial z^2} . \tag{11.119}$$

Let us now consider, for simplicity, a plane wave propagating in x-direction in a dielectric medium. Such waves are obtained if all the quantities, at any time, are constant over (y, z), including y and z coordinates extending to infinity. Under these conditions $\partial/\partial y = \partial/\partial z = 0$ and (11.110–11.113) reduce to

$$\mu \frac{\partial}{\partial t} \left\{ \begin{array}{c} H_x \\ H_y \\ H_z \end{array} \right\} = \left\{ \begin{array}{c} 0 \\ -\frac{\partial E_z}{\partial x} \\ -\frac{\partial E_y}{\partial x} \end{array} \right\} \tag{11.120}$$

$$\epsilon \frac{\partial}{\partial t} \left\{ \begin{array}{c} E_x \\ E_y \\ E_z \end{array} \right\} = \left\{ \begin{array}{c} 0 \\ -\frac{\partial H_z}{\partial x} \\ \frac{\partial H_y}{\partial x} \end{array} \right\} \tag{11.121}$$

$$\frac{\partial E_x}{\partial x} = 0 \tag{11.122}$$

$$\frac{\partial H_x}{\partial x} = 0 \tag{11.123}$$

Now we first write the wave equation for only $\mathbf{E}$, (11.116) and write

$$\frac{1}{c_o^2} \frac{\partial^2 E_y}{\partial t^2} - \frac{\partial^2 E_y}{\partial x^2} = 0 \tag{11.124}$$

$$\frac{1}{c_o^2} \frac{\partial^2 E_z}{\partial t^2} - \frac{\partial^2 E_z}{\partial x^2} = 0 \tag{11.125}$$

The two equations represent propagation of E_y and E_z waves in the x direction. For simplicity we assume further that the electromagnetic waves are so polarized that $\mathbf{E}$ is contained only within $\{x, y\}$ plane, that is $E_z = 0$; thus (11.125) need not be considered. The general solution of (11.124) is

$$E_y = f_E(x - c_o t) + g_E(x + c_o t) \tag{11.126}$$

where f provides wave propagation in the $+x$ direction and g provides in $-x$ direction. For the present we consider propagation only in the $+x$ direction and write

$$E_y = f_E(x - c_o t) \equiv A_E \exp^{i\omega(t - x/c_o)} \ . \tag{11.127}$$

Accompanying the E_y wave is a companion wave component of the magnetic field) for which we write from (11.117) for plane polarized wave as

$$\frac{1}{c_o^2} \frac{\partial^2 H_z}{\partial t^2} - \frac{\partial^2 H_z}{\partial x^2} = 0 \tag{11.128}$$

and the magnetic wave propagating in $+x$ direction is given by

$$H_z = f_H(x - c_o t) \equiv A_M \exp^{i\omega(t - x/c_o)} \ . \tag{11.129}$$

Both these waves propagate in the $+x$ direction in dielectric medium with undiminished amplitude.

In this connection we introduce a simple *refractive index, n,* defined as the ratio of the wave speed in vacuum or dielectric medium, c_o, to the wave speed $c = 1/\sqrt{\epsilon\mu}$ in an arbitrary medium including a non-dielectric medium ($\sigma \neq 0$). For such a medium the equivalent wave equation of (11.116, 11.117) are

$$\frac{1}{c_o^2} \frac{\partial^2 \mathbf{E}}{\partial t^2} = \nabla^2 \mathbf{E} - \sigma\mu \frac{\partial \mathbf{E}}{\partial t} \tag{11.130}$$

$$\frac{1}{c_o^2} \frac{\partial^2 \mathbf{H}}{\partial t^2} = \nabla^2 \mathbf{H} - \sigma\mu \frac{\partial \mathbf{H}}{\partial t} \tag{11.131}$$

It can be shown again for the polarized plane wave of E_y (and similarly for the companion wave of H_z), the governing equation is

$$\frac{1}{c_o^2}\frac{\partial^2 E_y}{\partial t^2} - \frac{\partial^2 E_y}{\partial x^2} = 0 \tag{11.132}$$

and (11.130) is satisfied if we consider the trial function

$$E_y \equiv A_E \exp^{i\omega(t-x/c_o)} \exp^{-\omega\kappa x/c_o} \tag{11.133}$$

where κ is called the *extinction coefficient* of the medium; the additional factor is an attenuation term indicating the absorption of the wave energy as it propagates inside the medium. For a dielectric medium obviously, $\kappa = 0$ and $n = 1$.

A comparison of (11.133) with (11.127) shows that a simple *refractive index* is replaced by *complex refractive index* $\bar{n} = n - i\kappa$. Substituting (11.133) into (11.132) and separating into the real and the imaginary parts we get the real and imaginary components of the *complex refractive index*([19]).

Now the energy carried per unit time and unit area (energy flux) by an electromagnetic wave is given by the *Pointing vector* $\mathbf{S} = \mathbf{E} \times \mathbf{H}$ [Wm^{-2}]. For a plane polarized wave the magnitude is $| \mathbf{S} | = E_y H_z$. Now since the interaction between the electric and magnetic waves in a plane polarized wave is given by

$$\mu\frac{\partial H_z}{\partial t} = -\frac{\partial E_y}{\partial x} \tag{11.134}$$

and for E_y we put the trial function (11.133). Thus we get

$$H_z = \frac{\bar{n}}{\mu c_o} E_y \tag{11.135}$$

and the magnitude of the *Pointing vector* is

$$| \mathbf{S} |= \frac{\bar{n}}{\mu c_o} E_y^2 \equiv \frac{\bar{n}}{\mu c_o} | E |^2 , \ \mathrm{W} \tag{11.136}$$

which decays in the direction of propagation with the factor $\exp^{-2\omega\kappa x/c_o}$. This may now be compared with the decay of thermal radiation factor in an emitting-absorbing medium discussed in Chap. 4 as $\exp^{-a_\nu x}$, where a_ν is the absorption coefficient of intensity of thermal radiation at frequency ν (or equivalent wave length λ) and we get

$$a_\nu = \frac{2\omega\kappa}{c_o} = \frac{4\pi\nu\kappa}{c_o}, \ \mathrm{m}^{-1} . \tag{11.137}$$

As a summary for this section, it is, therefore, concluded, that in an ionized gas (plasma) it is not possible to have a sustained electro-magnetic wave passing through it. The cases discussed in the next chapter will, therefore, show all without electromagnetic waves, although the wave like *Alfven wave* is possible.

12 Some Practical Examples

In this chapter we investigate now some of the plasma dynamic and magneto-gas-dynamic problems of practical interest. First we discuss the *"one-dimenional"* electromagneto-gas-dynamic nozzle and channel flow and the flow through a convergent-divergent nozzle of a rectangular shape, followed by a discussion on arc plasma flow in a tube, impinging plasma jet and particle-plasma interaction. Further, the interaction between the electromagnetic fields and cross-gas flow fields, magnetohydrodynamic power generation and flow interaction, and application of plasma in manufacturing and processing are discussed. Weakly ionized plasma, because of its aerospace application, is just emerging as a hot subject, without being fully understood at present; the final section on this has, therefore, a very tentative discussion on this.

12.1 Magneto-gas-dynamic Waves and Shocks

We investigate the interactions in perturbations of magnetic and velocity fields in a channel. This is the classical problem studied by *Alfven* ([34]). Let there be a flow in a rectangular channel with the flow velocity u in the x-direction, and a constant externally applied magnetic field H_o in the z-direction (Fig. 12.1). As such the problem is quite different from the electromagnetic wave propagation, discussed in the previous chapter; there is no externally applied electric field. However, there will be an induced electric field in the y-direction resulting in an electric current density $j_y = -\sigma_o \mu_o H_o u$, if the opposite walls are short-circuited. Because of this current flow, there is a non-zero value of the curl of the vector H in the y-direction. Noting H_x to be independent of x, this gives a variation of H_x in the z-coordinate direction, and due to an interaction between the magnetic field component H_x with the electric current density j_y, a force is exerted in the z-direction. The latter, for a fully developed flow, is responsible for u to be a function of z. Thus for the velocity and magnetic fields it is assumed that the velocity vector is $\mathbf{V} = \{u, 0, 0\}$ and all derivatives of the velocity field except $\partial u/\partial z$ are zero. Further assumptions are: (1) the displacement current is neglected, (2) the density variation is small, $\rho = $ constant, (3) the effect of the Hall current and ion-slip are neglected, and (4) the shear stress terms in momentum and

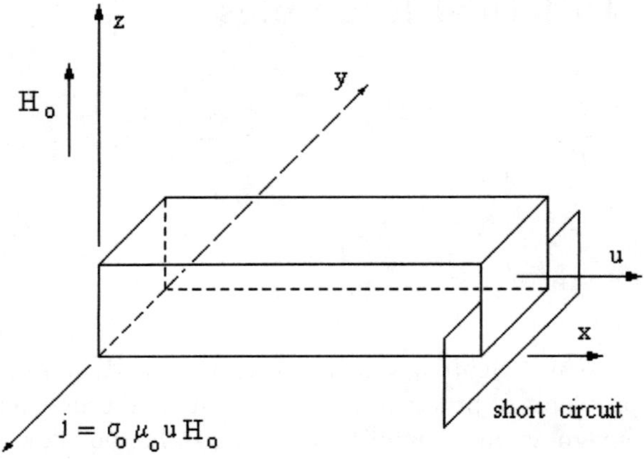

Fig. 12.1. Production of Alfven wave

energy equations are neglected. Under these circumstances the relevant basic equations are:

Maxwell equations:

$$\nabla \times \mathbf{E} = -\frac{1}{\mu_o}\frac{\partial \mathbf{H}}{\partial t} \tag{12.1}$$

$$\nabla \times \mathbf{H} = \mathbf{j} \tag{12.2}$$

$$\nabla \cdot \mathbf{H} = 0 \tag{12.3}$$

Ohm's law (neglecting the *Hall term*):

$$\mathbf{j} = \sigma_o(\mathbf{E}' + \mu_o \mathbf{V} \times \mathbf{H}) \tag{12.4}$$

Momentum equation:

$$\rho\frac{D\mathbf{V}}{Dt} = -\nabla p + \mu_o(\mathbf{j} \times \mathbf{H}) \tag{12.5}$$

From (12.4), an expression for $\mathbf{E}'$ is obtained, the curl of which, taking (12.2) into account, from (12.1) and after some rearrangement gives the electromagneto-gas-dynamic equation

$$\frac{\partial \mathbf{H}}{\partial t} = \nabla \times (\mathbf{V} \times \mathbf{H}) - \frac{1}{\mu_o \sigma_o}\nabla \times (\nabla \times \mathbf{H}) \ . \tag{12.6}$$

Further from (12.2) and (12.5)

$$\rho\frac{D\mathbf{V}}{Dt} = -\nabla p + \mu_o[(\nabla \times \mathbf{H}) \times \mathbf{H}] \ . \tag{12.7}$$

Now,

$$\nabla \times (\mathbf{V} \times \mathbf{H}) = (\mathbf{H} \cdot \nabla)\mathbf{V} - (\mathbf{V} \cdot \nabla)\mathbf{H} - \mathbf{H}(\nabla \cdot \mathbf{V}) + \mathbf{V}(\nabla \cdot \mathbf{H}) \tag{12.8}$$

and

$$\nabla \times (\nabla \times \mathbf{H}) = \nabla(\nabla \cdot \mathbf{H}) - \nabla^2 \mathbf{H} \ . \tag{12.9}$$

Since from (12.3), $\nabla \cdot \mathbf{H} = 0$, the fourth term in the right hand side of (12.8) and the first term in the right hand side of (12.9) can be neglected. In the cartesian coordinate system in x-direction, note the following expressions:

$$[\nabla \times (\mathbf{V} \times \mathbf{H})]_x = \left[H_x \frac{\partial u}{\partial x} + H_y \frac{\partial u}{\partial y} + H_z \frac{\partial u}{\partial z} \right]$$
$$- \left[u \frac{\partial H_x}{\partial x} + v \frac{\partial H_x}{\partial y} + w \frac{\partial H_x}{\partial z} \right]$$
$$- H_x \left[\frac{\partial u}{\partial x} + \frac{\partial v}{\partial y} + \frac{\partial w}{\partial z} \right] \tag{12.10}$$

$$[\nabla \times (\nabla \times \mathbf{H})]_x = - \left[\frac{\partial^2 H_x}{\partial x^2} + \frac{\partial^2 H_x}{\partial y^2} + \frac{\partial^2 H_x}{\partial z^2} \right] \tag{12.11}$$

where u, v, and w are velocity components in the three cartesian coordinates, $H_z = H_o$, $\partial u/\partial z \neq 0$ and $\partial H_x/\partial z \neq 0$. Thus from (12.6) and noting (12.8, 12.9), we get

$$\frac{\partial H_x}{\partial t} - H_o \frac{\partial u}{\partial z} = 0 \ . \tag{12.12}$$

Further for $\nabla p = 0$, (12.7) becomes

$$\rho \frac{\partial u}{\partial t} - \mu_o H_o \frac{\partial H_x}{\partial z} = 0 \ . \tag{12.13}$$

Now taking the derivative of (12.12) with respect to t, and that of (12.13) with respect to z, and subtracting one from the other we get

$$\frac{\partial^2 H_x}{\partial t^2} - c^2 \frac{\partial^2 H_x}{\partial z^2} = 0 \tag{12.14}$$

where $c = H_o \sqrt{\mu_o/\rho}$. Equation (12.14) is the same as the wave equation with c as the *wave speed*, which is named the "*Alfven speed*" after its discoverer. Similar to the definition of the gas dynamic Mach number an *Alfven Mach number* M_A is defined with the help of the following relation,

$$M_A = \frac{u}{H_o} \sqrt{\frac{\rho}{\mu_o}} \ . \tag{12.15}$$

A general solution of (12.15) is

$$H_x = f_1(z + ct) + f_2(z - ct) \ . \tag{12.16}$$

Let us consider simple harmonic type functions

$$H_x = A \exp^{i(z-ct)} \ \text{ and } \ u = C \exp^{i(z-ct)} \tag{12.17}$$

where $i = \sqrt{-1}$ is the imaginary unit number and A and C are constant amplitudes. By substituting above functions into (12.12) or (12.13), it is evident that $C = -cA/H_o$. Thus,

$$u = -\sqrt{\mu/\rho}\,A \exp^{i(z-ct)}$$

(12.18)

and from the two functions follows that

$$\frac{H_x}{u} = -\sqrt{\frac{\rho}{\mu_o}}\ .$$

(12.19)

Therefore, it may be concluded that the fluctuating magnetic field and the velocity both have components in the flow direction, and the waves of these propagate in the direction of the external magnetic field with these exactly out of phase from each other. This result has led to several investigations in which the magnetic field and the velocity fields interact in such a manner that only small perturbations occur which vanish at infinity. Further assumptions are: quasi-neutrality exists; the displacement current is neglected; the shear stress, heat conduction, dissipation function and the radiation terms are neglected; and the product of the perturbed quantities is neglected. It has been found that very interesting changes in the flow character occur when the magnetic and the velocity fields are not only both uniform at infinity, but are also parallel (or anti-parallel). Thus the total velocity vector $\mathbf{V}$ consists of the unperturbed velocity vector $\mathbf{U} = \{U, 0, 0\}$ and the perturbed velocity vector $\mathbf{V} = \{V'_x, V'_y, V'_z\}$; similarly the total magnetic field $\mathbf{H}$ consists of the unperturbed magnetic field $\mathbf{H}_o = \{H_o, 0, 0\}$ and the perturbed magnetic field $\mathbf{H}' = \{H'_x, H'_y, H'_z\}$. Now in a two-dimensional steady state case, there will be perturbation of the current density component in the z-direction, which will interact with the unperturbed velocity field to modify the magnetic field.

For the special case of the aligned field, $\mathbf{U}\mathbf{X}\mathbf{H}_o = 0$, *McCune* and *Resler* [82] derived the stream function equation

$$m^2\frac{\partial^2\psi}{\partial x^2} + \frac{\partial^2\psi}{\partial y^2} = 0$$

(12.20)

where ψ is the stream function defined in the usual manner in fluid mechanics and m^2 is given by the relation

$$m^2 = \frac{\beta^2(1-\alpha^2)}{(1-\beta^2\alpha^2)}$$

(12.21)

where

$$\beta^2 = 1 - M_\infty^2$$

(12.22)

and

$$\alpha = \frac{H_o}{U}\sqrt{\frac{\mu_o}{\rho}}\ .$$

(12.23)

In (12.22) M_∞ is the usual Mach number, that is the ratio of the flow velocity to the sonic speed of the approaching flow. Thus from (12.15) and (12.23) it can be seen that α^2 is an inverse of the Alfven Mach number, and α is proportional to the square root of the externally applied magnetic field. Therefore $\alpha = 0$, and consequently $m^2 = \beta^2$ is the case of the small perturbation theory for the classical two-dimensional gas dynamics. For this case it is well known that in the subsonic flow, $\beta^2 > 0$, and the flow is described by the elliptic partial differential equation, which, for arbitrary subsonic Mach numbers, allows Prandtl-Glauert coordinate transformations to relate the results in incompressible flows. In the supersonic flow, $\beta^2 < 0$, and the flow is described by the hyperbolic partial differential type of equation, in which the solution gives the flow dependent on $\arctan \sqrt{(-\beta^2)} - 1$. In principle a very similar situation exists for the magneto-gas-dynamic flows in aligned fields. It now depends very much on the sign of m^2; so long $m^2 > 0$, (12.20) is of an elliptic type and is amenable to *Prandtl-Glauert coordinate transformation,* whereas if $m^2 < 0$, then the flow variables depend on $\arctan \sqrt{(-m^2)} - 1$. These facts in different flow regimes and the types of the equation are shown in Table 12.1.

In Table 12.1 the Hyperbolic II region is analogous to the usual supersonic gas dynamic flow having Mach lines inclined rearwards to the flow direction; they are steeper (larger Mach angle) in comparison to those if there is no magnetic field. However, by keeping α^2 constant, an increase in Mach number results in a lesser steepening of the Mach lines. On the other hand, in the subsonic Hyperbolic I region and constant α^2, an increase in the Mach number causes steepening of the Mach lines. The implications of this was investigated by *McCune* and *Resler* in terms of the moving disturbances in acoustic coordinates, and they conclude that the correct Mach lines must be forward facing.

It is well known from gas dynamics that large disturbances with converging characteristics develop into shocks. Whereas in gas dynamics for normal shocks, the motion of the gas by definition is normal to the shock front, this is not necessarily so in magneto-gas-dynamics. In magnetogasdynamic flows the direction of propagation of the shock is perpendicular or parallel to the

Table 12.1. Different flow regimes in magneto-gas-dynamics

Flow Regime	β^2	α^2	m^2	Eq. Type
Incompressible $(0<M_\infty)$	1	-	1	Elliptic I
Subsonic $(0< M_\infty <1)$	$0< \beta^2 <1$	$0< \alpha^2 <1$	>0	Elliptic II
		$1< \alpha^2 < \beta^{-2}$	<0	Hyperbolic I
		$\beta^{-2} < \alpha^2 < \infty$	>0	Elliptic III
Supersonic $(M_\infty >1)$	$\beta^2 < 0$	$0< \alpha^2 <1$	>0	Elliptic IV
	$-$	$\alpha^2 >1$	<0	Hyperbolic II

magnetic induction lines. For such cases the solutions of the shock equation are obtained under the assumption of all motions parallel to the shock, which are called "*longitudinal shocks*". In addition there can be "*transverse shocks*" in which the motion of the material is along the surface of the discontinuity and the magnetic induction lines in general have different directions before and after the shock has passed.

For a normal longitudinal shock, if the gas velocity is u, the density is ρ, the pressure is p, the magnetic induction is B and the enthalpy is h, and the index 1 and 2 denote conditions before and after the shock, the conservation equations of mass, momentum and energy in magneto-gas-dynamic approximation are

$$\rho_1 u_1 = \rho_2 u_2 \tag{12.24}$$

$$\rho_1 u_1^2 + p_1 + \frac{B_1^2}{2\mu} = \rho_2 u_2^2 + p_2 + \frac{B_2^2}{2\mu} \tag{12.25}$$

$$h_1 + \frac{u_1^2}{2} + \frac{B_1^2}{\rho_1 \mu} = h_2 + \frac{u_2^2}{2} + \frac{B_2^2}{\rho_2 \mu} \tag{12.26}$$

where μ is the magnetic permittivity of the gas. Introducing the two auxiliary variables

$$p^* = p + \frac{B^2}{2\mu} \text{ and } h^* = h + \frac{B^2}{\rho \mu} \tag{12.27}$$

the conservation equations reduce to those for an ordinary gas dynamic shock. Thus the magneto-gas-dynamic equivalent of the *Rankine-Hugoniot equation*, (9.13), will be valid if p and h are replaced by the above p^* and h^*.

12.2 Arc Plasma Flow in a Tube

Model of a fully-developed arc plasma in a plasma tube was studied originally by *Elenbaas* [60] and *Heller* [68]; the details are given by *Maecker* [81]. Subsequently the subject of thermal non-equilibrium in such arc plasma tubes, which causes enhancement of plasma transport properties at moderate temperatures, was studied by various authors also. Motivation and justification for discussion of a sixty-year old model is because these results are of importance as boundary conditions for a number of plasma flow field calculations like impinging plasma jet, with electric current (transferred arc) or without, and plasma particles spraying. Since among the noble gases helium has the highest ionization potential and xenon the smallest, one would expect also that for a given arc tube radius, gas pressure and arc current, helium would have the highest arc center line temperature and xenon the lowest. However, as it turns out, the situation is more complex than just consideration of the ionization potential. Method of analysis presently is by numerical integration of coupled differential equations of mass, energy and current-conservation with appropriate boundary conditions, for which the required thermophysical and transport properties were computed with the help of the method given in Chap. 7. At least for the two-temperature argon plasma the temperature distribution results and the wall electron temperature are in reasonable agreement with experimental results from other authors.

We consider a fully-developed arc plasma flow in a plasma tube of diameter d (Fig. 12.2), in which an external electric field $\mathbf{E}$ is applied in the axial direction; fully-developed arc plasma being considered is to be found near the anode if the arc tube is quite long. The resultant electric current I is, therefore, in the axial direction also, whereas the temperature and the corresponding plasma thermophysical and transport properties change only in the radial direction for this fully-developed arc model. Only boundary condition, that is prescribed, is the wall temperature T_w, which for the single temperature model is equal to the gas temperature at the wall and for the two-temperature model is equal to the heavy particles temperature at the wall, T_{hw}. The common center line temperature of the electrons and the heavy particles is determined after integration of the energy equation.

For the single temperature model the energy equation can be written as

$$\frac{1}{r}\frac{\mathrm{d}}{\mathrm{d}r}\left(rk\frac{\mathrm{d}T}{\mathrm{d}r}\right) + \sigma E^2 - P_{rad} = 0 \tag{12.28}$$

where k is the thermal conductivity (including conduction and diffusion), T is the temperature, σ is the electrical conductivity, E is the electric field (in axial direction), P_{rad} is the emitted radiative power and r is the radial co-ordinate.

For the two-temperature model the energy equation for the electrons and the heavy particles (subscripts e and h, respectively) can be written as

$$\frac{\mathrm{d}}{\mathrm{d}r}\left(rk_e\frac{\mathrm{d}T_e}{\mathrm{d}r}\right) - r\left(3\frac{m_e}{m_h}k_B(T_e - T_h)\Gamma_{eh} - \sigma E^2 + P_{rad}\right) = 0 \tag{12.29}$$

$$\frac{\mathrm{d}}{\mathrm{d}r}\left(rk_h\frac{\mathrm{d}T_h}{\mathrm{d}r}\right) + 3\frac{m_e}{m_h}k_B(T_e - T_h)\Gamma_{eh}r = 0 \tag{12.30}$$

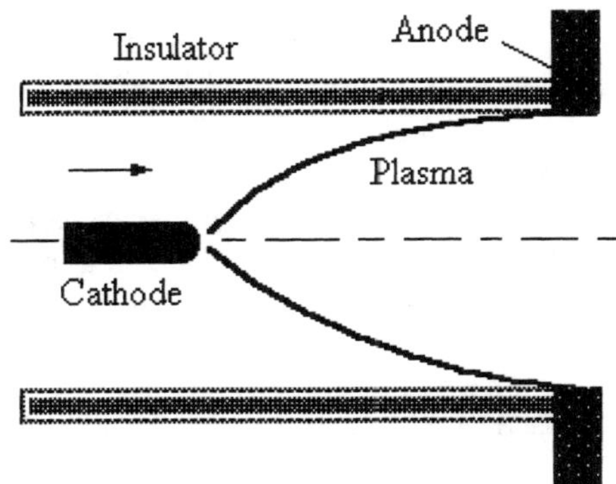

Fig. 12.2. Schematic of a plasma flow in a tube

Kruger [75] in his analysis has, in electron continuity, electron energy and global energy equations, terms corresponding to electron production. In the present case equilibrium composition at elevated electron temperature, if needed, has been assumed to eliminate the electron continuity equation, and in the energy equation, the effect of diffusion and recombination has been taken care of in computing the energy transport by ambipolar diffusion under equilibrium condition, and has been added to the pure thermal conductivity coefficient to obtain a total conductivity coefficient for electrons and heavy particles.

For laboratory plasmas radiation can be considered as *"optically thin"*, and its overall effect on the temperature profile may be assumed to be small. In any case, in absence of reliable data or expression for P_{rad} for all the five noble gases, this has been neglected.

For both the models the electric current conservation requirement gives the relation

$$I = 2\pi E \int_0^{d/2} \sigma r \mathrm{d}r \tag{12.31}$$

where I is the (prescribed) electric current. E is the (axial) electric field computed from (12.31) and is substituted in (12.28) or (12.29, 12.30), and d is the tube diameter.

The center line boundary condition requires that the center line temperature gradient is zero on the axis. For the two-temperature plasma with electrically insulated wall (at negative potential with respect to the plasma) the electron temperature gradient is determined from a sheath analysis, which has been discussed in some detail in Chap. 8. However from extensive numerical analysis it is shown that one could also assume that the gradient of electron temperature at the wall is approximately equal to zero. Therefore, for the two-temperature model, the wall electron temperature is much higher that the wall heavy particles temperature.

In actual calculation, first guess temperature profiles are used to compute k_e, k_h, σ and Γ_{eh}. Subsequently (12.31) is used to obtain the value of the electric field, which, in turn, is used for solution of Eqs. (12.28) or (12.29, 12.30) to determine the temperature profiles of the electrons and the heavy particles. Steady-state solution in temperature profiles is obtained in an iterative manner by using a relaxation factor.

In order to understand the interaction between various arc and gas parameters, we examine the case of a single temperature arc and in which we neglect the radiation term. We define the following non-dimensional variables, in which the superscript (*) refers to a non-dimensional variable, and the subscripts o and w refer to center line and wall values, respectively:

$$r^* = 2r/d; \sigma^* = \sigma/\sigma_o; k^* = k/k_o; T^* = (T - T_w)/(T_o - T_w)$$

$$\Omega_1 = \frac{I^2}{\pi^2 \sigma_o d^2 k_o (T_o - T_w)}; \Omega_2 = \frac{2I}{\pi E \sigma_o d^2}$$

By simple integration of (12.31) from $r^* = 0$ to 1, we can show, that

$$\Omega_2 = \int_0^1 \sigma^* r^* \, \mathrm{d}r^* \ . \tag{12.32}$$

Since σ^* can be between 0 (at wall) and 1 (center-line), Ω_2 can have values between 0 and 1/2; a smaller value shows that the electric current conducting core is concentrated more near the axis. Similarly, the value of Ω_1 can change only in the limited range, since a higher current I and smaller tube diameter causes increase in the center line temperature and decrease in the electric field. For an arc tube diameter of 1 cm, the gas pressure of 1 bar, the wall temperature of heavy particles, $T_{hw} = 1000$ K, and different arc currents, results of computation for different noble gases and two different temperature models were carried out. However, only for argon plasma at 1 bar and tube diameter 1 cm, but for different arc currents, the electron and heavy particles temperature are plotted in Fig. 12.3. Numerical results for electron temperature only for argon plasma are also plotted in Fig. 12.4, and are compared for argon plasma with experimental results (*Kruger* [75], *Bott* [48], *Giannaris* and *Incropera* [65]), wherever possible; the agreement is reasonable in view of the discussion that follows. *Kruger* [75] compared the temperature profile, measured from the free-free and free-bound continuum spectral distribution

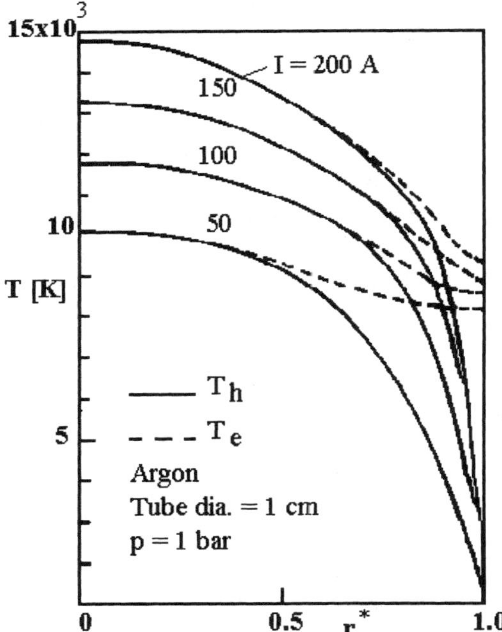

Fig. 12.3. Temperature distribution for two-temperature argon plasma for different arc currents

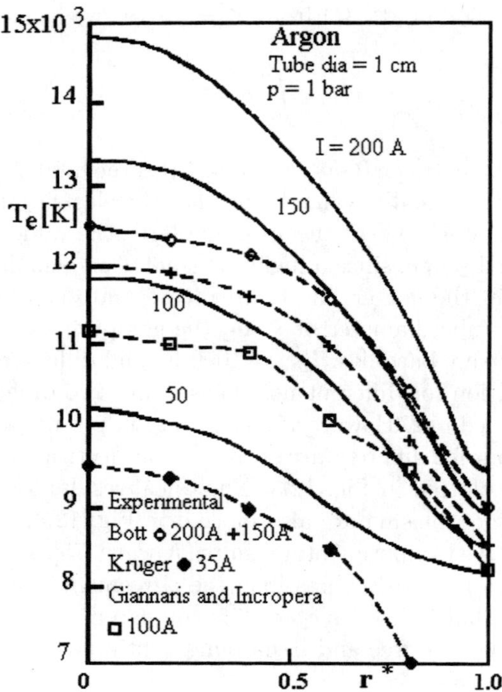

Fig. 12.4. Comparison of calculated and experimental electron temperature for two-temperature argon plasma

and the temperature deduced by calculating plasma composition from the single-temperature model for a one cm diameter arc at 1 bar, and two arc currents 150 and 200 Amps. The respective *"experimental"* wall temperature were 8500 and 9000 K, whereas from the present calculations the same are 8930 and 9430, respectively (Fig. 12.4). Further, *Bose* and *Seeniraj* [41] compared experimental results evaluated from spectroscopical measurements (*Giannaris* and *Incropera* [65]) but evaluated with two-temperature model for arc tube radius 1 cm and pressure 1 bar; for argon the results are given for 35 amps arc current and for helium the results are for 100, 160 and 210 amps. However the Helium experimental results are suspect since higher electric current is not shown to give higher center line temperature. For example, the respective center line temperature for helium deduced from experiments (*Giannaris* and *Incropera* [65]) at the three currents of 100, 160 and 210 amps are 16,000, 17,000 and 14,600 K. In addition the electron temperature profile maximum is not on the axis and the wall electron temperature for helium plasma is given for higher electric current. Further, Fig. 12.5 presents results of the present calculation of the center line temperature for the five noble gases. It shows that the center line temperature for helium, among the five noble gases, is maximum for all currents being considered.

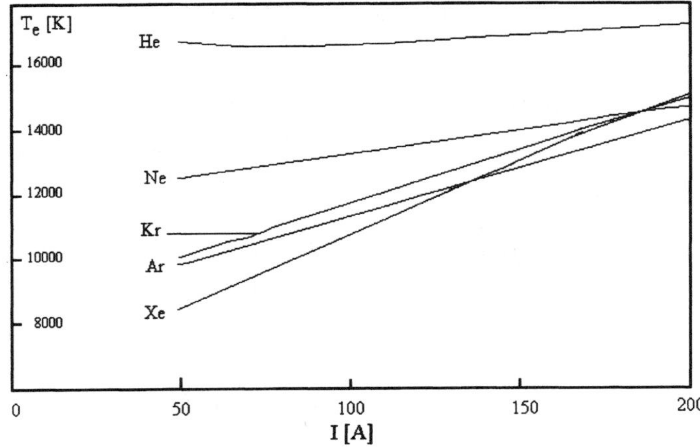

Fig. 12.5. Centerline temperature vs. arc current for two-temperature Elenbaas-Heller model for different noble gases

Experimental measurements and computed results are reported for plasma jets emanating into free atmosphere (*Dilawari*, et al. [59], *Chyou* and *Pfender* [57, 58]). Data were operated with the torch of 12.7 mm diameter in an argon environment operating at the power level between 4.8 and 15.3 kW, and the current level between 250 and 750 A. The mathematical formulation of the problem is by writing down the continuity, momentum and energy equations in axi-symmetric system without electro-magnetic fields outside the jet and adding a radiation loss term. Along the entrainment boundary, which is placed sufficiently far away from the axis of symmetry, the static pressure is assumed constant and the axial velocity zero. The entrainment velocity is computed by setting the radial gradient of the radial flux equal to zero and the entraining region is assigned an enthalpy value corresponding to a temperature 300K. The position of the entrainment boundary is obtained in such a way that moving further away from the axis produced no significant difference in the computed fields of velocities and temperature; the assumption of zero axial velocity at the entrainment boundary is a *"standard"* assumption for such calculations. Results of such calculations as a sample is taken from (*Chyou* and *Pfender* [57, 58]), and is shown in Fig. 12.6 where there is only qualitative agreement. Further such results of calculated temperature field are also available elsewhere and a detailed description of the solution procedure is given in (*Lee* [78]). In view of the results of calculation, carried out by this author for five noble gases – helium, neon, argon, krypton and xenon – the following conclusions can be made:

(a) For a given electric current the arc center line temperature is maximum for helium.

(b) The maximum arc center line temperature with single-temperature model

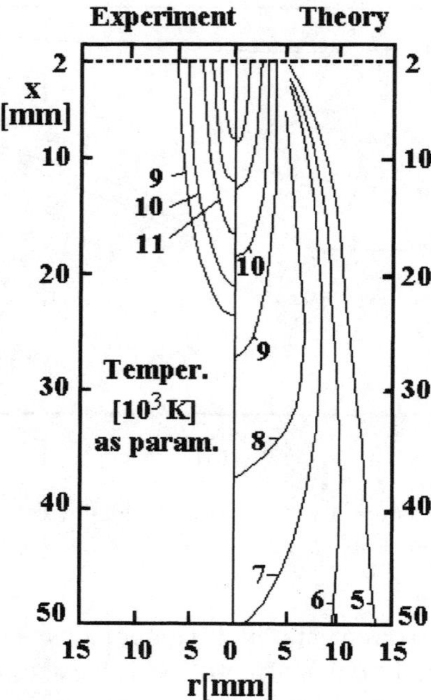

Fig. 12.6. Comparison of experimental and computational isotherms in a plasma jet

is higher than the center line electron temperature in a two-temperature model, which is again higher than the center line heavy particles temperature; the latter difference is maximum for helium;

(c) For a given arc current, the diameter of the electric current conducting core is minimum for helium and maximum for xenon; the core diameter is larger for the two-temperature model in comparison to the single temperature model, the reason being that the electrical conductivity distribution profile in a two-temperature plasma is flatter than in a single-temperature case;

(d) Diameter of electric current conducting core increases with increasing electric current; and

(e) The wall electron temperature at the wall determined with two-temperature model can have a quite high value; for neon with arc current 50 amps, it is even higher than the center line temperature of the heavy particles.

12.3 Impinging Plasma Jet

Now we discuss modelling of a plasma jet impinging on a flat plate (Fig. 12.7). Such a model is of considerable practical interest in plasma welding (with or without electric field outside the plasma tube), plasma spraying (by putting solid particles into the plasma by introducing it, hopefully uniformly at the exit plane of the plasma tube), etc. The present analysis is done without any electric field outside the plasma tube, and it is assumed that the plasma state (pressure, temperature, gas velocity) is known at the exit plane of the tube by the method described in the previous section.

The equations for an impinging plasma jet in cylindrical (x, r) co-ordinate system with thin shear layer approximation, where (u, v) are the respective velocity components, p is the pressure, T is the temperature and h is the specific enthalpy (per unit mass), are given with the help of a matrix equation as follows:

$$
\begin{bmatrix} \rho r \\ \rho u r \\ \rho v r \\ (\rho h - p)r \end{bmatrix}_t
+ \begin{bmatrix} \rho u r \\ (p + \rho u^2)r \\ (\rho u v - \mu v_x)r \\ (\rho u h - k T_x)r \end{bmatrix}_x
+ \begin{bmatrix} \rho v r \\ (\rho u v - \mu u_r)r \\ (p + \rho v^2)r \\ (\rho v h - k T_r)r \end{bmatrix}_r = 0
\qquad (12.33)
$$

The boundary conditions are:

$$
r = 0 : v = 0, \frac{\partial}{\partial r} = 0; x = 0 : u = v = 0, T = T_w
$$
$$
x = H, r < R_j : \text{ flow prescribed } ; x = H, r > R_j; v = 0 .
$$

The outer boundary condition at $r = R_b$ are obtained by extrapolation. In addition at $x = H$, $r > R_{tube}$, various alternate options like $\nabla u = 0$ have

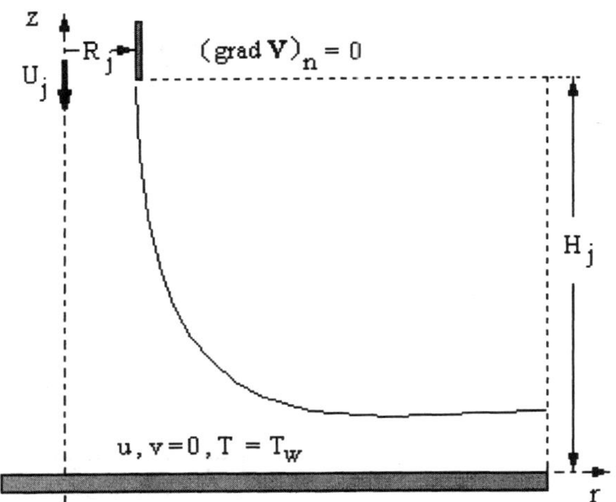

Fig. 12.7. Model of an impinging [lasma jet

been tried out. Since we do not consider any current flow outside the plasma tube, a single-temperature plasma model is sufficient for understanding of the problem. Equation (12.33) was solved by the *Finite Volume Method*, developed specifically for study of impinging plasma jets, and which was found to be numerically very stable.

Solution is obtained by prescribing initial prescription of fields of various variables, like pressure, temperature, density, velocity components, and carrying out numerical integration in time. Numerical calculation of the flow field was done for a large number of impinging flow data for argon and nitrogen plasma. However, only the sample results are presented for $I = 200$ A, tube height $H = 5$ cm, tube diameter $d_j = 1$ cm (corresponding to a non-dimensional height $H^* = 2H/d_j = 10$) and mass flow rate in the tube is about 0.5 g.s^{-1}. Figure 12.8 shows a comparative heat flux results (in kW.m^{-2}) vs non-dimensional radius ($r^* = 2r/d_j$) for nitrogen and argon. It can be seen that between argon and nitrogen plasma, the latter gives a somewhat higher heat flux to the wall than the former, but in both cases the maximum heat flux was on the centerline. The entire level of heat flux increases with increasing current. In these calculations the radiative heat flux has not been taken into account.

At the outset, we intend to indicate at this stage what happens if there is an externally applied electric field outside the plasma tube (transferred arc) (Fig. 12.9). In order, that the electric current flows between the exit plane of the plasma tube and the impinging plate, a two-temperature plasma model is needed near the (cold) plate and in the outer regions of the impinging plasma. Further, the heat flux to the wall depends not only on the gradient of temperature at the wall (convective heat flux), but also on the energy that

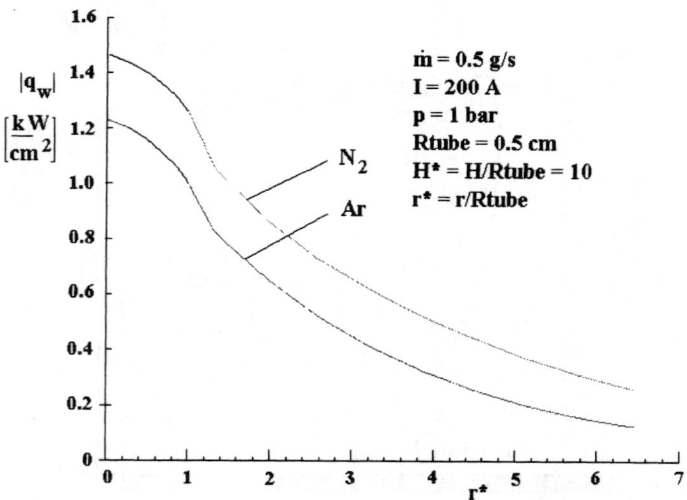

Fig. 12.8. Impinging jet heat flux for argon and nitrogen

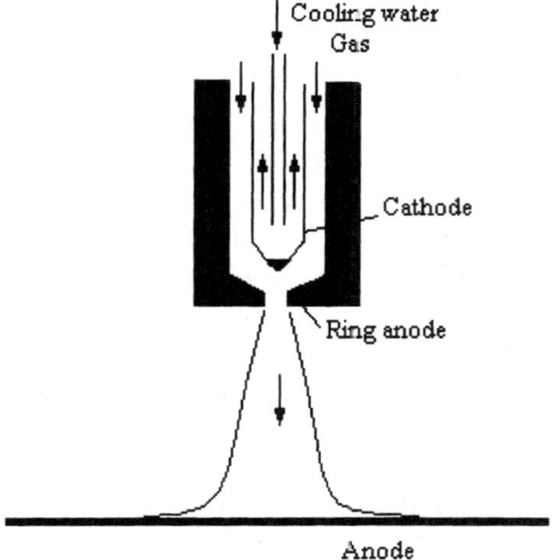

Fig. 12.9. Schematic diagram of a transferred arc

is carried by the charged particles to the wall and ionize or recombine there. If the plate is anode, the second energy transfer mechanism can be shown to be about an order of magnitude larger than due to pure convection.

12.4 Particle-Plasma Interaction

We would now discuss the problems in the application of thermal plasmas for physical and chemical processing of fine powders, particularly with reference to introduction of fine powders in an impinging plasma jet (*Bourdin*, et al. [50], *Lee*, et al. [78], 1985; *Gokhale* and *Bose* [67], and *Pfender* [101]). The high temperature of the plasma leads to rapid increase in the temperature of the particles, melting of the particle and part evaporation. For the purpose of good adhesion of the particles on to the impinging plate, it is not only necessary that the powders are, at the time of impingement in melted form, but also the impinging plate be cooled rapidly, so that the molten droplets stick to the plate. This leads not only to modelling of the mechanism of heating of the powder at and inside the plasma, but also to the heat transfer inside the plate. However, in spite of intensive efforts in recent years, our understanding of interaction of fine particles with thermal plasmas, and further with the flow, remains incomplete.

A particle injected into a thermal plasma will usually undergo the following processes: (1) heating of the particle in the solid phase from an initial temperature to the melting point of the material; (2) melting of the solid phase at a constant melting (or sublimation) temperature; (3) heating up to the evaporation point; and (4) evaporation of the liquid phase. Of course, for a solid material like graphite, only processes (1) and (4) are relevant. On the other hand, only processes (3) to (4) are relevant for liquid droplets into the plasma. Due to the final thermal conductivity of the particle during the heating process, temperature differences may develop inside the particle, although the assumption of an infinite thermal conductivity of the particle can considerably simplify the analysis.

For estimation of the role of internal conduction in a particle immersed in a hot plasma for the purpose of deciding about the difference between the particle surface temperature and the center temperature of the particle, the value of the so-called *Bio-number*, defined as the ratio of the convective to conductive heat transfer in a particle is an important criterion. If $Bi < 0.1$, internal conduction is relatively high and the temperature variation within a particle is negligible. Since the condition is generally met, the discussion in this paper is mainly restricted to particles with uniform temperature. A simple analysis is considered about the drag, heat and mass transfer for a single particle, and the relations for different characteristic times are given: for example, relaxation times for momentum and energy transfer, characteristic times to bring the particles to melting temperature, for melting, to bring to the evaporation temperature and for evaporation.

Let us consider now a gas-particle mixture, in which the particles are of spherical shape and let the gas (assumed to be an ideal gas) and solid particles properties be considered as given: For gas: density $\tilde{\rho}_g$, temperature T_g, pressure $p_g = \tilde{\rho} R_g T_g$, velocity u_g, gas mass flow rate $\dot{m}_g$, gas constant R_g, isobaric specific heat c_{pg}, volume flow rate $\dot{V}_g = \dot{m}_g/\tilde{\rho}_g$ and specific heat ratio γ.

For particles: (average) mass of a particle M_p, number density n_p, (average) diameter d_p, volume of a single particle $V_p^1 = (\pi/6)d_p^3$, particle volume flow rate $\dot{V}_p = \dot{m}_p/\tilde{\rho}_p$, elemental particle mass density $\tilde{\rho}_p = 6M_p/(\pi d_p^3) = (6\rho_p)/(\pi d_p^3 n_p)$, (average) mass density $\delta_p = n_p M_p$, particles mass flow rate $\dot{m}_p$, and specific heat c_p. For the gas-particle mixture, therefore, the following expressions can be derived: Total volume flow rate $= \dot{V}_p + \dot{V}_g = (\dot{m}_p/\tilde{\rho}_p) + (\dot{m}_g/\tilde{\rho}_g)$, volume fraction of particles $\varphi = (\dot{m}_p/\tilde{\rho}_p)/(\dot{m}_p/\tilde{\rho}_p + \dot{m}_g/\tilde{\rho}_g)$, mixture mass density $\rho = \rho_g + \rho_p = \tilde{\rho}_p\varphi + \tilde{\rho}_g(1 - \varphi)$, loading ratio $\dot{m}^* = \dot{m}_p/\dot{m}_g$ and mass fraction of particles $= \dot{m}_p/(\dot{m}_p + \dot{m}_g) = \tilde{\rho}_p\varphi/\rho$.

One can now show that even for $\tilde{\rho}_p/\tilde{\rho}_g = 10^3$ to 10^4, the volume fraction is between 2 to 0.2 percent for ρ_p/ρ_g as high as 0.95.

Solids can alter the flow in following ways: 1. penetration of solids through inner layer of the turbulent boundary layer causing thinning of the layer; 2. presence of solids may cause a damping of the convection eddies and a reduction of turbulent transport energy; 3. slip between particles and gas may enhance the turbulent mixing of the carrier gas; and 4. radial motion of particles enhancing transfer of heat. In general, solid particles of smallest size retain the slope of the thin initial crystal, whereas melted particles are of spherical shape. Now from fluid mechanics one can estimate the circulation (line integral of velocity component along the line in a closed path) for a rotating sphere or in a shear flow and the *"lift"* force on the particle. The phenomenon is important to understand the interaction between particles (solid or liquid) and gas in a flow situation.

12.4.1 Drag and Heat Transfer

Further, the underlying assumptions are: 1. the particle is at a uniform temperature (infinite heat conductivity); 2. the flow process is steady; and 3. the influence of vapor from evaporating or sublimating particles on the plasma properties is neglected. In addition, we consider first only the interaction of the plasma with a single particle with mass M_p, diameter $d_p = 2.r_p$, material mass density $\tilde{\rho}_p$, the respective particle and gas velocity, u_p and u_g, the respective specific heat of particle and gas c_p and c_{pg}, the respective specific heat conductivity coefficient of particle and gas, k_p and k_g, and the gas viscosity μ_g. At low Reynolds numbers, $\mathrm{Re} = \rho_g(u_g - u_p)d_p/\mu_g$, where ρ_g is the gas mass density, and all gas properties are to be evaluated at an average temperature between the gas and the particle. The drag coefficient c_{Do} in the Stokes region ($\mathrm{Re} < 1$) is given by the relation

$$c_{Do} = \frac{24}{\mathrm{Re}} \; . \tag{12.34}$$

It must, however, be noted, that a correction to the above relation for the drag coefficient can be found easily if we define a drag coefficient ratio, $f_D = c_D/c_{Do}$, and the equations for calculating f_D for various ranges of Reynolds number are as follows (*Gokhale* and Bose [67]):

$$\mathrm{Re} < 1000 : f_D = 1 + 0.15\mathrm{Re}^{0.687} \tag{12.35}$$

$$\mathrm{Re} > 1000 : f_D = 0.01833\mathrm{Re} \; . \tag{12.36}$$

Noting that the mass of a single particle is $M_p = \pi\tilde{\rho}_p d_p^3/6$, drag on a particle is $c_D\pi d_p^2\rho_g(u_g - u_p)^2/8$, and writing a force balance between the inertia and drag, we get

$$M_p\frac{\mathrm{d}u_p}{\mathrm{d}t} = \frac{\pi}{6}\tilde{\rho}_p \mathrm{d}_p^3\frac{\mathrm{d}u_p}{\mathrm{d}t} = c_{Do}f_D\pi d_p^2\rho_g(u_g - u_p)^2/8 \; . \tag{12.37}$$

By putting the relation between c_{Do} and Re, and in the event of a step change in the gas speed, we can define a velocity relaxation time $\tau_u = \tilde{\rho}_p d_p^2/(18\mu_g)$, and the above differential equation becomes

$$\frac{\mathrm{d}u_p}{\mathrm{d}t} = \frac{(u_g - u_p)}{\tau_u}f_D \; . \tag{12.38}$$

By integration we can show that for a step change of the gas speed u_g, the particle speed at any time will be

$$\frac{u_g - u_p}{(u_g - u_p)_{t=0}} = \exp^{-\int_0^\infty (f_D/\tau_u)\mathrm{d}t} \; . \tag{12.39}$$

Strictly speaking, the drag term in the momentum equation need be supplemented by *Basset history term* representing the time-dependent nature of the boundary layer around the particle (*El-Kaddah*, et al. [63], *Boulos* and *Gauvin* [49]). The history term is necessary, "since for plasma systems particle residence times are comparable with boundary layer relaxation times (approx. 1 ms)". *Boulos* and *Gauvin* [49] report that the history term should be retained for particle diameters greater than 30μ in a non-homogeneous velocity field in plasma. Unfortunately, insufficient information on this effect is only available and therefore, it is not being considered any further.

For the gas side heat transfer without any gas flow velocity, the energy conservation equation is

$$\dot{Q} = \dot{Q}_p = 4\pi r^2 k_g \frac{\mathrm{d}T}{\mathrm{d}r} \ , \ \mathrm{W} \ . \tag{12.40}$$

By introducing the heat conductivity potential defined as (*Chen* and *Pfender* [53])

$$\phi = \int_{T_o}^{T} k_g \mathrm{d}T \tag{12.41}$$

where $\phi = 0$ if $T = T_o$, (12.40) can be integrated from a large radial distance to the wall to get the heat flux relation on the wall

$$\dot{q} = \frac{\dot{Q}_p}{4\pi r_p^2} = \frac{\phi_g - \phi_{gp}}{r_p} \tag{12.42}$$

where ϕ_g is the gas (plasma) heat conductivity potential and ϕ_{gp} is the same at the particles surface temperature. The relationship for the *Nusselt number* relation (without flow) is given in literatures as

$$\mathrm{Nu} = \frac{\dot{q}d_p}{(\phi_g - \phi_{gp})} = 2 \ . \tag{12.43}$$

Under flow condition a correction to the *Nusselt number* $f_N = \mathrm{Nu}/\mathrm{Nu}_o$ is required, which is given by the relation valid for a quite large range of *Reynolds number*, as

$$f_N = 1 + 0.2295\mathrm{Re}^{0.55}\mathrm{Pr}^{0.33} \tag{12.44}$$

where $\mathrm{Pr} = \mu c_p/k$ is the Prandtl number of the gas. It may be noted that all the gas properties in Re and Pr are to be evaluated at an average temperature between the gas and the particle.

Now as a result of heat flowing from the gas to a particle, there will be a rise in the particle temperature, and from the energy balance we may write the equation

$$M_p c_p \frac{\mathrm{d}T_p}{\mathrm{d}t} = \pi d_p^2 \dot{q}_w = 2\pi d_p(\phi_g - \phi_{gp})f_N = 2\pi \bar{k}_g d_p(T_g - T_p)f_N \tag{12.45}$$

where $\bar{k}_g$ is the average heat conductivity at an average temperature of the gas and the particle,

$$\bar{k}_g = \frac{\phi_g - \phi_{gp}}{T_g - T_p}$$

(12.46)

and ϕ_{gp} is the gas heat conductivity potential at the particle temperature.

The particle temperature can be described by the relation

$$\frac{\mathrm{d}T_p}{\mathrm{d}t} = \frac{T_g - T_p}{\tau_T} f_N$$

(12.47)

where a particle temperature relaxation time (analogous to velocity relaxation time) is

$$\tau_T = \frac{3}{2} \mathrm{Pr}_g \tau_u \frac{c_g}{c_{gp}} \ .$$

(12.48)

The temperature can, therefore, be given (without phase change) by the relation

$$\frac{T_g - T_p}{(T_g - T_p)_{t=0}} = \exp^{-\int_c^t (f_N/\tau_T)dt} \ .$$

(12.49)

Note that τ_u and τ_T are directly proportional to d_p^2 and hence we can reduce these characteristic temperatures by dividing with d_p^2 to get reduced times $\tau_u^* = \tau_u/d_p^2$ and $\tau_T^* = \tau_T/d_p^2$. For an analysis of a typical particle immersed in argon plasma at 1 bar and 12,000 K with initial temperatures of 300 K and melting temperature around 1,800 K, the following data are taken:

$$\tilde{\rho}_p = 4,000 \mathrm{kgm}^{-3}, \quad \bar{c}_p = 500 \mathrm{Jkg}^{-1}\mathrm{K}^{-1},$$
$$\bar{c}_{pg} = 520 \mathrm{Jkg}^{-1}\mathrm{K}^{-1}, \quad \bar{\mu}_g = 2.2 \times 10^{-4}\mathrm{kgm}^{-1}\mathrm{s}^{-1},$$
$$\bar{k}_g = 0.11 \mathrm{Wm}^{-1}\mathrm{K}^{-1} \text{ and } \bar{\mathrm{Pr}}_g = 1.04$$

Hence the reduced relaxation time $\tau_u^* = 177$ Msm^{-2} and $\tau_T^* = 25.7$ Msm^{-2}. Further, considering a time-averaged mean value of $\bar{f}_N$ from the initial ($t = 0$) particle temperature to the particle melting temperature, T_M, we get

$$t_1 = \frac{\tau_T}{\bar{f}_N} \ln \left[\frac{(T_g - T_p)_{t=0}}{(T_g - T_M)} \right] \ .$$

(12.50)

Further, after reaching the melting temperature the time to melt a particle, t_2, is given by the relation

$$t_2 = \frac{M_p L_M}{\dot{Q}_p} = \frac{\tilde{\rho}_p d_p^2 L_M}{12(\phi_g - \phi_{gM})} = \tau_T \frac{\bar{k}_g}{\bar{c}_p} \frac{L_M}{(\phi_g - \phi_{gM})}$$

(12.51)

where ϕ_{gM} is the *heat conductivity potential* of the gas at the melting temperature of the particle and L_M is the *latent heat of melting of the particle* (*Chen* and *Pfender* [53, 54]). Further, the time required for the particle to teach the evaporation temperature of the particle, T_V, from its melting temperature, in analogy to (12.50), can now be written as

$$t_3 = \frac{\tau_T}{\bar{f}_N} \ln \left[\frac{(T_g - T_M)}{(T_g - T_V)} \right] \, .$$
(12.52)

Now let $\dot{M}_p$ be the *mass flow rate* $[\mathrm{kgs}^{-1}]$ of the particle material from each particle due to evaporation or sublimation. From the energy balance in the gas side, we write

$$4\pi r_p^2 \frac{k_g}{c_{pg}} \frac{\mathrm{d}h}{\mathrm{d}r} = \dot{M}_p (h - h_V + L_V)$$
(12.53)

where h_V is the enthalpy of the gas at evaporation (or sublimation) temperature and L_V is the *latent heat of evaporation*. By simple manipulation and integration we write

$$\int_{h_V}^{h} \frac{k_g}{c_{pg}} \frac{\mathrm{d}h}{(h - h_V + L_V)} = \frac{\dot{M}_p}{2\pi d_p}$$
(12.54)

from which we get for the particle mass evaporation rate from each particle as

$$\dot{M}_p = 2\pi d_p \int_{h_V}^{h} \frac{\mathrm{d}h}{(h - h_V + L_V)} \, , \, \mathrm{kgs}^{-1} \, .$$
(12.55)

Now the heating rate of a particle is

$$\dot{Q}_p = \dot{M}_p L_V$$
(12.56)

and the corresponding heat flux is

$$\dot{q}_p = \dot{Q}_p / (\pi d_p^2) = 2F L_V / d_p$$
(12.57)

where

$$F = \int_{h_V}^{h} \frac{k_g}{c_{pg}} \frac{\mathrm{d}h}{(h - h_V + L_V)} = F(h_g, h_p, L_V) \, .$$
(12.58)

While (12.57) gives the heat flux with evaporation or sublimation, (12.42) is the one without phase change. The ratio between the two is

$$\frac{\dot{q}_p}{\dot{q}} = \frac{F L_V}{\phi_g - \phi_{gp}} = F_1(h_g, h_p, L_V) \, .$$
(12.59)

Simplification of the above equation is possible, if (k_g/c_{pg}) can be replaced by their average value, and we get from (12.58)

$$F = \frac{\bar{k}_g}{\bar{c}_{pg}} \ln \left[1 + \frac{h_g - h_{gp}}{L_V} \right] \, .$$
(12.60)

We would now evaluate the evaporation time of the particle. For quasi-steady evaporation

$$\dot{M}_p = \tilde{\rho}_p \dot{V}_p = -4\pi \tilde{\rho}_p r_p^2 \dot{r}_p \tag{12.61}$$

and hence,

$$\dot{q}_p = \frac{FL_V}{r_p} = \frac{\dot{M}_p L_V}{4\pi r_p^2} = -\tilde{\rho}_p L_V \dot{r}_p \ . \tag{12.62}$$

Thus further,

$$r_p \dot{r}_p = \frac{1}{2}\frac{d}{dt}(r_p^2) = -\frac{F}{\tilde{\rho}_p} = -\frac{K}{2} \tag{12.63}$$

which on integration becomes

$$r_{po}^2 - r_p^2 = Kt \tag{12.64}$$

and hence the time required for a complete evaporation is

$$t_4 = \frac{r_{po}^2}{K} = \frac{\tilde{\rho}_p r_{po}^2}{2F} \ . \tag{12.65}$$

Evaporation constant K depends on the particle material, and the gas and the particle temperature. The typical value of the evaporation constant $[\mathrm{m^2 s^{-1}}]$ in argon and nitrogen plasma at 12,000 K is 4×10^{-8} and 1.5×10^{-7}, respectively.

An estimate of these times have been done by *Chen* and *Pfender* ([53, 54]) for argon plasma and different particle substances, for which the properties data given in Table 12.2 are used. Further, the following data of specific heats and thermal conductivity of solid particles were also used by them:

$$\text{Tungsten:} c_p = 125.46 + 4.632 \times 10^{13} T^{-2} \exp^{-38400/T}, \mathrm{Jkg^{-1}K^{-1}}$$
$$k_p = 178\mathrm{Wm^{-1}K^{-1}} \text{ at 300 K to 30 } \mathrm{Wm^{-1}K^{-1}} \text{ at } T_V$$
$$\text{Graphite:} c_p = 837 + 0.523(T - 300), \mathrm{Jkg^{-1}K^{-1}}$$
$$k_p = 192\mathrm{Wm^{-1}K^{-1}} \text{ at 300 K to 19 } \mathrm{Wm^{-1}K^{-1}} \text{ at 3,800 K}$$
$$\text{Alumina:} c_p = 1088 + 0.1151(T - 300), \mathrm{Jkg^{-1}K^{-1}}$$
$$k_p = 36\mathrm{Wm^{-1}K^{-1}} \text{ at 300 K to 6 } \mathrm{Wm^{-1}K^{-1}} \text{ at 2,000 K}$$

Using these properties data, *Chen* and *Pfender*([53, 54]) calculated times t_1 to t_4 for the above substances in argon plasma at various temperatures. These

Table 12.2. Different flow regimes in magneto-gas-dynamics

Properties	Water	Alumina	Graphite	Tungsten
$\tilde{\rho}_{p,solid}$, $\mathrm{kg^{-3}}$	–	4005	2145	19,350
T_M, K	273	2345	–	3653
L_M, $\mathrm{Jkg^{-1}}$	3.35e3	1.06e8	–	1.84e5
T_V, K	373	3800	4100	5950
L_V, $\mathrm{Jkg^{-1}}$	2.26e6	2.47e8	5.97e7	4.62e6

Table 12.3. Typical reduced times for selected particle substances

Substance	Reduced Time. Ms^{-2}			
	t_1^*	t_2^*	t_3^*	t_1^*
Alumina	1.00	0.40	1.20	20.00
Tungsten	1.20	0.30	1.40	22.00
Graphite	1.05	–	–	20.00

times are all proportional to the square of the particle diameter and hence reduced times t_1^* to t_4^* are obtained by dividing t_1 to t_4 by d_p^2 to obtain values as given in Table 12.3.

Importance of these characteristic times are apparent when compared with typical flow times, that should occur for an application. For plasma spraying applications, the particles may be introduced at the exit of the jet, where the gas jet speed may be order of several hundred meters per second, and downstream lengths may be several tens of centimeter per second. Taking a jet speed of 200 ms^{-1} and length of the exhaust of about 0.2 m, the characteristic flow time (residence time) available for the particles is 1 ms, which for an average particle diameter of 100μ gives a reduced flow time of 0.1 Msm^{-2}. On the other hand for a particle diameter of 10μ the reduced flow time is 10 Msm^{-2}. In comparison, the characteristic reduced relaxation times are $\tau_u^* = 2.22$ Msm^{-2} and $\tau_T^* = 3.33$ Msm^{-2}, respectively. These can be compared with the typical reduced time data given in Table 12.3, from which it can be seen that only for particles of diameter of a few microns the reduced flow time is small enough to have complete evaporation. On the other hand for spray of particles it is necessary to have a comparatively large solid particles. As pointed out by *Chen* and *Pfender* ([53, 54]), "it is interesting to notice that the evaporation constant of a water droplet is of the same order of magnitude as that of a fuel droplet in the combustion chamber of a jet engine; but the evaporation constant for W, C and Al_2O_3 are one order of magnitude lower. Therefore, the residence time for particles ($d_p = 100$ μ) of such materials in a plasma reactor, if complete evaporation is desired, should be in the range from 50 to 100 ms (milli-second)".

Vaporization and evaporation, although related to each other, are dependent on the surface temperature with respect to the boiling temperature. According to *Etemadi* and *Mostaghimi* [8], while "vaporization is defined as a mass transfer process driven by concentration gradients existing between the free stream and the particle surface, ... evaporation accounts for large amounts of mass transfer as the surface temperature reaches the boiling point. For the former, mass transport due to diffusion need be considered".

12.4.2 Internal Conduction

The above analysis was done under the assumption of infinite heat conductivity of the particle, velocity of which will now be considered. For this purpose, the *Biot number*, Bi, defined as the ratio of the convective to conductive heat transfer, serves as a criterion for determining the relative importance of heat conduction within a particle. Because of similarity in definition between Bi and Nu, we can evaluate the one from the other from the relation

$$\mathrm{Bi} = \mathrm{Nu}(\bar{k}_g/k_p) = 2(\bar{k}_g/k_p)f_N \ . \tag{12.66}$$

As such, under a typical plasma condition, Bi has the value around 0.01, with the result, that the non-uniformity in the particle temperature may be neglected (*Pfender* [101], *Bourdin*, et al. [50]).

12.4.3 Low Pressure Effects

The mean free path lengths of the plasma constituents in thermal plasmas are in the order of microns at atmospheric pressure *Pfender* [101]. Particles used in thermal plasma synthesis or plasma spraying are almost of the same order of magnitude (1–10 μ). Therefore the continuum approach is no longer valid and modifications become necessary. For this purpose the Knudsen number (ratio of mean free path to particle diameter) is defined; the higher *Knudsen numbers* are reached at reduced gas pressures or for smaller particle diameters (*Rizk* and *Elghobashi* [102]), and there is a strong influence on heat transfer with a temperature jump on the particle surface. Further, a correction term for the drag coefficient is proposed as follows *Pfender*([101]):

$$c_{D,slip} = c_{D,cont.} \left[\cfrac{1}{1 + \left(\cfrac{2-a}{a}\right)\left(\cfrac{\gamma}{\gamma+1}\right)\cfrac{4}{\mathrm{Pr}_w}\mathrm{Kn}^*} \right]^{0.45} \tag{12.67}$$

where a is the thermal accommodation coefficient, γ is the specific heat ratio, Pr_w is the *Prandtl number* of the gas at the surface temperature of the sphere and Kn^* is the *Knudsen number* based on an effective mean free path length.

12.4.4 Particle Charging Effect

A particle injected into a thermal plasma is always negatively charged due to different thermal velocities and mobilities of electrons and ions by retarding electron bombardment and increasing ion-flux.

12.4.5 Fluctuating Velocity and Temperature

While we have analyzed so far the interaction of the particles in a laminar flow, we consider now the situation that may arise in a turbulent flow. For this purpose we simplify the turbulence model by considering a fluctuation in the flow properties of a simple harmonic nature. For uniform particle and gas density, the gas and particle momentum equations are:

$$\rho_g \frac{du_g}{dt} = -\rho_p \frac{(u_g - u_p)}{\tau_u} \tag{12.68}$$

$$\rho_p \frac{du_p}{dt} = \rho_p \frac{(u_g - u_p)}{\tau_u} \tag{12.69}$$

where the particle density $\rho_p = n_p M_p$ and n_p is the particle number density.

Let $u_g = A_g \exp^{i\omega t}$ and $u_p = A_p \exp^{i(\omega t - \phi)}$, where A_g and A_p are amplitude of gas and particle fluctuation, respectively, ω is the radian frequency of harmonic oscillation of the gas and the particles, $i = \sqrt{-1}$ is the unit imaginary number and ϕ is the phase shift of oscillation between the particle and the gas. Substituting these into the particle momentum equation and after some manipulation we get

$$A_p = A_g \frac{1 - i\omega\tau_u}{\sqrt{1 + \omega^2\tau_u^2}} \exp^{i\phi} \ . \tag{12.70}$$

The middle term in the right hand side of the above equation can be represented in a complex plane, in which the real component is 1 and the imaginary component is $-\omega\tau_u$. This is equivalent of an angle θ in the complex plane, so that,

$$\exp^{-i\theta} = \frac{1 - i\omega\tau_u}{\sqrt{1 + \omega^2\tau_u^2}} \tag{12.71}$$

and thus further, $A_p = A_g \exp^{i(\phi - \theta)}$. Assuming the phase shift $\phi = \theta = \arctan(\omega\tau_u)$, we get $A_p = A_g/\sqrt{1 + \omega^2\tau_u^2}$ and the following limiting conditions:

$$\omega\tau_u = 0 : \phi = 0 \text{ and } A_p = A_g, \omega\tau_u \to \infty : \phi = \pi/2 \text{ and } A_p = 0. \tag{12.72}$$

While the first limiting case is for small particles and the particles follow the gas completely without any phase shift, the second limiting case is for a very heavy particle, where there is a phase shift of 90° between the particles and the gas and the amplitude of the particle tends to zero.

In case the turbulence velocity in a fluid can be described by the amplitude and the frequency of any one velocity component, then the average turbulence kinetic energy of the particle and the gas is equal to the half of the square of the amplitude. Hence the ratio of *kinetic turbulence energy* of the gas, k_g, to that of the particles, k_p, is given by the relation

$$\frac{k_g}{k_p} = \frac{A_p^2}{A_g^2} = \frac{1}{1 + \omega^2\tau^2} \ . \tag{12.73}$$

The above expression gives, therefore, a simple method by which by using two turbulent kinetic energy equations for the particles and the gas, the typical value of the radian frequency can be obtained; there are similar relations given in literatures (*Soo* [20], *Rizk* and *Elgobashi* [102]).

$$\frac{k_g}{k_p} = \frac{1 + \gamma^2 + \sqrt{6}\gamma^{3/2} + 3\gamma + \sqrt{6\gamma}}{1 + (\gamma/\beta)^2 + \sqrt{6}\gamma^{3/2} + 3\gamma + \sqrt{6\gamma}} \qquad (12.74)$$

where $\alpha = 3\mu_g/(\rho_g d_p^2)$, $\beta = 3\rho_g/(\rho_g + \rho_p)$ and $\gamma = \beta/\alpha$. Once again one can see that,

$$d_p \to 0, \alpha \to \infty, \gamma \to 0 : k_p/k_g \to 1 \text{ and}$$
$$d_p \to \infty, \alpha \to 0, \gamma \to \infty : k_p/k_g \to \beta^2 \to 0 \text{ if } \rho_g \ll \rho_p. \ .$$

Thus once again, a small particle will follow the turbulence motion of the gas, while for large particles the ratio of the turbulent kinetic energy of the particles to that for the gas is dependent on the mass fraction of the gas.

12.5 A Transverse Blown Arc

Interaction between an arc and a cross-flowing gas is known since the beginning of invention of the arc. In fact the name *"arc"* was given to the discharge struck between two horizontal electrodes which bent due to the free-convection of the hot gas in upwards direction. There are several hundred studies, both analytical and experimental, to predict the maximum arc deflection of such an arc. This author ([38, 40]) also studied such interactions between the gas and the electromagnetic fields which is being described in this section.

For this purpose a model of an arc is considered, which is of circular shape of diameter D in which a given electric current I is flowing. Let, along the arc D be varying, and a characteristic diameter D_o be chosen to normalize the arc diameter. Further, for simplificaticns of the analysis, the following assumptions are made: (1) the gas properties inside the arc are constant, which are assumed to be the same as at the maximum temperature at the center line of the arc. Thus it is assumed that inside the arc the electrical conductivity $\sigma = \sigma_o$ and the mobility coefficient of the electrons $b_e = b_{eo}$, but outside the arc $\sigma = 0$ and $b_e = 0$; (2) the mass velocities of all particles due to the electromagnetic forces are independent of those due to the collisional forces; (3) the terms due to the electron partial pressure and the ion slip in the momentum equation are neglected; and (4) the electric current is mainly carried by the electrons (the electron current density $\mathbf{j}_e$ = overall current density $\mathbf{j}$).

Thus, one may write the following set of equations:

$$\text{electron conduction current: } \mathbf{j}_{ce} = -en_e\mathbf{V} + \mathbf{j}_e \qquad (12.75)$$

$$\text{Ohm's law: } \mathbf{j}_e = \sigma[\mathbf{E}' + \mathbf{V}\times\mathbf{B} - \frac{1}{en_e}\mathbf{j}_e\times\mathbf{B}] \qquad (12.76)$$

Because of the assumption (3), it follows that, inside the arc, $\mathbf{j}_e$ is uniform in the arc cross-section, which can be determined easily from the equation

$$\mathbf{I} = \frac{\pi D^2}{4}\mathbf{j}_e \;. \qquad (12.77)$$

Let us introduce the following dimensionless variables and vectors:

$$\mathbf{k} = \mathbf{I}/I, \mathbf{k}_c = \pi D^2\mathbf{j}_{ce}/(4I), \mathbf{E}^* = \pi D_o^2\sigma\mathbf{E}'/(4I)$$
$$\mathbf{V}^* = \mathbf{V}/V_o, \mathbf{B}^* = \mathbf{B}/B_o, D^* = D/D_o$$

where B_o is the characteristic magnetic induction, V_o is the characteristic cross-flow gas velocity and D_o is the characteristic maximum arc diameter. It may be noted that $\mathbf{k}$ is a unit vector, which does not change the magnitude but the direction. Further, the following two new non-dimensional parameters are defined:

$$\textit{blowing parameter} : B_p = \frac{en_eV_o\pi D_o^2}{4I} \qquad (12.78)$$

$$\textit{magnetic blowing parameter} : M_p = \frac{V_oB_o\pi D_o^2\sigma_o}{4I} \qquad (12.79)$$

In these two relations B_p gives the ratio of the gas velocity to the velocity of the electrons in the arc, and M_p gives the ratio of the velocity of the electrons due to the magnetic field interaction to the velocity of the electrons in the arc. Thus (12.75, 12.76) become

$$\mathbf{k}_c = -B_p\mathbf{V}^*D^{*2} + \mathbf{k} \qquad (12.80)$$

$$\mathbf{k} = D^{*2}\mathbf{E}^* + M_pD^{*2}\mathbf{V}^*\times\mathbf{B}^* - \frac{M_p}{B_p}\mathbf{k}\times\mathbf{B}^* \;. \qquad (12.81)$$

Now let the magnetic induction $\mathbf{B}^*$ be in the z-direction only. The electric field $\mathbf{E}^*$ is in the plane parallel to the gas-crossflow field and the electrodes; the gas-crossflow is in the y-direction (Fig. 12.10). At any point P on the arc, the different terms in (12.80, 12.81) are shown in Fig. 12.11. For many cases we may assume a uniform velocity field ($\mathbf{V}^* = 1$), uniform magnetic induction ($\mathbf{B}^* = 1$) and uniform arc diameter ($D^* = 1$). Further, as per the definition, $\mathbf{k} = 1$. It is now possible, from geometric considerations and under the assumptions just detailed, to write down the following set of equations:

$$\cos\psi = k_c\cos\theta = E^*\cos\phi - \frac{M_p}{B_p}\sin\psi + M_p \qquad (12.82)$$

$$\sin\psi = k_c\sin\theta + B_p = E^*\sin\phi + \frac{M_p}{B_p}\cos\psi \qquad (12.83)$$

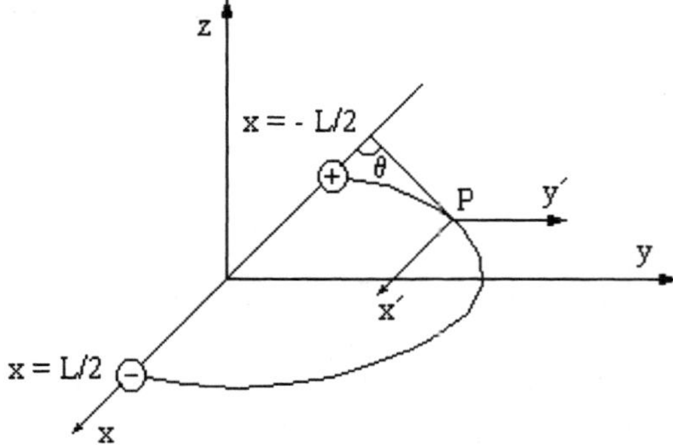

Fig. 12.10. Gas flow and electromagnetic fields

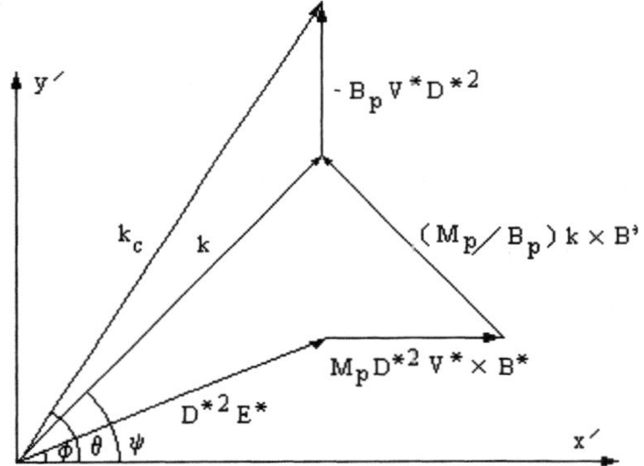

Fig. 12.11. Velocity diagram for interaction between gas flow and electromagnetic fields

It may be noted that the angles θ, ϕ and ψ refer to the angles, which are made on the x-axis by $\mathbf{k}_c$, $\mathbf{E}^*$ and $\mathbf{k}$, respectively. Whereas $\mathbf{k}$ is the direction in which the electrons should move without the effect of the flowing gas, $\mathbf{k}_c$ is the direction of movement of the electrons due to the interaction of the gas-crossflow and the electromagnetic fields. From equations (12.82, 12.83), one gets the equation

$$\tan\psi = \tan\theta + \frac{B_p}{\cos\psi} \text{ and} \tag{12.84}$$

$$\tan\phi = \frac{\sin\psi - (M_p/B_p)\cos\psi}{\cos\psi - (M_p/B_p)\sin\psi + M_p}. \tag{12.85}$$

Now the above relations are applied to two different cases. The first case is that of an arc, balanced gas dynamically, between two point electrodes. Since there is no externally applied magnetic field for this case, $M_p = 0$, and thus $\phi = \psi$. In the second case the arc is struck between two parallel electrodes, and is magnetically balanced, which means that $\phi = 0$ and $\psi = \tan^{-1}(M_p/B_p)$. Thus the shape of the arc are given in the respective two cases by the relations,

$$\tan\theta = \tan\phi - \frac{B_p}{\cos\phi} \quad \text{and} \tag{12.86}$$

$$\tan\theta + \frac{1}{\sqrt{1+\tan^2\theta}} = \frac{M_p}{B_p} \ . \tag{12.87}$$

From the latter equation it is, therefore, concluded that for the parallel rail electrodes θ is negative, so long as the electric, magnetic and gas flow field directions are as given in Fig. 12.11. For the gas dynamically balanced arc between the two point electrodes and without an externally applied magnetic field ($M_p = 0$), the electric field direction is given, for two point electrodes, by the relation

$$\tan\phi = -\frac{4xy}{2(y^2 - x^2 + 1)} \tag{12.88}$$

where x and y are non-dimensionalized with the semi-width between the two electrodes. Now the method of evaluation is as follows: For every point in the (x, y) plane, the slope of the electric field lines, ϕ, are obtained from (12.88). Further for a given B_p, the arc slope is obtained from (12.86). These slopes may be plotted on a graph sheet and a smooth curve (without wrinkles) is drawn between the two electrodes to obtain the arc shape. The maximum deflection, given in Fig. 12.12, was obtained in this manner and showed good agreement with experimental results.

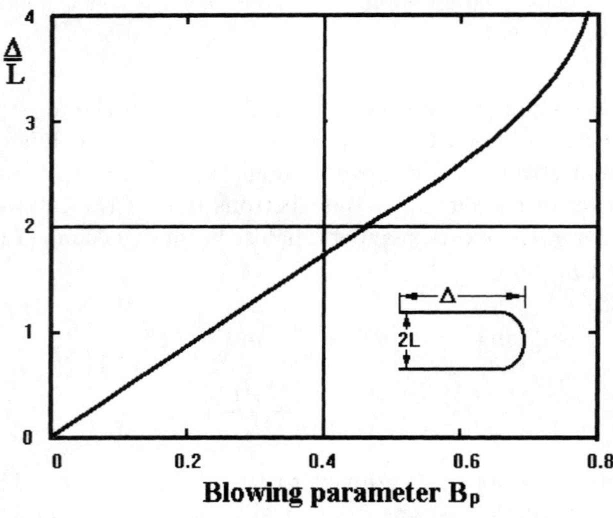

Fig. 12.12. Theoretical maximum deflection in blown arc between point electrodes

For blown arcs, there is the question, whether the so-called *Steenbeck's minimization principle* is violated. According to this principle, for a given electrode geometry and electric current without any transverse gas blowing , the arc current follows the *path of the least resistance* (the shortest distance), that is the potential difference between the two electrodes becomes minimum. Since the heat produced due to passing of the electric current through a small volume $\mathrm{d}\Omega = \pi D^2 \mathrm{d}s$ is $\mathbf{j} \cdot \mathbf{E}'\mathrm{d}s$, it is possible to define an integral

$$Q = \int (\mathbf{j} \cdot \mathbf{E}')\mathrm{d}s \tag{12.89}$$

where the arc line element $\mathrm{d}s$ is non-dimensionalized with semi-width between the electrodes. This equation, for constant arc diameter (non-dimensionalized $D^* = 1$), is now examined. For the first case with $M_p = 0$ and $\mathbf{k} = \mathbf{E} = 1$. For an arc without cross-flow blowing the minimization requirement for the heat (minimum entropy production in unit time) should lead to a straight arc. For the blown arc the same minimization of energy production (minimum of entropy production) allows determination of the arc shape, although, for various values of B_p, and as per (12.89) the voltage difference between the electrodes increases with increasing blowing parameter. This is observed experimentally also.

12.6 Magneto-gas-dynamic Flow Inside Ducts

We consider now a duct flow in x-direction in a channel of arbitrary cross-section placed in a permanent, uniform, transverse (in y-direction) magnetic induction B, as shown in Fig. 12.13. This has a number of applications in MGD channel flows and two-dimensional applications like Hartmann flow and MHD Couette flow, and they were studied analytically by many people (*Pai* [18], *Moreau* [16]). Here there will be an induced electric field in the z-direction, causing an electric current to flow in the z-direction also if the

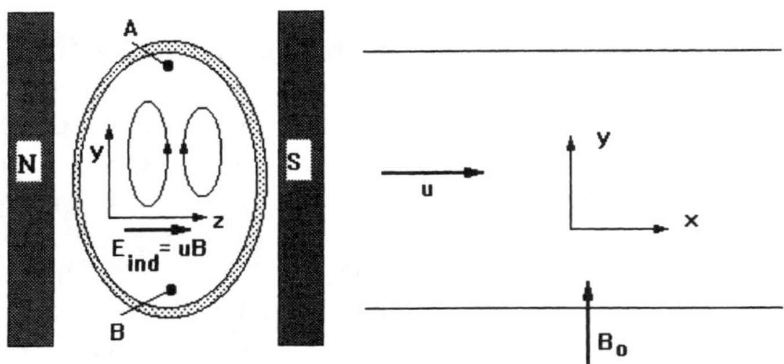

Fig. 12.13. Duct flow in presence of transverse magnetic induction

channel flow had been a fully two-dimensional flow extending infinitely in x and y-directions. The induced electric field has only one non-zero component in z-direction, which has the value uB, but it is distributed in a non-uniform way in the cross-section since both u and temperature are functions of (y, z). If the magnet and the duct are very long, one can assume that the current density in z-direction remains in the plane of cross-section (y, z). Now the conservation of charge requires that $\nabla \cdot \mathbf{j} = 0$ and the lines of the electric current be closed surfaces; if the wall is conducting and the gas is also sufficiently conducting near the wall, part of the current closes up in the wall, or else the current loop will close within the fluid only. The only way to reconcile this is to accept that the positive charges are concentrated in the "A" region and the negative charges in the "B" region, until the electric field created by the distribution of the surface charge (in an electrically insulated surface) globally balances the electromotive force. In the (y, z) plane, therefore, the electric field E_z appears so that the total current in any segment in (x, y) plane is zero. Thus we may write

$$\int_{y_{\min}}^{y_{\max}} j_z \mathrm{d}y = \int_{y_{\min}}^{y_{\max}} \sigma(-E_z + uB_y)\mathrm{d}y = 0 \ . \tag{12.90}$$

The electric potential will be maximum around "A" and minimum around "B". In the central region where u is strongest and also for plasmas it is the region of maximum electrical conductivity, the electromotive field dominates and j is directed from "B" to "A". Thus the electrical circuit is made up of two opposing "coils" which induce B_x component of the magnetic field. Identifying various components of the current density in expressions drawn separately from Ampere and Ohm's law, we write

$$j_x = -\frac{1}{\mu_o}\frac{\partial B_z}{\partial x} = 0; \tag{12.91}$$

$$j_y = -\frac{1}{\mu_o}\frac{\partial B_x}{\partial x} = \sigma E_y; \ \text{and} \tag{12.92}$$

$$j_z = -\frac{1}{\mu_o}\frac{\partial B_x}{\partial y} = \sigma(E_z + uB_y) \ . \tag{12.93}$$

which shows immediately that B_y component is uniform and is equal to B_o (because $\nabla \cdot \mathbf{B} = 0$).

We discuss now several typical magnetohydrodynamic cases for which analytical solutions are available, at least when the properties like density, viscosity coefficient and electrical conductivity are constant. Among the simplest cases amenable to analytical solutions are *Hartmann flow* and *MHD Couette flow*, both being *"fully developed"* (is valid far away from inlet and magnetic field extending infinitely both in x- and z-direction) and an externally applied magnetic induction B_o is in the direction normal to the parallel plates (y-direction). The flow velocity in x-direction is $u = u(y)$ and there is an induced electric field and an associated electric current in z-direction, as a result of which there is an electro-magnetic volumetric force in the direction opposing the flow and there is an induced magnetic induction gradient $\partial B_x/\partial y$. We write the equations in non-dimensional form as follows. We non-dimensionalize all lengths by dividing with L (distance between the plates is $2L$), all velocities by dividing with U (the characteristic velocity), all properties like density, viscosity coefficient and electrical conductivity by dividing with characteristic respective values $\hat{\rho}$, $\hat{\mu}$ and $\hat{\sigma}$, pressure by dividing with ρU^2 and magnetic induction by dividing with B_o. The non-dimensionalized equations are now

$$x - \text{momentum} : \frac{1}{\text{Re}}\frac{\mathrm{d}}{\mathrm{d}y}\left(\mu\frac{\mathrm{d}u}{\mathrm{d}y}\right) + \text{R}_H\frac{\mathrm{d}B_x}{\mathrm{d}y} = \frac{1}{\rho}\frac{\partial p}{\partial x} \tag{12.94}$$

$$y - \text{momentum} : \text{R}_H B_x \frac{\mathrm{d}B_x}{\mathrm{d}y} = -\frac{1}{\rho}\frac{\partial p}{\partial y} \tag{12.95}$$

$$x - \text{induction} : \frac{\mathrm{d}u}{\mathrm{d}y} + \frac{1}{\text{R}_\sigma}\frac{\partial^2 B_x}{\partial y^2} = 0 \tag{12.96}$$

where $\text{Re} = \hat{\rho}UL/\hat{\mu}$ is the Reynolds number, $\text{R}_\sigma = \mu_o\hat{\sigma}UL$ is the magnetic Reynolds number, $\text{R}_H = B_o^2/(\mu_o L\hat{\rho}_o U^2)$ is the ratio of magnetic force to dynamic pressure, and μ_o is the magnetic permeability (kindly note the difference of the latter with the non-dimensional viscosity μ). Now by making derivative of (12.94) with respect to y and derivative of (12.95) with respect to x, and taking (12.96) into account, we obtain a third order differential equation of u, which may be solved under appropriate boundary condition. However, we would do further analysis for *constant properties*, for which analytical expressions are available, and we get the following expression:

$$\frac{\mathrm{d}^3 u}{\mathrm{d}y^3} - \text{Rh}^2\frac{\mathrm{d}u}{\mathrm{d}y} = 0 \ . \tag{12.97}$$

Herein $\text{Rh} = \sqrt{\text{Re}.\text{R}_H.\text{R}_\sigma}$ is the *Hartmann number* and (12.97) has the general solution (*Pai* [18], *Moreau* [16])

$$u = \frac{1}{\text{Rh}}[A\cosh(\text{Rh}y) + B\sinh(\text{Rh}y) + C] \tag{12.98}$$

and the special solution (between $y = -1$ and $y = +1$) for flow between stationary parallel plates (*Hartmann flow*) is:

$$u = \frac{\cosh(\text{Rh}) - \mathrm{ccsh}(\text{Rh}y)}{\cosh(\text{Rh}) - 1} \tag{12.99}$$

and for flow between plates moving in opposite directions (*MHD Couette flow*) is:

$$u = \frac{\sinh(\text{Rh}y)}{\sinh(\text{Rh})} \ . \tag{12.100}$$

For the magnetic induction the distribution of B_x depends on whether the walls are electrically conducting or insulated; the wall magnetic induction of electrically insulated (non-conducting) wall is $B_x = 0$ and for conducting wall, $\partial B_x/\partial y = 0$. Finally the relation for the pressure distribution is obtained in closed form. For insulated wall, the results are:

$$B_x = \mathrm{R}_\sigma \left[\frac{y+1}{2} \int_{-1}^{1} u dy - \int_{-1}^{1} u dy \right] \tag{12.101}$$

$$p = -\frac{1}{2}\mathrm{R}_H B_x^2 + A_p x + B_p \tag{12.102}$$

In the latter equation, the constant B_p is the pressure at $x = 0$ and where $B_x = 0$, while $(A_p.\mathrm{Re})$ is obtained from (12.94).

12.7 MGD Power Generation or Gas Acceleration

Since 1960s, there has been proposals for magneto-gas-dynamic power generation cycles to top the conventional cycles. The basic idea arises out of the thermodynamic principle that improvement in the thermodynamic efficiency in a process depends on the ratio of the highest to the lowest temperature in the cycle. In this section therefore, we deal with the thermodynamic aspects of electric power generation and the irreversibilities occurring for the seeded combustion plasmas in magneto-gas-dynamic power generators by considering the enthalpy-entropy chart and the electrical conductivity of the plasma. For countries with large coal reserve such power generators operating with seeded combustion products of the coal or various coal gasification products are of interest. For operational reasons these gasification products of coal, like the water gas, the Lurgi gas, the Kopper-Totzek gas and the producer gas are to be preferred against direct combustion of coal, to which various alkali seeds are added. For comparison purpose these are compared with methane as the fuel gas. As oxidizer, in addition to the natural air, which gives a comparatively low adiabatic flame temperature, also the oxygen enrichment is considered. Among the gasification products of coal, the water gas, burning with atleast fifty percent oxygen enriched air and one to three percent seed to fuel gas mass ratio seems to give adequate electrical conductivity for further study of the magneto-gas-dynamic process.

There are basically two types of magneto-gas-dynamic generators. In both of these two types the induced voltage (due to the integrated electric field through the gas) is the driving force for the electric current through the external electric circuit, and hence the external electric current direction is in the opposite direction of that in the gas. In a Faraday type magneto-gas-dynamic channel flow, as it is well known, a hot electrical conducting gas is allowed through a rectangular channel, in which a magnetic field is applied in one direction and an induced electric field is generated in the

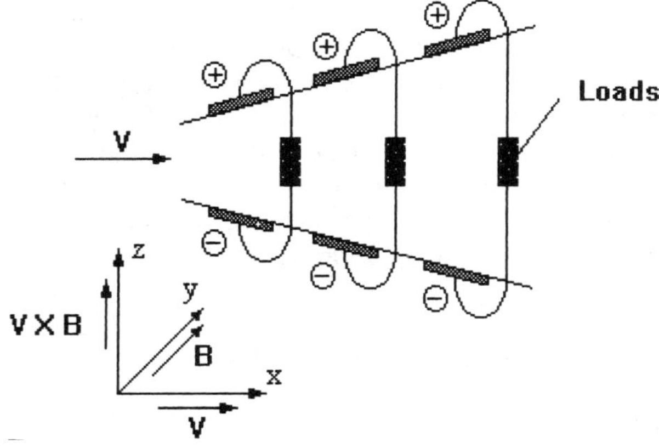

Fig. 12.14. Schematic sketch of a Faraday generator

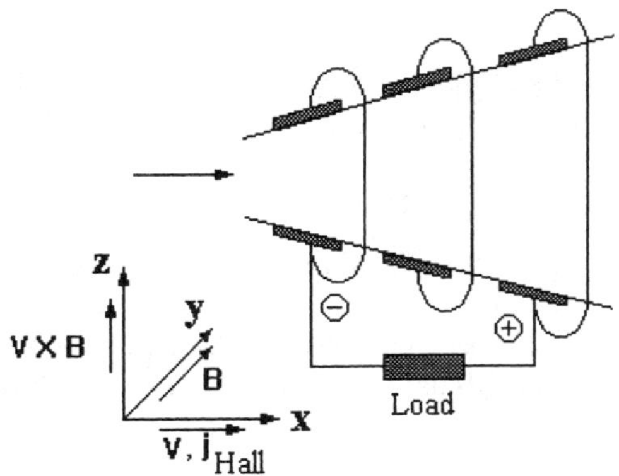

Fig. 12.15. Schematic sketch of a Hall generator

direction perpendicular to both the gas flow and the magnetic field directions
(Fig. 12.14). In the Hall type generator the opposite electrodes are short-
circuited to enable a current flow, which interacts with the magnetic field to
induce an electric field in the flow direction (Fig. 12.15). Thus in Hall type
of generator the electric power can be extracted from the potential difference
between the first and the last electrode.

The extracted electric power per unit length of the channel in a *Faraday
generator* is dependent on the gas flow velocity, externally applied magnetic
induction and the electrical conductivity of the flowing gas. Generation of
high gas velocity in the magneto-gas-dynamic channel is of no consequence,
if during the expansion process to generate high gas velocity the gas is cooled
and the electrical conductivity is reduced beyond an acceptable limit. For this
purpose, we consider the one-dimensional channel flow process, as shown in

Fig. 12.14 for Faraday generators, in which the gas flows in x direction with velocity $\mathbf{V} = U$, and the externally applied magnetic induction B is in the y-direction. Thus an electric field UB is induced in the z- direction, and for short-circuited electrodes a current density $j = \sigma UB$ flows in the z-direction (the electrons move in the $-z$ direction). In case, however, the electrodes are not connected (open circuit), an induced electric field E is produced in the $-z$ direction to retard the electron flow. Thus if the current is allowed to flow over a load, which is symbolically represented in Fig. 12.14 by a resistance, the electric current density is given by the relation

$$j = \sigma(UB_E) = \sigma UB(1 - K) \tag{12.103}$$

where the *load factor*

$$K = \frac{E}{UB} \ . \tag{12.104}$$

is the ratio of the actual electric field to the maximum induced electric field, and it is less than or equal to one. Note that $K = 0$ is for short-circuited electrodes and $K = 1$ is if there is an open-circuit situation. The induced voltage is the line integral inside the gas of $UB(1 - K)$ and the electric power extracted from the gas per unit volume of the gas through the external circuit is then

$$P = \mathbf{j} \cdot \mathbf{E} = \sigma U^2 B^2 K(1 - K) \tag{12.105}$$

which shows that the maximum extraction of the electric power is when $K = 1/2$. An electromagnetodynamic volume force in x direction, $F = \mathbf{j} \times \mathbf{B} = \sigma UB^2 K(1 - K)$, will oppose the flow. Further, the state change in an one-dimensional magneto-gas-dynamic channel flow is obtained from the equations of momentum and energy as (*Bose* [44]),

$$\rho\frac{\mathrm{d}U}{\mathrm{d}z} = -\frac{\mathrm{d}p}{\mathrm{d}z} - \sigma UB^2(1 - K); \tag{12.106}$$

$$\rho U\frac{\mathrm{d}h}{\mathrm{d}z} = U\frac{\mathrm{d}p}{\mathrm{d}z} + \sigma U^2 B^2(1 - K)^2 \text{ and} \tag{12.107}$$

$$\rho U\frac{\mathrm{d}h^o}{\mathrm{d}z} = -\sigma U^2 B^2 K(1 - K) = -P \ . \tag{12.108}$$

Note that the *stagnation enthalpy* h^o is related to the *static enthalpy* h and the *gas velocity* U by the relation

$$h^o = h + \frac{U^2}{2} \tag{12.109}$$

and, therefore, we get (12.108) by multiplying (12.106) with U and add the resultant equation with (12.107). Further from (12.107) and the *laws of thermodynamics* we write down a relation for entropy s by

$$T\frac{\mathrm{d}s}{\mathrm{d}z} = \frac{\mathrm{d}h}{\mathrm{d}z} - \frac{1}{\rho}\frac{\mathrm{d}p}{\mathrm{d}z} = \sigma UB^2(1 - K)^2/\rho \ . \tag{12.110}$$

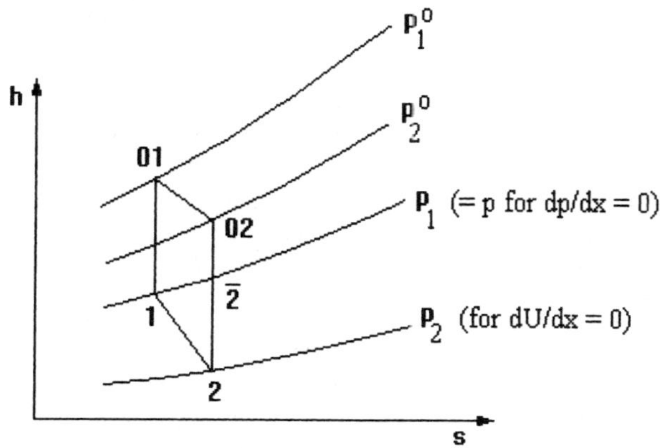

Fig. 12.16. Schematic Sketch of a Mollier (h, s) chart for Faraday generator

From Equations (12.108, 12.110), we can, therefore, write

$$\frac{\mathrm{d}s}{\mathrm{d}h^o} = \frac{K-1}{KT} \tag{12.111}$$

which is dependent on K; it is 0 if externally open-circuited ($K = 1$) and is $-1/T$, if $K = 1/2$. This slope of the entropy change with respect to the total enthalpy change for $K = 1/2$ is now plotted schematically into the enthalpy-entropy chart (Fig. 12.16) for proper understanding of the direction in which the magneto-gas-dynamic power generation process takes place. It can also be shown from (12.108, 12.111), that for both the extraction of the electric power per unit channel volume and the reduction of the relative entropy change with respect to the total enthalpy change (and less loss due to irreversibility in the thermodynamic sense) the temperature must be kept as high as possible. An estimate of the change in the thermodynamic properties can now be done on the basis of the above equations. For an open-circuit situation, $K = 1$ and $\mathrm{d}s = 0$, which is evident, since no current flows. On the other hand for short-circuit, $K = 0$ and $\mathrm{d}h^o = 0$, and $\mathrm{d}s/\mathrm{d}z$ can be calculated from (12.110). This entropy change is always positive, as expected.

Further to the general statement made above regarding the entropy change, the two special cases are now considered, namely U remains constant along the channel and p remains constant. For the first case, that is, $\mathrm{d}U/\mathrm{d}z = 0$,

$$\frac{\mathrm{d}p}{\mathrm{d}z} = -\sigma U B^2 (1 - K) \tag{12.112}$$

and

$$\rho U \frac{\mathrm{d}h}{\mathrm{d}z} = -\sigma U^2 B^2 K (1 - K) = \rho \frac{\mathrm{d}h^o}{\mathrm{d}z} \tag{12.113}$$

and hence, both the static pressure and the enthalpy (both static and stagnation) decrease in the flow direction depending on the value of the electrical conductivity, if the other parameters are kept constant. Similarly for the case p remaining constant in the flow direction, we get

$$\rho U \frac{dU}{dz} = -\sigma U B^2 (1 - K) \tag{12.114}$$

and

$$\rho U \frac{dh}{dz} = -\sigma U^2 B^2 (1 - K)^2 \tag{12.115}$$

and hence, while the flow velocity decreases in the flow direction, the static enthalpy increases. These two special cases for an ideal gas are shown in the schematic sketch of an enthalpy-entropy chart in Fig. 12.16. It can be seen that the exit static pressure is higher than the constant velocity case with the result that the plasma in the channel has adequate electrical conductivity; the maximum kinetic energy of the flow is, however, restricted. On the other hand for the constant velocity case, the temperature and thus, the electrical conductivity may fall fast, so that no electro-magnetic interaction is possible. Thus the comparison brings to focus the dilemma of using the two limiting cases for a seeded combustion plasma in a magneto-gas-dynamic generator. While in the constant pressure case the extracted power per unit channel length becomes smaller and smaller as U tends to zero, in the constant velocity case this happens due to the electrical conductivity becoming smaller and smaller inspite of decreasing pressure as the static temperature is decreased. For such studies it is, therefore, necessary to plot the electrical conductivity into the enthalpy-entropy chart of the flowing gas, as shown in Fig. 12.17 for equilibrium water gas (typically, 49.5 percent H_2, 37.5 percent CO, 5.5 percent N_2 and 7.0 percent CO_2 as volume fraction) burnt with 25 percent oxygen enriched air (adiabatic flame temperature 2510 ^oK at one bar pressure) to which one percent potassium (per mass fraction) is added to generate sufficient electrical conductivity in the plasma.

We would now discuss the case of the Hall generator. While, in the *Faraday generator*, separate and parallel load current is obtained in z direction, in the Hall generator the opposite electrodes in z direction are short-circuited, but the load is applied between the electrodes at extreme ends of the generator in x direction. For this let us consider the full Ohms's law equation, without the third pressure gradient term, as follows:

$$\frac{\mathbf{j}}{\sigma} = \mathbf{E} + \mathbf{V} \times \mathbf{B} - \frac{1}{en_e} \mathbf{j} \times \mathbf{B} . \tag{12.116}$$

For the short-circuited opposite electrodes in the z-direction, there will be electric current density, $j_z = \sigma U B$ and if the electrodes at the opposite ends in z-direction are short-circuited, there will be electric current density in x direction,

$$j_x = -\frac{\sigma}{en_e} \mathbf{j} \times \mathbf{B} = \frac{\sigma^2}{en_e} U B^2 = \sigma b_e U B^2 \tag{12.117}$$

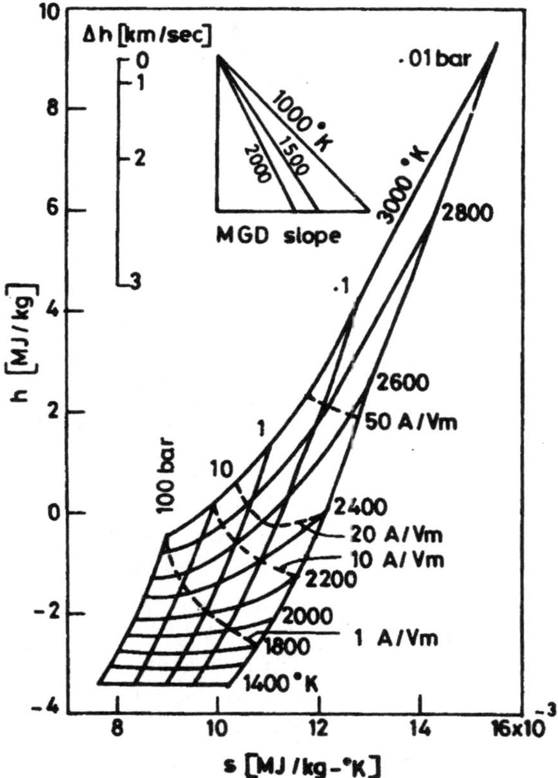

Fig. 12.17. (h,s) Chart of Equilibrium Water Gas Burnt With 25 Percent Oxygen Enriched Air

where $b_e \approx \sigma/(en_e)$ is the mobility coefficient of the electrons. Once again if additionally the electrodes in the opposite sides in the x direction are not closed (open circuit), there will be an induced electric field

$$E_x = -\frac{\sigma U B^2}{en_e} = -b_e U B^2 \; . \tag{12.118}$$

Between these two extremes the load will be replaced by actual electric field, $K = en_e E_x/(\sigma U B^2)$; the electric current density in x direction and the extracted electric power per unit volume of the gas are:

$$j_x = \frac{\sigma^2 U B^2}{en_e}(1 - K) \tag{12.119}$$

$$P = \sigma(b_e U B)^2(1 - K)K \tag{12.120}$$

For optimum power again $K = 1/2$ and further analysis follows the Faraday generator.

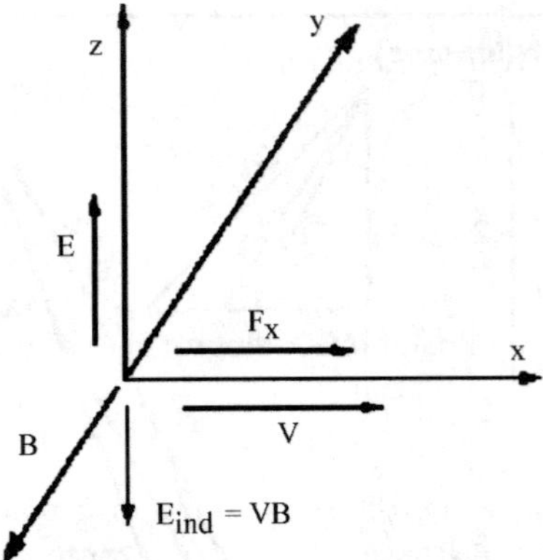

Fig. 12.18. Schematic sketch of a Faraday accelerator

Contrary to two generators discussed already, where there is no externally applied electric field but the induced electric field depends on the extracted load, in *Faraday accelerator* (Bose [47]) an external electric field is imposed in z direction and the external magnetic induction is imposed in $-y$ direction (Fig. 12.18). Hence an induced electric field, $E_{ind} = -UB$ is obtained in $-z$ direction. An electric current density in $+y$ direction, $j = \sigma(E - UB) = \sigma UB(K - 1)$ flows so long $K = E/(UB) > 1$, where K is a computed property changing from point to point, and there is an accelerating volumetric electromagnetic force in x direction, $\sigma U^2 B^2 (K - 1)$. The corresponding equations of (12.106–12.108) remain valid and obviously, due to flow acceleration (energy introduced), stagnation enthalpy of the flow will increase.

12.8 Plasma Manufacturing and Processing

High temperature gases of tens of thousand degrees Celsius have exotic properties required for manufacturing and other processes, which are required for various strategically important defence and aerospace applications. The present section is only an enumeration of such processes, for which various high intensity arcs of several megawatt power have been developed, like free burning arcs, wall-stabilized arcs, electrode stabilized arcs, self-stabilized arcs, vortex-stabilized arcs, etc. In addition, there are electrodeless high frequency discharges, basically of two types.The first one, due to a capacitive coupling in a very high frequency electric field is mainly for non-conducting materials (dielectric heating), for example in a microwave oven, but also to generate plasma. The second one, due to inductive coupling of a magnetic field, works on the principle that the required magnetic field is produced by circulating a high frequency current through a primary winding coil of a transformer.

Use of plasma in manufacturing processes is found when materials with very high melting points are required. Such materials carry names like *"refractory materials"*, but also as *"ceramics"* and, in general, are carbides, nitrides or oxides of various metals. These are very important materials for a number of applications in aerospace and defence. For example, the turbine blades of jet engines which propel our jet aircrafts, are to be sprayed with such oxides, the main component of which is the oxide of zirconium. Similarly one can use ceramic materials at the nose cone of missiles or reentry bodies. These ceramic materials are often in fine powder form, and they are introduced from outside to the gas coming out of an arc plasma jet forming a spray coating of ceramic materials. It is used in a number of industrial products as follows:

(a) in aerospace: landing gear components, gas path seals and turbine blades;
(b) in mechanics: as coating of bearing surfaces or coating hard ceramics like chromium oxide for dry low friction surfaces, as reinforcement of bearing surfaces on axles and shafts, as protection against corrosive fluids, coating for conveyor belts against wear and corrosion, and coatings of high temperature surfaces and dies;
(c) in electrical engineering: for coating heaters and contacts; and
(d) in nuclear reactors: protection with nickel based alloys, or with special hafnium carbide powder with large neutron cross-section.

Nearly all binary refracting carbides and some refractory nitrides and oxides are synthesized in plasmas by gas phase reactions in a high frequency plasma generators, where at temperatures around 10 to 20 thousand degrees Celsius, reactions are extremely fast. Some of the reactions are as follows:

(a) titanium micro-crystals formed from a mixture of titanium tetrachloride and hydrogen in argon plasma;
(b) aluminium nitride and silicon nitride syntheses from elements;
(c) ultrafine refractory carbides;
(d) boronnitride with special properties produced by arc vaporization of solution of boron oxide;
(e) silicon carbide from silicontetrachloride and methanated silicontrichloride or methylchloridilane;
(f) oxides of uranium, aluminium and zirconium; and
(g) carbides of uranium, boron, niobium, tantalum and tungsten.

As an example, commercially produced siliconcarbide powder, which, with its hardness and resistance to wear between diamond and corundum, and an outstanding resistance to chemical corrosion and thermal shocks, is attractive for grinding and cutting purposes, and also used as coatings for components to handle abrasive and corrosive products. This is an important raw material produced in a special furnace by carbothermic reaction of silica at temperatures above 1500 °C with diameter in the range of 1 to 85 microns (one-thousandth part of one millimeter), whereas such powders produced by plasma manufacturing processes can be one-thousandth time smaller in diameter. For the purpose of producing refractory powders, the starting material may be some suitable solution of a salt, which is sprayed into an atmosphere of nitrogen, oxygen or carbon, as the case may be. Because of the high temperature and cooling rate downstream of a plasma jet, rapid nucleation is obtained. The submicron products of various refractory materials thus produced can be directly sintered in a plasma equipment at about 80 percent of the melting temperature.

Rapid sintering of refractory material powders with the help of plasma has been demonstrated by various investigators. Enhanced densification up to 96 percent of theoretical density value and fine grain surface are observed for oxides of aluminium, beryllium, hafnium and magnesium after sintering for a few minutes, and in sintering of cold pressed oxide of uranium. One typical feature of plasma sintering is that the run-away growth, which is typical in conventional pressureless sintering, is restrained, and thus the grain growth after plasma sintering is quite uniform and overall fine structure is retained. Further, decomposition of metallic ores and oxides is an important application of plasma processing, for example, in decomposition of molybdenum sulphide to metallic molybdenum. Use of arc furnaces for extraction of iron from iron ore and making of steel is, of course, well known. Incidentally in Germany, a process to produce acetylene was developed, in which a natural gas is fed directly into an electric arc struck between a graphite cathode and a copper anode, and quenching is achieved by injecting water (or a hydro carbon to increase ethylene content).

Destruction of toxic wastes by allowing reaction in a plasma reactor either in the presence of an oxidizing or a reducing agent, as the case may be, is an important application of plasma processing, for example, in handling polychlorinated biphenyl wastes. Arc welding and arc cutting are two examples of use of plasma for fabrication of structures for a long time. Even for these, the theory behind the mechanism of transfer of energy seems to be not well understood, which would be necessary, for example, to minimize the electrical energy requirement but maximize the weldment speed and spot size. Another important application of plasma, although not as a manufacturing process, is when a plasma is generated to simulate very high speeds in wind tunnels (*arc-heated tunnels*) for testing of reentry vehicles or very high speed flow fields. When plasma for such purposes is generated electrically, we have a comparatively clean gas at high temperatures, which can be easily controlled.

Diamond-plating with the help of plasma in electrical discharges has an important industrial application. Formerly high pressure presses were used to squeeze carbon into diamonds. The presses were expensive and produced with great difficulty industrial diamonds of less than one carat. In recent years, the scientists have found a vapor deposition process, which is cost effective, by heating a mixture of methane and hydrogen without air in an induction plasma reactor, where the carbon of methane is deposited into the primary carbon of graphite or diamond structure depending on the substrate temperature. Diamond layer is formed on substrates of fused silica, tungsten carbide, graphite and silicon, if the substrate temperature is around 1000 °C. Diamond film produced this way is very hard and wear resistant, has a thermal conductivity five times that of copper and an electrical insulator property on par with the best insulators like quartz. Although generally a good thermal insulator is also a good electrical insulator, but there are a few materials like diamond or boronnitride, for which the general rule is not applicable. In addition, contrary to the boron nitride, which is very soft, a diamond film, apart being very hard and wear resistant, can transmit light from ultra-violet to far infrared. Thus a coating of diamond on easily damageable optical windows made of germanium, zincselenide or zincsulfate for heat seeking missiles is ideal when missiles move through ice particles and rain.

12.9 Weakly Ionized Plasma

This section, as the previous one, is only a description of the problems concerning weakly ionized plasma, but actual development of the subject should require lot of work in the future. Some of the properties of weakly ionized plasma being considered here are the following: pressure about 10 to 30 Torr, electron temperature 3 to 20 eV, heavy particles temperature 1,000 to 5,000 K. Such plasmas are generated in a variety of ways, like glow discharge, microwave, r.f. discharge, etc. In a supersonic flow of such a plasma, one finds the shock stand-off distance and other shock properties to change so much that effectively sonic Mach number (ratio of local gas speed to local sonic speed) is reduced (*Mishin* [84], *Bedin* and *Mishin* [35]); there is also considerable reduction in the drag of the body. Currently, the anomalous gas dynamic behavior of such weakly ionized plasma is not well understood, and remains among the most important topics of current research. Some of the solutions, as discussed by *Tidman* and *Krall* [23] postulate considerable charge separation near the shock, which, according to Poisson equation, can generate strong electric field, with the possibility of trapping of charge particles near the shock. Interested readers may look into the references cited and other references.

12.10 Exercise

12.10.1 Mercury, as a liquid metal at room temperature, is attractive as a MHD flow medium for which the following properties are estimated at 300K: density = 13510 kgm^{-3}, dynamic viscosity = 0.00204

$\mathrm{kgm^{-1}s^{-1}}$, specific heat = 140 $\mathrm{Jkg^{-1}K^{-1}}$, electrical conductivity = 1.09e6 $\mathrm{AV^{-1}m^{-1}}$, and thermal conductivity = 8.0 $\mathrm{Wm^{-1}K^{-1}}$.

Let us consider a MHD channel of 2 m long, 0.1 m width and 0.1 m depth, and channel inlet fluid velocity = 20 $\mathrm{ms^{-1}}$. The externally applied magnetic induction is B_o = 10,000 Gauss = 1 $\mathrm{Vsm^{-2}}$.

For Faraday generator and constant fluid velocity compute (a) Δp in flow direction (drop in pressure is due to conversion into electrical power), (b) voltage across electrodes, (c) electric power generated, and (d) total current. (Ans: (a) Δp = -218 bar, (b) 1 V, (c) 2.18 MW and (d) 2.18e6 A. Show further that the constant pressure case is not feasible.

12.10.2 In a plasma tube of tube radius = 5 mm and wall temperature = 500K with fully developed arc plasma (Elenbaas-Heller problem) the following plasma properties are given: average heat conductivity coefficient = 0.1 $\mathrm{Wm^{-1}K^{-1}}$ and electrical conductivity σ = 0 if $T <$ 10,000K and σ = 10,000 $\mathrm{AV^{-1}m^{-1}}$ if $T >$ 10,000K. Compute (by using a computer, if necessary) the centerline temperature and axial electric field for electric currents 100A and 200A. (Ans.: I = 100 A: centerline temperature = 21785 K, $\mathbf{E}$ = 279.5 $\mathrm{Vm^{-1}}$; I = 200 A: centerline temperature = 40736K, $\mathbf{E}$ = 348.3 $\mathrm{Vm^{-1}}$).

12.10.3 For a Hartmann flow with semi-channel width L = 5 mm and the rest of the data from Exercise 12.10.1, compute Re, R_H, R_σ and Rh. (Ans: Re = 6.622e5, R_H = 0.1473, R_σ = 0.137, Rh = 115.6)

12.10.4 Compute the Alfven speed in mercury flow in a channel with magnetic induction 10,000 Gauss. (Ans: 7.6748 $\mathrm{ms^{-1}}$)

12.10.5 For the MHD channel of Exercise 12.10.1, but for a Faraday accelerator and K = 2, compute the change of $p + \rho U^2/2$. (Ans: 218 bar)

12.10.6 Calculate the typical residence time in a thermal plasma of alumina particle of 10 micron diameter to enable complete melting of the particles. (Ans: 1.4e-4 s)

A Statistical Weights and Energy (cm^{-1}) for Selected Atoms and Molecules

g_i	E_i	g_i	E_i	g_i	E_i	g_i	E_i
He-I							
4	0.000000e0	3	1.598500e5	13	1.693380e5	36	1.856980e5
122	1.927810e5	252	1.962160e5	117	1.973040e5	142	1.976860e5
159	1.979860e5	inf	1.983050e5				
He-II							
2	0.000000e5	8	3.291820e5	22	4.114780e5	32	4.114780e5
32	4.213530e5	44	4.267170e5	58	4.299520e5	74	4.320510e5
92	4.334900e5	112	4.345200e5	34	4.352810e5	158	4.358610e5
184	4.363170e5	212	4.366690e5	242	4.569570e5	inf	4.389087e5
He-III							
1	0.000000e0						
Ne-I							
1	0.000000e0	12	1.346750e5	36	1.501980e5	72	1.613990e5
36	1.631800e5	52	1.668110e5	88	1.673410e5	36	1.678590e5
140	1.696770e5	36	1.700790e5	72	1.710710e5	84	1.711420e5
36	1.713110e5	72	1.719080e5	84	1.719510e5	136	1.720530e5
72	1.724470e5	84	1.724760e5	36	1.725510e5	120	1.727200e5
96	1.729670e5	124	1.731890e5	20	1.732800e5	inf	1.739317e5
Ne-II							
4	0.000000e0	2	7.820000e2	2	2.170500e5	18	2.210590e5
42	2.478110e5	22	2.524940e5	52	2.771340e5	88	2.811120e5
64	3.036350e5	76	3.043050e5	36	3.071100e5	10	3.279600e5
inf	3.331350e5						
Ne-III							
5	0.000000e0	3	6.500000e2	1	9.270000e2	5	2.584100e5
1	5.574700e4	9	2.045890e5	11	3.088650e5	44	3.542910e5
9	3.744480e5	39	3.896440e5	15	3.982020e5	33	3.988360e5
15	4.098480e5	12	4.117700e5	21	4.355810e5	27	4.365830e5
15	4.368410e5	12	4.397510e5	inf	5.141480e5		

g_i	E_i	g_i	E_i	g_i	E_i	g_i	E_i
Ne-IV							
4	0.000000e0	10	4.096000e3	6	6.216370e4	12	1.842220e5
10	2.538130e5	8	3.148260e5	12	4.793020e5	6	4.849440e5
6	4.886880e5	26	5.209080e5	18	5.738420e5	26	5.796470e5
14	5.871670e5	32	6.086390e5	18	6.347890e5	36	6.435590e5
56	6.705850e5	10	6.735230e5	12	6.939330e5	40	7.017030e5
34	7.400400e5	inf	7.838800e5				
Ne-V							
1	0.000000e0	3	4.140000e2	5	1.112000e3	5	3.029400e4
1	6.390000e4	5	8.834200e4	15	1.758760e5	9	2.081610e5
11	2.820320e5	9	4.130670e5	12	5.991050e5	35	6.972460e5
12	7.020620e5	16	7.154800e5	24	7.977590e5	69	8.417367e5
inf	1.019950e6						
Ne-VI							
2	0.000000e0	4	1.816000e3	12	9.930000e4	10	1.790070e5
8	2.455260e5	2	7.226100e5	6	7.639820e5	22	8.260640e5
18	8.965370e5	20	9.247910e5				
Ar-I							
1	0.000000e0	12	8.481040e4	36	1.062775e5	60	1.136065e5
12	1.140062e5	36	1.175092e5	60	1.197577e5	12	1.201898e5
84	1.206991e5	36	1.216811e5	60	1.226248e5	12	1.229299e5
84	1.231810e5	36	1.237068e5	60	1.242093e5	12	1.243933e5
84	1.245278e5	36	1.248225e5	60	1.251242e5	12	1.252537e5
68	1.252621e5	92	1.253397e5	56	1.253903e5	24	1.255242e5
56	1.257525e5	12	1.258542e5	48	1.258998e5	4	1.260799e5
60	1.261767e5	4	1.262700e5	60	1.263878e5	48	1.265466e5
60	1.266650e5	48	1.267666e5	12	1.268067e5	inf	1.271099e5
Ar-II							
4	0.000000e0	2	1.432000e3	2	1.087220e5	20	1.324761e5
12	1.347507e5	30	1.358289e3	6	1.385827e5	28	1.427057e5
12	1.476447e5	10	1.487544e5	36	1.572430e5	18	1.595845e5
2	1.673087e5	30	1.718305e5	14	1.786369e5	12	1.820306e5
6	1.833664e5	60	1.852312e5	30	1.896173e5	18	1.906258e5
6	1.920948e5	48	1.932663e5	10	1.958665e5	16	1.963616e5
12	1.991381e5	50	1.996970e5	6	2.003631e5	16	2.045355e5
10	2.129337e5	inf	2.228200e5				
Ar-III							
5	0.000000e0	3	1.112400e3	1	1.570200e3	5	1.401000e4
1	3.326700e4	9	1.143027e5	28	1.447980e5	15	1.569427e5
5	1.743750e5	3	1.806790e5	48	1.885667e5	30	2.006705e5

g_i	E_i	g_i	E_i	g_i	E_i	g_i	E_i
9	2.073817e5	9	2.091452e5	24	2.120236e5	45	2.270996e5
27	2.411425e5	48	2.489451e5	75	2.684358e5	15	2.722054e5
45	2.824358e5	9	2.859470e5	inf	3.299650e5		

Ar-IV

g_i	E_i	g_i	E_i	g_i	E_i	g_i	E_i
4	0.000000e0	4	2.109000e4	6	2.121900e4	2	3.485400e4
4	3.503500e4	12	1.181268e5	18	1.564257e5	18	2.531922e5
10	2.681594e5	36	2.880277e5	18	2.935964e5	24	3.050657e5
inf	4.824000e5						

Ar-V

g_i	E_i	g_i	E_i	g_i	E_i	g_i	E_i
1	0.000000e0	3	7.650000e2	5	2.032000e3	5	1.630100e4
15	1.217304e5	12	1.542103e5	3	1.953560e5	24	2.220659e5
9	2.971060e5	3	3.013000e5	inf	6.051000e5		

Ar-VI

g_i	E_i	g_i	E_i	g_i	E_i	g_i	E_i
2	0.000000e0	4	2.210000e3	12	1.012843e5	8	1.797843e5
10	2.186300e5	36	2.853321e5	2	3.422860e5	24	2.220659e5
10	4.547900e5	10	5.555650e5	inf	7.366000e5		

Ar-VII

g_i	E_i	g_i	E_i	g_i	E_i	g_i	E_i
1	0.000000e0	9	1.147444e5	3	1.707200e5	9	2.716566e5
15	3.241506e5	3	5.140830e5	3	5.663620e5	15	6.346494e5
21	6.600920e5	15	7.723340e5	inf	1.000400e6		

Ar-VIII

g_i	E_i	g_i	E_i	g_i	E_i	g_i	E_i
2	0.000000e0	6	1.418700e5	10	3.326666e5	2	5.759100e5
6	6.289047e5	10	6.975172e5	14	7.168374e5	2	8.124220e5
6	8.325423e5	10	8.651002e5	14	8.752646e5	10	9.555600e5
inf	1.157400e6						

Ar-IX

g_i	E_i	g_i	E_i	g_i	E_i	g_i	E_i
1	0.000000e0	8	2.033500e6	4	2.052120e6	inf	3.395360e6

Kr-I

g_i	E_i	g_i	E_i	g_i	E_i	g_i	E_i
1	0.000000e0	12	8.191260e4	36	9.407450e4	60	9.984360e4
12	1.012830e5	36	1.047200e5	60	1.064220e5	24	1.069070e5
12	1.072030e5	68	1.079560e5	56	1.079580e5	48	1.085660e5
24	1.088630e5	56	1.095830e5	12	1.098650e5	24	1.099530e5
48	1.101340e5	56	1.103950e5	12	1.106390e5	40	1.109200e5
48	1.110950e5	12	1.112800e5	35	1.112820e5	40	1.115410e5
45	1.117460e5	49	1.120600e5	12	1.121220e5	48	1.123540e5
12	1.126750e5	12	1.130470e5	24	1.134670e5	24	1.136220e5
3	1.176250e5	inf	1.182847e5				

g_i	E_i	g_i	E_i	g_i	E_i	g_i	E_i
Kr-II							
4	0.000000e0	42	5.364590e3	20	1.150470e5	20	1.203010e5
10	1.274500e5	28	1.275720e5	42	1.326720e5	42	1.356380e5
24	1.409390e5	62	1.487310e5	16	1.517810e5	80	1.614430e5
40	1.665800e5	86	1.680380e5	34	1.707010e5	58	1.727250e5
54	1.768020e5	36	1.784000e5	26	1.824380e5	inf	1.981820e5
Kr-III							
5	0.000000e0	4	4.739250e3	5	1.464400e4	1	3.307900e4
9	1.177050e5	25	1.385430e5	47	1.508900e5	82	1.655580e5
36	1.754570e5	43	1.809950e5	63	1.932540e5	49	2.073440e5
30	2.163860e5	18	2.197570e5	42	2.352310e5	12	2.507180e5
inf	2.980200e5						
Kr-IV							
4	0.000000e0	12	1.208730e5	76	1.638600e5	32	1.722200e5
14	1.791580e5	36	2.080790e5	30	2.111740e5		
Xe-I							
1	0.000000e1	8	4.678729e4	28	5.431762e4	52	5.759861e4
28	6.240280e4	76	6.266208e4	68	6.348980e4	72	6.500350e4
68	6.523037e4	164	6.617834e4	68	6.711069e4	48	6.735241e4
133	6.743130e4	72	6.765255e4	111	6.768198e4	inf	6.787909e4
Xe-II							
4	0.000000e1	2	7.310949e3	14	6.555224e4	48	6.813448e4
34	7.395317e4	52	7.827079e4	48	8.346817e4	38	9.022247e4
26	9.254537e4	70	9.486895e4	64	9.590760e4	36	9.801819e4
26	1.000878e5	30	1.023199e5	30	1.034626e5	22	1.075589e5
inf	1.186900e5						
Xe-III							
5	0.000000e1	4	6.507302e3	5	1.186426e4	1	2.594734e4
9	7.026693e4	25	7.756865e4	23	8.453387e4	31	8.798005e4
44	9.176344e4	15	9.560232e4	20	9.886395e4	13	1.014921e5
22	1.030497e5	26	1.055232e5	77	1.137331e5	23	1.214824e5
45	1.238871e5	28	1.355536e5	3	1.387979e5	inf	1.797601e5
Xe-IV							
4	0.000000e1	10	1.097229e4	6	2.585666e4	10	6.914850e4
12	7.445480e4	14	8.635375e4	24	9.386362e4	40	1.052249e5
14	1.142916e5	54	1.221894e5				
Li-I							
2	0.000000e0	6	1.490390e4	2	2.790612e4	6	3.092588e4
10	3.128310e4	2	3.501210e4	30	3.659580e4	32	3.903450e4

g_i	E_i	g_i	E_i	g_i	E_i	g_i	E_i
20	4.043150e4	44	4.150130e4	44	4.234350e4	40	4.286020e4
156	4.332940e4	inf	4.348720e4				

Li-II

g_i	E_i	g_i	E_i	g_i	E_i	g_i	E_i
1	0.000000e0	3	4.760460e5	1	4.900790e5	9	4.942730e5
3	5.018160e5	13	5.583516e5	23	5.613175e5	64	5.823981e5
64	5.923997e5	61	5.978077e5	inf	6.100790e5		

Li-III

g_i	E_i	g_i	E_i	g_i	E_i	g_i	E_i
2	0.000000e0	8	7.407466e5	14	8.779235e5	22	9.259335e5
32	9.481546e5	44	9.602247e5	58	9.675033e5	inf	9.876578e5

Li-IV

g_i	E_i
1	0.000000e0

Na-I

g_i	E_i	g_i	E_i	g_i	E_i	g_i	E_i
2	0.000000e0	6	1.696760e4	2	2.573990e4	10	2.917290e4
6	3.027060e4	2	3.320000e4	10	3.454880e4	14	3.458860e4
6	3.504200e4	2	3.637260e4	10	3.703680e4	32	3.705910e4
6	3.729730e4	2	3.801210e4	10	3.838730e4	44	3.844680e4
56	3.945440e4	40	4.005790e4	202	4.083170e4	186	4.138230e4
inf	4.144960e4						

Na-II

g_i	E_i	g_i	E_i	g_i	E_i	g_i	E_i
1	0.000000e0	12	2.661921e5	32	2.981160e5	13	3.220644e5
31	3.314210e5	32	3.326090e5	15	3.537466e5	12	3.669370e5
inf	3.815280e5						

Na-III

g_i	E_i	g_i	E_i	g_i	E_i	g_i	E_i
4	0.000000e0	2	1.364000e3	2	2.644490e5	12	3.669370e5
6	3.739825e5	10	3.991807e5	12	4.064338e5	20	4.114726e5
22	4.163931e5	2	4.350310e5	30	4.424186e5	84	4.628204e5
24	4.686671e5	50	4.938069e5	26	5.135455e5	10	5.294782e5
16	5.445451e5	inf	5.780330e5				

Na-IV

g_i	E_i	g_i	E_i	g_i	E_i	g_i	E_i
5	0.000000e0	3	1.106000e3	1	1.576000e3	5	3.111800e4
1	6.678000e4	9	2.441902e5	3	3.439720e5	3	4.866480e5
15	5.251130e5	5	5.316960e5	9	5.501760e5	3	5.570810e5
15	5.949171e5	15	6.388750e5	3	6.414680e5	9	6.431614e5
18	6.451852e5	39	6.653351e5	35	6.878936e5	12	7.153960e5
42	7.322296e5	42	7.554009e5	15	7.724150e5	inf	7.977410e5

Na-V

g_i	E_i	g_i	E_i	g_i	E_i	g_i	E_i
4	0.000000e0	10	4.758000e4	6	7.248000e4	12	2.164687e5
10	2.971296e5	2	3.499870e5	6	3.721670e5	6	5.678543e5
12	6.721645e5	6	6.832720e5	10	7.092770e5	2	7.486400e5

g_i	E_i	g_i	E_i	g_i	E_i	g_i	E_i
2	7.925077e5	26	7.971288e5	20	7.984764e5	4	8.019500e5
24	8.203303e5	16	8.342139e5	14	8.467573e5	10	8.667800e5
20	8.782880e5	22	8.936323e5	26	9.059545e5	22	9.190700e5
10	9.280530e5	6	9.376690e5	26	9.394225e5	44	9.445159e5
24	9.736408e5	36	1.007071e6	14	1.010292e6	24	1.038473e6
inf	1.118170e6						

Na-VI

g_i	E_i	g_i	E_i	g_i	E_i	g_i	E_i
1	0.000000e0	3	6.980000e2	5	1.160000e3	5	3.535800e4
1	7.427400e4	5	1.035080e5	15	2.041870e5	9	2.413140e5
11	3.248350e5	9	4.779260e5	5	5.393100e5	12	8.105050e5
9	8.729710e5	56	9.236850e5	28	9.436380e5	47	1.006740e6
55	1.042526e6	9	1.047717e6	21	1.054678e6	22	1.069282e6
5	1.077752e6	9	1.090756e6	42	1.127008e6	29	1.132548e6
49	1.140464e6	15	1.205485e6	9	1.214191e6	15	1.228205e6
9	1.228882e6	7	1.230972e6	51	1.254288e6	16	1.271887e6
60	1.337564e6	inf	1.390558e6				

Na-VII

g_i	E_i	g_i	E_i	g_i	E_i	g_i	E_i
2	0.000000e0	4	2.139000e3	12	1.163310e5	10	2.054260e5
2	2.644000e5	6	2.847490e5	4	3.674810e5	10	4.123450e5
6	4.650800e5	2	9.513470e5	16	1.041064e6	12	1.078733e6
22	1.133839e6	44	1.187701e6	26	1.209374e6	18	1.252500e6
44	1.296564e6	62	1.339932e6	64	1.387741e6	22	1.419680e6
42	1.437607e6	72	1.465991e6	122	1.576964e6	52	1.655867e6
inf	1.681679e6						

Na-VIII

g_i	E_i	g_i	E_i	g_i	E_i	g_i	E_i
1	0.000000e0	9	1.275930e5	3	2.432330e5	9	3.291830e5
5	3.610460e5	1	4.460990e5	7	1.266601e6	15	1.327485e6
5	1.347756e6	12	1.407657e6	21	1.441520e6	35	1.475264e6
19	1.501436e6	7	1.660863e6	20	1.685157e6	94	1.832896e6
119	2.015301e6	inf	2.131139e6				

Na-IV

g_i	E_i	g_i	E_i	g_i	E_i	g_i	E_i
1	0.000000e0	4	1.460260e5	1	1.375944e6	12	1.425393e6
13	1.858963e6	13	2.061031e6	12	2.170878e6	12	2.236732e6
inf	2.418520e6						

K-I

g_i	E_i	g_i	E_i	g_i	E_i	g_i	E_i
2	0.000000e0	6	1.302370e4	12	2.145060e4	6	2.471390e4
12	2.740360e4	14	2.812770e4	6	2.900490e4	44	3.050080e4
6	3.107300e4	66	3.191300e4	18	3.248090e4	20	3.281760e4
42	3.334460e4	40	3.377100e4	42	3.427860e4	90	3.478430e4
150	3.494590e4	132	3.498450e4	inf	3.500980e4		

g_i	E_i	g_i	E_i	g_i	E_i	g_i	E_i
K-II							
1	0.000000e0	12	1.638983e4	40	1.692018e5	36	1.878370e5
57	2.166564e5	8	2.231241e5	inf	2.566370e5	inf	2.566370e5
K-III							
4	0.000000e0	2	2.162000e3	2	1.306090e5	14	2.011650e5
18	2.098651e5	10	2.250634e5	12	2.378026e5	28	2.414857e5
32	2.449523e5	6	2.631420e5	10	2.894460e5	10	3.030032e5
inf	3.690000e5						
K-IV							
5	0.000000e0	3	1.673000e3	1	2.324000e3	5	1.638600e4
1	3.854800e4	9	1.349261e5	3	1.711400e5	15	1.906595e5
19	2.261404e5	9	2.565948e5	21	2.623971e5	20	2.767794e5
5	2.823730e5	12	2.947338e5	3	3.678900e5	inf	4.913000e5
K-V							
4	0.000000e0	10	2.414220e4	6	3.995770e4	12	1.374678e5
36	2.050615e5	20	2.224702e5	12	2.586455e5	30	2.628215e5
10	2.810240e5	34	3.787764e5	12	3.382390e5	6	3.449260e5
10	3.570090e5	inf	6.423190e5				
K-VI							
1	0.000000e0	3	1.131000e3	5	2.924000e3	5	1.897300e4
15	1.408647e5	9	1.634340e5	3	2.183160e5	3	2.238400e5
9	2.529072e5	9	3.893587e5	inf	8.045130e5		
K-VII							
2	0.000000e0	4	3.129000e3	12	1.158172e5	10	1.519822e5
2	1.930790e5	6	2.077917e5	10	2.507394e5	4	3.074790e5
20	3.657026e5	2	4.392970e5	12	5.672208e5	10	5.709062e5
inf	9.502000e5						
K-VIII							
1	0.000000e0	9	1.302742e5	3	1.925402e5	14	3.043858e5
15	3.680824e5	3	6.316540e5	15	7.703068e5	21	8.015110e5
inf	1.247000e6						
K-IX							
2	0.000000e0	6	1.596697e5	10	3.749632e5	2	6.989020e5
6	7.591347e5	10	8.367978e5	14	8.608081e5	2	9.799010e5
18	1.044271e6	10	1.049150e6	14	1.061150e6	inf	1.419425e6
K-X							
1	0.000000e0	8	2.407300e6	4	2.430300e6	4	2.760200e6
6	2.813600e6	12	3.214200e6	9	3.219400e6	3	3.237600e6
6	3.368050e6	inf	4.064300e6				

g_i	E_i	g_i	E_i	g_i	E_i	g_i	E_i
K-XI							
4	0.000000e0	2	2.347500e4	12	2.646700e6	6	2.671300e6
10	2.727880e6	inf	5.000000e6				
Cs-I							
2	0.000000e0	6	1.154765e4	10	1.155804e4	2	1.853551e4
6	2.188632e4	10	2.261465e4	2	2.431717e4	14	2.447234e4
6	2.576423e4	10	2.606044e4	16	2.696388e4	18	2.467695e4
56	2.834454e4	16	2.880012e4	16	2.914591e4	16	2.945033e4
16	2.967734e4	16	2.988441e4	74	3.028814e4	100	3.077632e4
120	3.116480e4	228	3.135760e4	inf	3.140670e4		
Cs-II							
1	0.000000e0	8	1.075846e5	8	1.146494e5	71	1.232424e5
9	1.307725e5	20	1.307725e5	64	1.549105e5	74	1.648669e5
61	1.713786e5	46	1.937207e5	inf	2.022630e5		
Cs-III							
4	0.000000e0	2	1.388400e4	2	1.277860e5	4	1.372995e5
6	1.561890e5	6	1.843660e5	inf	2.788167e7		
H-I							
2	0.000000e0	8	8.225910e4	14	9.749230e4	22	1.028239e5
32	1.052916e5	44	1.066322e5	58	1.070404e5	74	1.079650e5
92	1.083247e5	112	1.085820e5	134	1.087723e5	158	1.089171e5
184	1.090298e5	212	1.091192e5	242	1.091913e5	274	1.092503e5
308	1.092992e5	344	1.093402e5	382	1.093749e5	422	1.094046e5
464	1.094301e5	508	1.094521e5	554	1.094714e5	602	1.094883e5
652	1.095033e5	704	1.095165e5	758	1.095283e5	814	1.095389e5
872	1.095483e5	932	1.095569e5	994	1.095646e5	1058	1.095716e5
1124	1.095780e5	1192	1.095839e5	1262	1.095892e5	1334	1.095941e5
1408	1.095986e5	1484	1.096028e5	1562	1.096066e5	1642	1.096102e5
inf	1.096788e5						
H-II							
1	0.000000e0						
N-I							
4	0.000000e0	10	1.922620e4	6	2.884000e4	12	8.333700e4
6	8.619260e4	12	8.813470e4	22	9.472480e4	12	9.551100e4
20	9.710540e4	10	9.966200e4	52	1.044129e5	56	1.049732e5
22	1.067937e5	16	1.071238e5	46	1.101350e5	40	1.103052e5
32	1.104516e5	6	1.123121e5	12	1.126391e5	6	1.127937e5
68	1.128310e5	22	1.129127e5	46	1.141432e5	32	1.140042e5
38	1.149550e5	42	1.150530e5	18	1.154820e5	62	1.155594e5
80	1.159091e5	58	1.161477e5	22	1.162504e5	58	1.163456e5
28	1.164448e5	18	1.165547e5	inf	1.173450e5		

g_i	E_i	g_i	E_i	g_i	E_i	g_i	E_i
N-II							
1	0.000000e0	3	4.910000e1	5	1.313000e2	5	1.531570e4
1	3.268710e4	5	4.716770e4	15	9.224537e4	9	9.224537e4
17	1.476253e5	21	1.669627e5	9	1.706371e5	6	1.748898e5
21	1.865926e5	15	1.874717e5	19	1.892460e5	12	1.969549e5
27	2.028721e5	24	2.056074e5	26	2.097929e5	59	2.108099e5
59	2.113811e5	18	2.130639e5	18	2.197355e5	64	2.211921e5
25	2.241139e5	20	2.270759e5	inf	2.388467e5		
N-III							
2	0.000000e0	4	1.745000e2	12	5.728260e4	10	1.010269e5
6	1.459497e5	10	2.030789e5	8	2.281310e5	6	2.456897e5
10	2.672420e5	12	2.876459e5	8	2.981911e5	26	3.096363e5
10	3.127143e5	36	3.186141e5	12	3.220348e5	50	3.314659e5
36	3.371626e5	72	3.434019e5	30	3.605828e5	36	3.755521e5
32	3.784746e5	inf	3.826255e5				
N-IV							
1	0.000000e0	9	6.727270e4	14	1.803433e5	4	3.801190e5
12	4.055754e5	20	4.222703e5	12	4.672948e5	21	4.843854e5
40	4.983909e5	34	5.043555e5	60	5.151320e5	60	5.589172e5
20	5.930232e5	inf	6.248510e5				
N-V							
2	0.000000e0	6	8.063720e4	16	4.819484e5	18	3.801190e5
18	6.785998e5	50	7.130835e5	66	7.334304e5	76	7.466041e5
inf	7.895329e5						
N-VI							
1	0.000000e0	3	3.385890e6	12	3.447277e6	3	4.016390e6
3	4.206810e6	inf	4.452800e6				
N-VII							
2	0.000000e0	8	4.034991e6	inf	5.379860e6		
N-VIII							
1	0.000000e0						
O-I							
5	0.000000e0	3	1.585000e2	1	2.265000e1	5	1.586770e4
1	3.379240e4	5	7.376781e4	3	7.679469e4	15	8.662870e4
9	8.863070e4	8	9.575730e4	45	9.744300e4	24	9.931370e4
15	1.011431e5	8	1.022270e5	45	1.028568e5	9	1.038694e5
57	1.054307e5	48	1.067266e5	48	1.075580e5	96	1.082809e5
48	1.087264e5	inf	1.098367e5				

g_i	E_i	g_i	E_i	g_i	E_i	g_i	E_i
O-II							
4	0.000000e0	10	2.681680e4	6	4.046740e4	12	1.199331e5
10	1.659910e5	18	1.866044e5	4	1.998263e5	30	2.069204e5
22	2.098877e5	16	2.131127e5	26	2.290604e5	46	2.315247e5
34	2.328051e5	34	2.368648e5	42	2.460057e5	82	2.529922e5
88	2.553483e5	132	2.566015e5	108	2.638625e5	124	2.661565e5
122	2.759224e5	inf	2.825509e5				
O-III							
1	0.000000e0	3	1.134000e2	5	3.068000e2	5	2.027100e4
1	4.318350e4	5	6.031210e5	15	1.200411e5	9	1.423838e5
11	1.961712e5	12	2.689073e5	12	2.856390e5	23	2.954467e5
15	3.033392e5	50	3.263474e5	26	3.363648e5	21	3.543195e5
61	3.667750e5	85	3.761343e5	19	3.807940e5	62	3.944139e5
40	3.982972e5	66	4.035813e5	39	4.197078e5	76	4.371319e5
inf	4.431935e5						
O-IV							
2	0.000000e0	4	3.865000e2	12	7.137880e4	10	1.269419e5
2	1.643669e5	6	1.806435e5	20	2.605456e5	18	4.028878e5
18	4.435446e5	42	4.717141e5	12	4.833212e5	50	4.966937e5
22	5.028312e5	36	5.126577e5	18	5.472778e5	26	5.524662e5
12	5.688740e5	58	5.743090e5	30	5.819077e5	104	5.944537e5
56	6.063844e5	inf	6.243965e5				
O-V							
1	0.000000e0	9	8.241280e4	3	1.587980e5	9	2.139294e5
5	2.317220e5	1	2.879090e5	3	5.471500e5	1	5.612780e5
12	5.825006e5	20	6.038618e5	12	6.561974e5	18	6.768152e5
12	6.883757e5	10	6.959080e5	25	7.059422e5	13	7.166614e5
13	7.361814e5	27	7.450587e5	26	8.052798e5	46	8.335356e5
65	8.424197e5	41	8.677600e5	59	9.023789e5	inf	9.187020e5
O-VI							
2	0.000000e0	6	9.673000e4	2	6.400398e5	6	6.662176e5
10	6.746564e5	2	8.526960e5	30	8.662785e5	18	9.544678e5
62	1.004011e6	168	1.043634e6	inf	1.113999e6		
O-VII							
1	0.000000e0	3	4.525340e6	12	4.597197e6	31	5.362835e6
3	5.628100e6	3	5.748450e6	3	5.813950e6	inf	5.963000e6
O-VIII							
2	0.000000e0	8	5.271141e6	inf	7.027970e6		

g_i	E_i	g_i	E_i	g_i	E_i	g_i	E_i
O-IX							
1	0.000000e0						
H$_2$							
1	0.000000e0	9	1.386150e5	54	1.590950e5	15	1.667350e5
inf	1.791000e5						
O$_2$							
1	0.000000e0	1	1.141000e4	1	1.901500e4	3	5.201400e4
3	7.176500e4	9	1.351400e5	inf	1.417000e5		
N$_2$							
1	0.000000e0	6	7.913400e4	1	9.984700e4	12	1.384700e5
8	1.522840e5	8	1.673790e5	inf	1.809000e5		
CO							
1	0.000000e0	6	7.535600e4	4	9.077300e4	12	1.320000e5
3	1.548800e5	inf	1.626000e5				
OH							
2	0.000000e0	2	4.710000e4				
NO							
2	0.000000e0	2	1.750000e2	10	7.371000e4		

B Enthalpy (MJ/kmol) for Different Gases (1MJ = 1 Mega Joule)

T[K]	H_2	H	O_2	O	H_2O	OH	N_2	N	NO	CO	CO_2	C	CN	C_{graph}	CH_4	NH_3	A	A^+	H^+	N^+	O^+	C^+	e^-
15000	550.28	556.75	569.61	607.36	810.09	550.42	615.80	909.17	669.02	556.96		1180.61					330.26	1837.49	1939.92	2213.01	1934.98	2126.23	311.79
14500	535.03	537.63	554.14	587.75	766.38	536.40	589.21	881.33	647.89	523.13		1138.22					313.97	1827.05	1829.53	2201.13	1919.71	2113.55	301.47
14000	519.45	520.80	538.43	570.43	722.52	522.17	563.03	856.92	626.84	490.16		1101.06					299.33	1816.62	1819.19	2189.22	1904.52	2101.04	291.01
13500	503.61	505.74	522.25	554.75	678.63	507.78	537.39	835.12	606.02	458.08		1069.00					285.98	1806.20	1808.74	2177.37	1889.31	2088.71	280.67
13000	487.30	492.18	505.79	540.46	634.86	493.13	512.26	815.42	585.38	427.16		1041.66					273.57	1795.77	1798.35	2165.50	1874.21	2076.53	270.22
12500	470.60	479.57	488.95	527.08	591.36	478.16	487.80	797.26	564.86	397.35		1018.06					261.86	1785.35	1787.96	2153.69	1859.17	2064.52	259.83
12000	453.53	467.67	471.70	514.26	548.06	463.00	463.87	780.03	544.40	368.77		997.73					250.63	1774.93	1777.61	2141.93	1844.32	2052.75	249.43
11500	436.03	456.37	454.07	501.91	505.17	447.49	440.60	763.64	524.07	341.49		979.95					239.73	1764.50	1767.17	2130.11	1829.53	2041.17	239.04
11000	418.04	445.37	435.99	489.86	462.78	431.69	417.90	747.64	503.86	315.48		974.05					229.03	1754.03	1756.78	2118.41	1814.99	2029.71	228.64
10500	399.61	434.68	417.40	487.02	421.08	415.05	395.70	731.94	483.76	290.60		949.63					218.48	1743.65	1746.38	2106.68	1800.80	2018.42	218.25
10000	380.68	424.12	398.41	466.29	380.16	398.94	374.16	716.46	463.77	265.37		916.21					208.00	1733.22	1736.00	2094.95	1786.71	2007.26	207.87
9500	361.29	413.61	378.98	454.65	340.06	382.02	353.08	701.17	443.90	243.22		923.47					197.56	1722.78	1725.61	2083.29	1773.00	1996.27	197.47
9000	341.49	403.15	359.11	443.07	300.81	364.71	332.45	686.05	424.13	221.76		911.15					187.14	1712.35	1715.21	2071.66	1759.60	1985.40	187.08
8500	321.25	392.73	338.81	431.54	262.54	346.99	312.21	671.13	404.46	200.89		899.12					176.74	1701.90	1704.82	2060.07	1746.58	1974.66	176.69
8000	300.67	382.33	316.13	420.05	225.21	328.89	292.26	656.47	384.88	180.49		887.24					166.34	1691.45	1694.43	2048.53	1733.90	1964.00	166.30
7500	279.81	371.93	279.12	408.63	188.83	310.39	272.58	642.14	365.40	160.49		841.20					155.94	1681.00	1684.03	2037.05	1721.62	1953.42	155.90
7000	258.78	361.54	275.87	397.27	153.33	291.55	253.11	628.18	346.01	140.78		861.75					145.55	1670.54	1673.64	2025.64	1709.69	1942.90	143.51
6500	237.69	351.14	254.41	385.98	118.73	272.40	233.79	614.68	326.71	121.31		852.08					135.15	1660.06	1663.24	2014.31	1698.12	1932.55	135.11
6000	216.66	340.75	232.87	374.80	84.94	253.02	214.59	601.65	307.48	101.99	-39.97	840.46	609.74		479.03	416.32	124.75	1649.58	1652.85	2003.06	1686.85	1922.00	124.72
5800	208.30	336.60	224.25	370.35	71.67	245.23	206.94	596.58	299.81	94.30	-52.93	835.83	600.50		457.77	396.98	120.60	1645.38	1648.70	1998.60	1682.43	1917.84	120.57
5600	199.97	332.44	215.64	365.92	58.52	237.41	199.30	591.60	292.15	86.63	-65.88	831.21	591.25		436.58	377.78	116.44	1641.18	1644.54	1994.15	1678.04	1913.67	116.41
5400	191.68	328.28	207.05	361.51	45.50	229.58	191.69	586.71	284.51	78.97	-78.97	826.60	581.95		415.39	358.79	112.28	1636.98	1640.38	1989.72	1673.70	1909.51	112.25
5200	183.44	324.12	198.45	357.12	32.60	221.76	184.03	581.90	276.88	71.34	-91.66	822.00	572.69		394.22	340.02	108.12	1632.77	1636.22	1985.32	1669.38	1905.35	108.09
5000	175.26	319.96	189.88	352.76	19.79	213.93	176.49	577.17	269.26	63.71	-104.51	817.42	563.46	119.00	373.09	321.46	103.96	1628.56	1632.06	1980.93	1665.10	1901.19	103.93
4800	167.14	315.89	181.34	348.40	7.13	206.12	168.90	572.51	261.66	56.10	-117.31	812.15	554.24	113.15	352.00	303.16	99.20	1624.35	1627.91	1976.56	1660.84	1897.03	99.78
4600	159.08	311.65	172.84	344.07	-5.42	198.33	161.34	567.93	254.07	48.51	-130.07	808.38	545.07	107.35	330.94	285.02	95.64	1620.13	1623.75	1972.21	1656.61	1892.87	95.62
4400	151.09	307.49	164.39	339.76	-17.83	190.56	153.79	563.42	246.48	40.93	-142.82	801.77	535.92	101.60	309.96	267.08	91.49	1615.91	1619.59	1967.89	1652.39	1888.71	91.46
4200	143.17	303.33	156.00	335.48	-30.13	182.82	146.25	558.97	238.91	33.36	-155.52	799.26	526.82	95.90	288.99	249.40	87.33	1611.68	1615.43	1963.58	1648.19	1884.55	87.30
4000	135.31	299.18	147.65	331.20	-42.34	175.11	138.73	554.58	231.36	25.82	-168.20	794.78	517.78	90.25	268.13	231.82	83.17	1607.45	1611.28	1959.30	1644.00	1880.39	83.15
3800	127.54	295.02	139.36	326.95	-54.39	167.46	131.22	550.25	223.82	18.28	-180.83	790.32	508.88	84.66	247.35	214.58	79.01						
3600	119.83	290.87	131.11	322.72	-66.31	159.84	123.74	545.95	216.29	10.76	-193.43	785.89	499.94	79.11	226.62	197.47	74.85						
3400	112.21	286.71	122.94	318.50	-78.10	152.29	116.27	541.70	208.78	3.26	-205.99	781.48	491.20	73.61	205.99	180.61	70.69						
3200	104.66	282.55	114.83	314.29	-89.76	144.78	108.83	537.47	201.29	-4.21	-218.51	777.11	482.60	68.16	185.51	163.94	66.54						
3000	97.20	278.39	106.80	310.10	-101.25	137.34	101.40	533.27	193.81	-11.67	-230.97	772.77	474.12	62.70	105.10	147.52	62.38						
2900	93.51	276.31	102.81	308.01	-106.95	133.65	97.70	531.17	190.08	-15.38	-237.19	770.61	469.93	60.09	155.05	139.37	60.30						
2800	89.83	274.23	98.84	305.92	-112.60	129.98	94.01	529.08	186.36	-19.10	-243.39	768.46	465.79	57.42	144.97	131.31	58.22						
2700	86.19	272.15	94.89	303.83	-118.20	126.32	90.32	527.00	182.64	-22.80	-249.59	766.32	461.70	54.77	134.95	123.39	56.14						
2600	82.56	270.07	90.97	301.74	-123.76	122.68	86.65	524.91	178.93	-26.50	-255.77	764.18	457.65	52.14	124.99	115.37	54.06						
2500	78.96	268.00	87.06	299.65	-129.25	119.06	82.98	522.83	175.22	-30.19	-261.92	762.05	453.65	49.51	115.06	107.51	51.98						
2400	75.40	265.92	83.18	297.57	-134.71	115.47	79.32	520.74	171.52	-33.87	-268.07	759.93	449.69	46.91	105.22	99.73	49.90						
2300	71.86	263.84	79.32	295.49	-140.11	111.90	75.68	518.66	167.83	-37.53	-274.20	757.82	445.78	44.32	95.44	92.04	47.82						
2200	68.35	261.76	73.49	293.41	-145.44	108.35	72.04	516.58	164.15	-41.19	-280.30	755.71	441.91	41.74	85.74	84.43	45.74						
2100	64.86	259.68	71.67	291.32	-150.72	104.83	68.42	514.50	160.48	-44.84	-286.37	753.61	438.09	39.18	76.15	76.91	43.60						
2000	61.42	257.60	67.88	289.24	-155.92	101.34	64.81	512.41	156.82	-48.47	-292.42	751.51	434.32	36.64	66.65	69.50	41.58						
1900	58.01	255.53	64.12	287.16	-161.06	97.88	61.22	510.35	153.17	-52.09	-298.45	749.42	430.58	34.12	57.28	62.21	39.51						
1800	54.63	253.45	60.37	285.08	-166.12	94.45	57.66	508.27	149.54	-55.68	-304.44	747.33	426.88	31.62	48.02	55.04	37.43						
1700	51.30	251.37	56.66	282.99	-171.09	91.06	54.10	506.19	145.92	-59.27	-310.39	745.24	423.22	29.15	38.92	48.00	35.35						
1600	48.01	249.29	52.96	280.91	-175.98	87.71	50.58	504.11	142.32	-62.83	-316.29	743.16	419.59	26.69	29.98	41.09	33.27						
1500	44.76	247.21	49.29	278.82	-180.76	84.39	47.08	502.03	138.73	-66.36	-322.16	741.07	416.01	24.27	21.22	34.32	31.19						
1400	41.55	245.13	45.65	276.74	-185.44	81.12	43.61	499.95	135.17	-69.87	-327.98	739.00	412.47	21.88	12.67	27.73	29.11						
1300	38.38	243.05	42.03	274.66	-190.01	77.89	40.17	497.87	131.75	-73.34	-333.73	736.92	408.96	19.54	4.37	21.32	27.03						
1200	35.27	240.98	38.45	272.57	-194.47	74.70	36.78	495.80	128.15	-76.78	-339.40	734.84	405.49	17.24	-3.66	15.11	24.95						
1100	32.19	238.90	34.89	270.48	-198.80	71.57	33.43	493.72	124.69	-80.17	-344.99	732.76	402.07	15.00	-11.38	9.12	22.87						
1000	29.15	236.82	31.39	263.39	-203.01	68.48	30.13	491.64	121.27	-83.52	-350.47	730.68	398.69	12.82	-18.75	3.36	20.79						
900	26.15	234.74	27.92	266.29	-207.08	65.43	26.89	489.56	117.90	-86.81	-355.83	728.60	395.38	10.70	-25.72	-2.15	18.71						
800	23.17	232.66	24.52	264.20	-211.02	62.42	23.72	487.48	114.59	-90.03	-361.06	726.52	392.12	8.67	-32.26	-7.39	16.63						
700	20.22	230.58	21.48	262.10	-214.84	59.44	20.61	485.40	111.35	-93.19	-366.11	724.44	388.98	6.73	-38.30	-12.36	14.55						
600	17.28	228.50	17.92	259.99	-218.54	56.48	17.56	483.32	108.19	-96.27	-370.95	722.36	385.83	4.94	-43.80	-17.02	12.48						
500	14.35	226.42	14.77	257.87	-222.11	53.54	14.58	481.24	105.11	-99.28	-375.56	720.28	382.80	3.34	-48.73	-21.37	10.40						
400	11.43	224.34	11.71	255.73	-225.59	50.58	11.64	479.16	102.09	-102.23	-379.86	718.20	379.84	2.00	-53.07	-25.39	8.32						
298.15	8.45	222.23	8.68	253.53	-229.04	47.56	8.67	477.05	99.05	-105.21	-383.86	716.08	376.87	0.98	-56.94	-29.18	6.24						
0	0	216.03	0	246.80	-238.95	38.74	0	470.85	89.87	-113.88	-393.23	709.54	368.19	0	-66.96	-39.22	0						

C Entropy (MJ/kmol) for Different Gases (1MJ = 1 Mega Joule)

T[K]	H_2	H	O_2	O	H_2O	OH	N_2	N	NO	CO	CO_2	C	CN	C_{graph}	CH_4	NH_3	A	A^+	H^+	N^+	O^+	C^+	e^-
15000	0.26519	0.19828	0.34875	0.24750	0.40192	0.31468	0.33263	0.24678	0.35076	0.34395		0.25310					0.23770	0.25125	0.19040	0.24429	0.24223	0.23763	0.10243
14500	0.26415	0.19699	0.34770	0.24616	0.39895	0.31373	0.33094	0.24489	0.34933	0.34165		0.25020					0.23659	0.25054	0.18970	0.24347	0.24120	0.23676	0.10172
14000	0.26306	0.19580	0.34660	0.24424	0.39588	0.31274	0.32910	0.24316	0.34785	0.33934		0.24758					0.23557	0.24981	0.18897	0.24265	0.24013	0.23589	0.10099
13500	0.26191	0.19471	0.34542	0.24381	0.39269	0.31169	0.32724	0.24158	0.34634	0.33700		0.24524					0.23460	0.24905	0.18821	0.24178	0.23903	0.23499	0.10024
13000	0.26068	0.19368	0.34418	0.24272	0.38938	0.31058	0.32534	0.24010	0.34478	0.33467		0.24317					0.23366	0.24827	0.18743	0.24090	0.23789	0.23407	0.09945
12500	0.25937	0.19270	0.34286	0.24168	0.38596	0.30941	0.32342	0.23587	0.34317	0.33233		0.24133					0.23274	0.24745	0.18661	0.23996	0.23671	0.23313	0.09863
12000	0.25797	0.19173	0.34145	0.24062	0.38243	0.30817	0.32147	0.23445	0.34150	0.33000		0.23966					0.23182	0.24660	0.18577	0.23900	0.23549	0.23217	0.09778
11500	0.25618	0.19076	0.33995	0.23958	0.37879	0.30686	0.31948	0.23298	0.33977	0.32768		0.23814					0.23089	0.24571	0.18488	0.23800	0.23424	0.23118	0.09690
11000	0.25489	0.18978	0.33834	0.23851	0.37502	0.30545	0.31747	0.23445	0.33796	0.32536		0.23672					0.22994	0.24478	0.18395	0.23695	0.23294	0.23016	0.09598
10500	0.25317	0.18879	0.33661	0.23741	0.37113	0.30393	0.31540	0.23298	0.33611	0.32305		0.23539					0.22896	0.24381	0.18299	0.23587	0.23162	0.22911	0.09501
10000	0.25132	0.18776	0.33476	0.23627	0.36715	0.30233	0.31330	0.23148	0.33416	0.32057		0.23407					0.22794	0.24280	0.18197	0.23472	0.23025	0.22803	0.09400
9500	0.24933	0.18668	0.33277	0.23506	0.36802	0.30058	0.31114	0.22900	0.33211	0.31830		0.23276					0.22687	0.24173	0.18091	0.23351	0.22884	0.22690	0.09293
9000	0.24719	0.18555	0.33062	0.23382	0.35879	0.29871	0.30891	0.22828	0.32998	0.31598		0.23144					0.22574	0.24060	0.17978	0.23227	0.22739	0.22572	0.09181
8500	0.24498	0.18436	0.32830	0.23250	0.35439	0.29669	0.30659	0.22657	0.32773	0.31359		0.23006					0.22455	0.23940	0.17859	0.23094	0.22590	0.22449	0.09062
8000	0.24238	0.18310	0.32580	0.23110	0.34989	0.29449	0.30417	0.22479	0.32536	0.31112		0.22862					0.22329	0.23814	0.17734	0.22954	0.22437	0.22320	0.08936
7500	0.23968	0.18176	0.32308	0.22963	0.34519	0.29211	0.30163	0.22291	0.32384	0.30854		0.22710					0.22195	0.23679	0.17599	0.22805	0.22278	0.22184	0.08802
7000	0.23679	0.18032	0.32015	0.22806	0.34029	0.28951	0.29895	0.22020	0.32017	0.30582		0.22548					0.22052	0.23534	0.17456	0.22648	0.22114	0.22039	0.08658
6500	0.23366	0.17878	0.31697	0.22638	0.33515	0.28667	0.29608	0.21902	0.31739	0.30293		0.22375					0.21897	0.23379	0.17302	0.22480	0.21942	0.21884	0.08504
6000	0.23029	0.17712	0.31552	0.22461	0.32976	0.28357	0.29301	0.21693	0.31423	0.29984	0.37815	0.22190	0.31150		0.41811	0.38450	0.21731	0.23211	0.17136	0.22301	0.21762	0.21716	0.08338
5800	0.22888	0.17642	0.31206	0.22386	0.32751	0.28225	0.29172	0.21607	0.31293	0.29854	0.37596	0.22111	0.30993		0.41450	0.38123	0.21659	0.23140	0.17065	0.22225	0.21687	0.21646	0.08267
5600	0.22742	0.17589	0.31055	0.22307	0.32521	0.28087	0.29038	0.21519	0.31158	0.29719	0.37368	0.22031	0.30831		0.41078	0.37785	0.21586	0.23066	0.16992	0.22147	0.21610	0.21573	0.08194
5400	0.22591	0.17493	0.30897	0.22227	0.32280	0.27946	0.28899	0.21431	0.31019	0.29580	0.37133	0.21948	0.30661		0.40693	0.37440	0.21512	0.22990	0.16917	0.22067	0.21531	0.21487	0.08119
5200	0.22435	0.17415	0.30736	0.22144	0.32040	0.27798	0.28755	0.21340	0.30875	0.29436	0.36891	0.21860	0.30487		0.40294	0.37085	0.21432	0.22911	0.16838	0.21983	0.21450	0.21418	0.08040
5000	0.22275	0.17333	0.30568	0.22059	0.31789	0.27645	0.28607	0.21248	0.30726	0.29286	0.36639	0.21770	0.30306	0.06540	0.39880	0.36722	0.21352	0.22828	0.16757	0.21897	0.21365	0.21337	0.07959
4800	0.22199	0.17248	0.30394	0.21970	0.31530	0.27485	0.28452	0.21152	0.30571	0.29131	0.36379	0.21676	0.30118	0.06420	0.39448	0.36348	0.21267	0.22742	0.16672	0.21808	0.21279	0.21252	0.07874
4600	0.21938	0.17160	0.30213	0.21877	0.31263	0.27318	0.28291	0.21054	0.30409	0.28970	0.36107	0.21580	0.29921	0.06297	0.39001	0.35963	0.21177	0.22652	0.16583	0.21715	0.21189	0.21164	0.07780
4400	0.21760	0.17067	0.30025	0.21782	0.30987	0.27146	0.28123	0.20955	0.30241	0.28801	0.35825	0.21479	0.29720	0.06169	0.38534	0.35565	0.21036	0.22558	0.16491	0.21619	0.21095	0.21071	0.07693
4200	0.21576	0.16970	0.29830	0.21682	0.30701	0.26906	0.27948	0.20851	0.30065	0.28625	0.35528	0.21375	0.29508	0.06037	0.38048	0.35152	0.20989	0.22460	0.16394	0.21519	0.20997	0.20974	0.07596
4000	0.21384	0.16869	0.29626	0.21578	0.30403	0.26777	0.27764	0.20744	0.29880	0.28411	0.35220	0.21265	0.29288	0.05899	0.37537	0.34725	0.20888	0.22357	0.16293	0.21414	0.20895	0.20873	0.07445
3800	0.21185	0.16763	0.29413	0.21469	0.30094	0.26582	0.27572	0.20632	0.29687	0.28248	0.34896	0.21150	0.29058	0.05755	0.37005	0.34282	0.20781						
3600	0.20977	0.16650	0.29191	0.21354	0.29773	0.26376	0.27369	0.20516	0.29484	0.28044	0.34555	0.21031	0.28818	0.05605	0.36445	0.33819	0.20669						
3400	0.20759	0.16531	0.28957	0.21234	0.29435	0.26160	0.27156	0.20395	0.29269	0.27830	0.34197	0.20905	0.28568	0.05448	0.35885	0.33338	0.20550						
3200	0.20530	0.16405	0.28711	0.21106	0.29081	0.25932	0.26930	0.20267	0.29042	0.27603	0.33818	0.20772	0.28308	0.05283	0.35235	0.32833	0.20424						
3000	0.20289	0.16271	0.28452	0.20971	0.28710	0.25692	0.26691	0.20131	0.28800	0.27363	0.33415	0.20632	0.28033	0.05109	0.34577	0.32301	0.20290						
2900	0.20164	0.16201	0.28317	0.20901	0.28518	0.25568	0.26565	0.20061	0.28674	0.27237	0.33204	0.20559	0.27891	0.05018	0.34234	0.32027	0.20219						
2800	0.20035	0.16128	0.28178	0.20827	0.28319	0.25439	0.26436	0.19987	0.28543	0.27107	0.32986	0.20484	0.27746	0.04925	0.33861	0.31743	0.20146						
2700	0.19902	0.16052	0.28034	0.20750	0.28116	0.25306	0.26302	0.19911	0.28408	0.26972	0.32762	0.20405	0.27597	0.04828	0.33517	0.31453	0.20071						
2600	0.19766	0.15974	0.27886	0.20672	0.27907	0.25168	0.26163	0.19833	0.28270	0.26832	0.32529	0.20325	0.27444	0.04729	0.33140	0.31153	0.19992						
2500	0.19625	0.15892	0.27733	0.20590	0.27691	0.25026	0.26019	0.19751	0.28123	0.26688	0.32286	0.20242	0.27288	0.04626	0.32752	0.30845	0.19911						
2400	0.19479	0.15807	0.27574	0.20506	0.27467	0.24879	0.25870	0.19666	0.27972	0.26538	0.32036	0.20156	0.27126	0.04520	0.32349	0.30528	0.19826						
2300	0.19328	0.15719	0.27410	0.20416	0.27238	0.24728	0.25715	0.19578	0.27815	0.26382	0.31776	0.20065	0.26959	0.04409	0.31934	0.30201	0.19737						
2200	0.19172	0.15626	0.27240	0.20323	0.27001	0.24570	0.25553	0.19485	0.27651	0.26219	0.31505	0.19972	0.26788	0.04295	0.31502	0.29852	0.19645						
2100	0.19010	0.15330	0.27062	0.20227	0.26756	0.24406	0.25385	0.19388	0.27480	0.26049	0.31221	0.19873	0.26610	0.04176	0.31056	0.29511	0.19548						
2000	0.18842	0.15428	0.26877	0.20185	0.26502	0.24236	0.25209	0.19286	0.27302	0.25872	0.30927	0.19772	0.26426	0.04052	0.30592	0.29150	0.19447						
1900	0.18667	0.15322	0.26684	0.20019	0.26249	0.24058	0.25025	0.19180	0.27115	0.25687	0.30618	0.19664	0.26234	0.03923	0.30113	0.28776	0.19340						
1800	0.18485	0.15209	0.26482	0.19905	0.25965	0.23872	0.24831	0.19068	0.26918	0.25492	0.30295	0.19552	0.26034	0.03788	0.29612	0.28389	0.19226						
1700	0.18294	0.15090	0.26269	0.19787	0.25680	0.23679	0.24629	0.18949	0.26711	0.25287	0.29954	0.19431	0.25825	0.03646	0.29001	0.27985	0.19169						
1600	0.18095	0.14964	0.26045	0.19661	0.25385	0.23476	0.24415	0.18823	0.26493	0.25072	0.29597	0.19305	0.25606	0.03497	0.28550	0.27568	0.18981						
1500	0.17885	0.14830	0.25808	0.19526	0.25076	0.23262	0.24189	0.18688	0.26262	0.24843	0.29219	0.19171	0.25374	0.03341	0.27984	0.27130	0.18850						
1400	0.17664	0.14687	0.25557	0.19382	0.24752	0.23036	0.23950	0.18545	0.26016	0.24601	0.28816	0.19028	0.25130	0.03176	0.27395	0.26676	0.18705						
1300	0.17429	0.14533	0.25289	0.19227	0.24413	0.22797	0.23695	0.18391	0.25755	0.24344	0.28691	0.18873	0.24870	0.03002	0.26779	0.26202	0.18551						
1200	0.17179	0.14366	0.25002	0.19061	0.24057	0.22542	0.23424	0.18224	0.25475	0.24069	0.27938	0.18707	0.24592	0.02819	0.26137	0.25705	0.18385						
1100	0.16912	0.14186	0.24693	0.18878	0.23680	0.22239	0.23132	0.18044	0.25173	0.23773	0.27450	0.18526	0.24294	0.02624	0.25465	0.25183	0.18204						
1000	0.16633	0.13987	0.24359	0.18600	0.23280	0.21974	0.22818	0.17846	0.24848	0.23455	0.26928	0.18328	0.23873	0.02416	0.24762	0.24635	0.18006						
900	0.16305	0.13768	0.23986	0.18459	0.22851	0.21653	0.22477	0.17626	0.24493	0.23108	0.26363	0.18108	0.23624	0.02103	0.24029	0.24054	0.17787						
800	0.15955	0.13524	0.23500	0.18213	0.22397	0.21299	0.22102	0.17381	0.24104	0.22729	0.25948	0.17864	0.23240	0.01954	0.23259	0.23436	0.17542						
700	0.15561	0.13246	0.23147	0.17931	0.21877	0.20901	0.21687	0.17104	0.23671	0.22308	0.25073	0.17587	0.22813	0.01696	0.22454	0.22773	0.17264						
600	0.15108	0.12926	0.22646	0.17606	0.21308	0.20445	0.21218	0.16784	0.23184	0.21833	0.24327	0.17265	0.22335	0.01420	0.21607	0.22055	0.16943						
500	0.14574	0.12547	0.22070	0.17221	0.20655	0.19908	0.20675	0.16405	0.22622	0.21248	0.23489	0.16886	0.21748	0.01129	0.20708	0.21263	0.16564						
400	0.13921	0.12083	0.21397	0.16744	0.19881	0.19250	0.20019	0.15941	0.21940	0.20625	0.22531	0.16421	0.21123	0.00833	0.19744	0.20367	0.16100						
298.15	0.13062	0.11472	0.20515	0.16109	0.18884	0.18375	0.19161	0.15330	0.21071	0.19766	0.21379	0.15810	0.20264	0.00540	0.18632	0.19277	0.15489						

References

1. Anderson, J.D.: *Hypersonic and High Temperature Gasdynamics*, 1st edn (McGraw-Hill,1989)
2. Bird, R.B., Stewart, W.E. and Lightfoot, E.N., *Transport Phenomena*, (Wiley, 5th print, 1965)
3. Bosnjakovic, F.: *Technische Thermodynamik*, **1**, (Verlag Theodor Steinkopff, Dresden, 1960)
4. Chandrasekhar, S.: *Radiative Transfer*, (Paperback ed.,Dover Publications, 1969)
5. Chapman, S. and Cowling, T.G.: *The Mathematical Theory of Nonuniform Gases*, (Cambridge University Press, 1970)
6. Chung, P.M., Talbot, L. and Touryan, K.J.: *Electric Probes in Stationary and Flowing Plasmas: Theory and Applications*, (Springer Verlag, New York, 1975)
7. Cobine, J.D.: *Gaseous Conductors*, (Dover Publ. New York, 1958)
8. Etemadi, K. and Mostaghimi, J. (Eds.) *Heat Transfer in Thermal Plasma Processing*, (HTD-Vol., **161** ASME, 1991)
9. Hasted, J.B.: *Physics of Atomic Collisions*, (Butterworths, London, 1964)
10. Hirschfelder, J.O., Curtiss, C.F. and Bird, R.B.: *Molecular Theory of Gases and Liquids*, (John Wiley and Sons, New York, 1967)
11. Hottel, H.C and Sarofim, A.F.: *Radiative Transfer*, (McGraw Hill, 1967)
12. Huddlestone, R.H. and Leonard, S.I.: *Plasma Diagnostic Techniques* (Academic Press., New York, 1965)
13. Kuo, K. K.: *Principles of Combustion*, Paperback, (John Wiley, New York, 1986)
14. Massey, H.S.W. and Burhop, E.H.S.: *Electronic and Ionic Impact Phenomena*, (Clarendon Press,1952), pp. 1694–1703.
15. Mitchener, H. and Kruger, jr., C.H.: *Partially Ionized Gases*, (John Wiley,1973)
16. Moreau, R.J.: *Magnetohydrodynamics* (translated from french by A.F. Wright), (Kluwer Acad. Publishers, 1990)
17. Moore, Ch.: *Atomic Energy Levels*, 3 vols., (National Bureau of Standards USA, 1949)
18. Pai, Shih-I.: *Magnetogasdynamics and Plasma Dynamics*, (Springer Verlag, Vienna, 1962)
19. Siegel, R. and Howell, J.R.: *Thermal Radiation Heat Transfer*, (Hemisphere, 1981)
20. Soo, S.L.: *Fluiddynamics of Multiphase System*, (Blaisdell Waltham, Mass, 1967)
21. Sutton, G.W. and Sherman, A.: *Engineering Magnetohydrodynamics*, (McGraw-Hill, New York, 1965)

22. Swift, J.D. and Schwar, M.J.R.: *Electrical Probes for Plasma Diagnostiques*, (American Elsevier, New York, 1971).

23. Tidman, D.A. and Krall, N.A.: *Shock Waves in Collisionless Plasma*, (Interscience, 1971)

24. Tranter, C.L.: *Integral Transforms in Mathematical Physics*, (Methuen London, 1966)

25. Unsoeld, A.: *Physik der Sternatmosphaeren*, (Springer Verlag, Berlin Heidelberg New York, 1955)

26. Finkelnburg, W. and Maecker, H.: Electric Arcs and Thermal Plasma (in german) In: *Handbuch der Physik*, **22**, (Springer Verlag, Berlin Heidelberg New York, 1956)

27. Finkelnburg, W. and Peters, Th.: Continuous Spectra (in german) In: *Handbuch der Physik*, **28**, (Springer Verlag, Berlin Heidelberg New York, 1957) pp. 79–204

28. Kerrebrock, J.L.: Conduction in gases with elevated electron temperature In: *Engineering Aspects of Magnetohydrodynamics*, (Columbia Press, 1962) pp. 327–46

29. Kollath, R.: Durchgang Langsamer Elektronen und Ionen Durch Gases In: *Handbuch der Physik*, **34**, (Springer Verlag, Berlin Heidelberg New York, 1958) pp. 1-52

30. Monti, R. and Napolitano, L.G.: Generalized Saha equation for non-equilibrium two-temperature plasma, XV Intern. Astron. Congress, Cannes (France), *AGARDOGRAPH 81*, 517 (1964)

31. Shahin, M.M.: Fundamental Definition and Relationships Pertinent to Plasmas: In *Reactions under Plasma Conditions* (Ed.: M. Venugopalan), **I**, (Willey Interscience, 1971)

32. Adamovich, I.V., Macheret, S.O., Rich, J.W. and Treanor, C.E.: Vibrational Relaxation and Dissociation Behind Shock Waves. Part 1: Linear Rate Models, *AIAA Jl.* **33**, 1064 (1995a)

33. Adamovich, I.V., Macheret, S.O., Rich, J.W. and Treanor, C.E.: Vibrational Relaxation and Dissociation Behind Shock Waves. Part 2: Master Equation Modeling, *AIAA Jl.* **33**, 1070 (1995b).

34. Alfven, H.: On the Existence of Electromagnetic-hydrodynamic Waves, *Arkiv Mat. Aston. Fysik*, **29B**, 1 (1943); reprinted in American Institute of Aeronautics and Astronautics (AIAA) selected reprint series (ed. W. Sears) on *Magnetofluiddynamics.*

35. Bedin, A.P. and Mishin, G.I.: Ballistic Studies of the Aerodynamic Drag on a Sphere in Ionized Air, *Tech. Phys. Lett.*, **21**, 5 (1995)

36. Beulens, J.J., Milojivic, D., Schram, D.C. and Vallinga, P.M.: A Two-dimensional Non-equilibrium Model of Confined Arc Plasma Flow, *Physics of Fluids*, **3** Part b, 2548 (1991)

37. Bose, T.K. and Pfender, E.: Direct and Indirect Measurements of the Anode Fall in a Coaxial Arc Configuration, *AIAA J.*, **7**, 1643 (1969)

38. Bose, T.K.: Cross-flow Blowing of a Two-dimensional Stationary Arc *AIAA J.*, **10**, 80 (1972a)

39. Bose, T.K.: Anode Heat Transfer for a Flowing Argon Plasma at Elevated Electron Temperature, *Intern. J. Heat and Mass Transfer*, **15**, 1745 (1972b)

40. Bose, T.K.: On Interaction of the Gas Cross-Flow with Electromagnetic Fields, *Plasma Physics*, **15**, 819 (1973)

41. Bose, T.K. and Seeniraj, R.V.: Two Temperature Elenbaas-Heller Problem with Argon Plasma, *Plasma Physics and Controlled Fusion*, **26**, 1163 (1984a)

42. Bose, T.K. and Seeniraj, R.V.: Laminar Stagnation Point Heat Transfer for a Two-temperature Argon Plasma, *AIAA Jl.*, **22**, 1080 (1984b)

43. Bose, T.K. and Seeniraj, R.V.: Electrostatic Probes for Dense Flowing Plasma, *Physics of Fluids*, **26**, 1561 (1984c)

44. Bose, T.K.: Thermodynamic Analysis of a Seeded Combustion Plasma, *AIAA Paper 86-1333* (1986)

45. Bose, T.K. Thermophysical and Transport Properties of Multi-component Gas Plasmas at Multiple Temperatures, *Progress in Aerospace Sciences*, **25**, 1 (1988)

46. Bose, T.K.: One-dimensional Analysis of the Wall Region for a Multiple-Temperature Argon Plasma, *Plasma Chemistry and Plasma Processing*, **10**, 189 (1990)

47. Bose, T.K.: Thermodynamic Analysis of Magnetogasdynamic Accelerator for Hypersonic Tunnels, *AIAA Paper 99-0870* (1999)

48. Bott, J.F.: Spectroscopic Measurement of Temperature in an Argon Plasma Arc, *Physics of Fluids*, **9**, 1540 (1966)

49. Boulos, H.I. and Gauvin, W.H.: Powder Processing in a Plasma Jet: A Proposed Model, *Can. J. Chem. Engg.*, **52**, 355 (1974)

50. Bourdin, E., Fauchais, P. and Boulos, M.I.: Transient Heat Conduction Under Plasma Conditions, *Intern. J. Heat Mass Transfer*, **26**, 567 (1983)

51. Brokaw, R.: Alignment Charts for Transport Properties, *NASA, Technical Report R-81* (1961)

52. Capitelli, M., Armenise, I. and Gorse, C.. State-to-State Approach in the Kinetics of Air Components Under Reentry Conditions, *J. Thermophysics and Heat Transfer*, **11**, 570 (1997)

53. Chen, Xi and Pfender, E.: Heat Transfer to a Single Particle Exposed to Thermal Plasma, *Plasma Chemistry and Plasma Processing*, **2**, 185 (1982a)

54. Chen, Xi and Pfender, E.: Unsteady Heating and Radiation Effects of Small Particles in a Thermal Plasma, *Plasma Chemistry and Plasma Processing*, **2**, 293 (1982b)

55. Chen, Xi and Pfender, E.: Effect of Knudsen Number on Heat Transfer to a Particle Immersed into a Plasma, *Plasma Chemistry and Plasma Processing*, **3**, 97 (1983)

56. Chen, Xi and Pfender, E.: Modeling of RF Plasma Torch with a Metallic Tube Inserted for Reactant Injection, *Plasma Chemistry and Plasma Processing*, **11**, 103 (1991)

57. Chyou, Y.P. and Pfender, E.: Behaviour of Particulates in Thermal Plasma Flows, *Plasma Chemistry and Plasma Processing*, **9**, 291 (1989a)

58. Chyou, Y.P. and Pfender, E.: Modeling of Plasma Jets with Superimposed Vortex Flow, *Plasma Chemistry and Plasma Processing*, **9**, 291 (1989b)

59. Dilawari, A.H., Szekely, J., Batdorf, J., Detering, R. and Shaw, C.B. : The Temperature Profiles in an Argon Plasma Issuing into an Argon Atmosphre: a Comparison of Measurements and Predictions, *Plasma Chemistry and Plasma Processing*, **10**, 321 (1990)

60. Elenbaas, W.: Die Quecksilber-Hochdruckentladung, *Physica*, **1**, 673 (1934)

61. Elwert, G.: Ueber die Ionisations und Rekombinations prozesse in eimem plasma und die Ionisationsformel der Sonnenkorona, *Zeitschrift für Naturforschung*, **7a**, 432 (1952)

62. Emmons, H.W.: Arc Measurement of High Temperature Gas Transport Properties, *Physics of Fluids*, **10**, 1135 (1967)

63. El-Kaddah, N., McKelliget, J. and Szekely, J.: Heat Transfer and Fluid Flow in Plasma Spraying, *Metallurgical Transactions*, **15B**, 59 (1984)

64. Gauvin, W.H.: Some Characteristics of Transferred-Arc Plasmas, *Plasma Chemistry and Plasma Processing*, **9**, 65S (1989)

65. Giannaris, R.J. and Incropera, F.P.: Radiation and Collisional Effects in a Cylindrically Confined Plasma. I. Optically Thin Consideration, *J. Quant. Spectroscopy and Radiation Transfer*, **13**, 167 (1971)

66. Gnoffo, R.N., Gupta, R.N. and Shina, J.L.: Conservation Equation and Physical Models for Hypersonic Air Flows in Thermal and Chemical Non-equilibrium, *NASA-TP-2867* (1989)

67. Gokhale, S.S. and Bose, T.K.: Reacting Solid Particles in One Dimensional Nozzle Flow, *Intern. J. Multiphase Flow*, **15**, 269 (1989)

68. Heller, G.: Dynamical Similarity Laws of the Mercury High Pressure Discharge, *Physics*, **6**, 389 (1935)

69. Hsu, K.C., Etemadi, K.C. and Pfender, E.: Studies of the Anode Region of a High Intensity Argon Arc, *J. Applied Physics*, **53**, 4136 (1982)

70. Igra, O. and Barcessat, M.: Supersonic Non-equilibrium Corner Expansion Flows of Ionized Argon, *Physics of Fluids*, **20**, 1449 (1977)

71. Incropera, F.P.: Procedure for Modelling Laminar Circular Arc Behaviour, *IEEE Plasma Sciences*, **PS-1**, 3 (1973)

72. Kannappan, D. and Bose, T.K.: Transport Properties of a Two-temperature Argon Plasma, *Physics of Fluids*, **20**, Part 1, 1668 (1977)

73. Kerrebrock, J.L.: Magnetohydrodynamic Generators with Non-equilibrium Ionization, *AIAA Jl.*, **3**, 591 (1965)

74. Knoche, K.F.: Lecture delivered at Verein Deutscher Ingenieure (VDI) Thermodynamics Colloquium, Constance (Germany) (1963)

75. Kruger, C.H.: Non-equilibrium in Confined Arc Plasma, *Physics of Fluids*, **13**, 1737 (1970)

76. Landrum, D.B. and Candler, G.V.: Vibration-Dissociation Coupling in Nonequilibrium Flows, *AIAA-91-0466*, (1991)

77. Leckner, B.: Spectral and Total Emissivity of Water Vapor and Carbon Dioxide, *Combustion and Flame*, **19**, 33 (1972)

78. Lee, Y.C., Chyou, Y.P. and Pfender, E.: Particle Dynamics and Particles Heat and Mass Transfer in Thermal Plasmas. Part II. Particles Heat and Mass Transfer in Thermal Plasmas, *Plasma Chem. and Plasma Processing*, **5**, 1391 (1985)

79. Liu, W.S., Whitten, B.T. and Glass, I.I.:Ionizing Argon Boundary Layers, Part 1. Quasi-steady Flat Plate Laminar Boundary Layer Flow, *J. Fluid Mech.*, **87**, Part 4, 609 (1978)

80. Lu, Z.P., Stachowicz, L., Kong, P., Heberlein, J. and Pfender, E.: Diamond Synthesis by D.C. Plasma CVD at 1 atm, *Plasma Chem. and Plasma Processing*, **11**, 387 (1991)

81. Maecker, H.: Ein Lichtbogen für hohe Leistungen, *Zeitschrift für Physik*, **129**, 108 (1951)

82. McCune, J.C. and Resler, Jr., E.L.: Compressibility Effects in Magnetohydrodynamic Flows Past Thin Bodies, *J. Aerospae Sciences*, **27**, 493 (1960)

83. Mehta, R.C. and Bose, T.K.: Energy Transfer in Cathode Region of an Arc, *Intern. J. Heat Mass Transfer*, **26**, 1959 (1983)

84. Mishin, G.I.: Total Pressure Behind a Shock Wave in Weakly Ionized Air, *Tech. Phys. Letters*, **21**, 857 (1994)

85. Monchik, L: Collision Integrals for the Exponential Repulsive Potential, *Physics of Fluids*, **2**, 695 (1959)

86. Morris, J.C. and Yos, J.M.: Radiation Studies of Arc-heated Plasmas, Aerospace Research Laboratories, *Report No. ARL 71-0317*, (1971)

87. Morris, J.C., Krey, R.U. and Bach, G.R.: Bremsstrahlung and Recombination Radiation of Neutral and Ionized Nitrogen, *Phys. Review*, **159**, 113 (1967)

88. Morris, J.C. and Garrison, R.L.: Bremsstrahlung and Recombination Radiation of Atomic and Ionic Oxygen. *J. Quant. Spectr. Radiative Transport*, **6**, p. 899 (1966)

89. Morro, A. and Romeo, M.: Thermodynamic Derivation of Saha's Equation for a Multi-temperature Plasma, *J. Plasma Physics* **39**, Part 1, 41 (1988)

90. Mostaghimi, J. and Boulos, M.I.: Two-dimensional Electromagnetic Field Effects in Induction Plasma Modelling, *Plasma Chemistry and Plasma Processing*, **9**, 25 (1989)

91. Mostaghimi, J., Proulx, P. and Boulos, M.I.: An Analysis of the Computer Modeling of the Flow and Temperature Fields in an Inductively Coupled Plasma, *Numerical Heat Transfer*, **8**, 187 (1985)

92. Nick, K.P., Richter, J. and Helbig, V.: Non-LTE Diagnostics of an Argon Arc Plasma, *J. Quant. Spectr. Radiative Transfer*, **32**, 1 (1984)

93. Oettinger, P.E. and Bershader, D.: A Unified Treatment of Relaxation Phenomenon in Radiating Argon Plasma Flows, *AIAA Jl.*, **5**, 1625 (1967)

94. Pai, Shih-I. and Speth, A.I.: Shock-waves in Radiation Magnetogasdynamics, *Physics of Fluids*, **4**, 1232 (1961).

95. Paik, S., Chen, Xi, Kong, P. and Pfender, E.: Modeling of a Counterflow Plasma Reactor, *Plasma Chemistry and Plasma Processing*, **11**, 229 (1991)

96. Paik, S.H. and Pfender, E.: Modeling of an Inductively Coupled Plasma at Reduced Pressures, *Plasma Chemistry and Plasma Processing*, **10**, 167 (1990)

97. Park, C.: On Convergence of Computation of Chemically Reacting Flows, *AIAA 85-0247* (1985)

98. Park, C.: Two-temperature Interpretation of Dissociation Rate Data for N_2 and O_2, *AIAA 88-0458* 1988

99. Park, C.: Assessment of the Two-temperature Kinetic Model for Ionizing Air, *J. Thermophysics and Heat Transfer*, **3**, 233 (1989)

100. Parsi, P.S. and Gauvin, W.H.: Heat Transfer from a Transferred-Arc Plasma to a Cylindrical Enclosure, *Plasma Chemistry and Plasma Processing*, **11**, 57 (1991)

101. Pfender, E: Particle Behavior in Thermal Plasmas, *Plasma Chemistry and Plasma Processing*, **9**, 167S 9 (1989 Supplement)

102. Rizk, M.A. and Elgobashi S.E.: A Two-Equation Turbulence Model for Dispersed Dilute Confined Two Phase Flows, *Intern. J. Multiphase Flow*, **15**, 119 (1989)

103. Sanders, N.A. and Pfender, E.: Measurement of Anode Falls and Anode Heat Transfer in Atmospheric Pressure High Intensity Arcs, *J. Applied Physics*, **55**, 714 (1984)

104. Seikel, G.R. and Reshotko, E.,: Hall Current Accelerator, *Bull. Amer. Phys. Soc.*, **7**, 414 (1962)

105. Shih, K.T., Pfender, E., Ibele, W.E. and Eckert, E.R.G.: Experimental Heat Transfer Studies in a Co-axial Arc Configuration, *AIAA Jl.*, **6**, 1482 (1968)

106. Spalding, D.B.: *Lecture delivered at Verein Deutscher Ingenieure (VDI) Thermodynamics Colloquium*, (Constance (Germany), 1963)

107. Sanden (Van de), M.C.M., Schramm, P.P.J.M., Peeters, A.G., Van der Mullen, J.A.M. and Kroesen, G.M.W.: Thermodynamic Generalization of the Saha Equation for a Two-temperature Plasma, *Physical Review A*, **40**, 5273 (1989)

108. Veis, S.: The Saha Equation and Lowering of the Ionization Energy for a Two Temperature Plasma, *Proc. Czechoslovak Conference on Electronics and Vacuum Physics*, 105 (1968)

109. Wilhelmi, H., Wimmer, W. and Pfender, F.: Modeling of Transport Phenomena in the Anode Region of High Current Arcs, *Numerical Heat Transfer*, **8**, 731 (1985)

110. Zhiguler, V.N., Rominshevski, Ye. A. and Ventushkin, V.K.: On the Role of Radiation in Modern Problems of Gasdynamics, *AIAA Jl.*, **1**, 1473 (Russian Supplement, 1963). Translated from *Inzhenernii Zhurnal*, **1**, 60 (1961)

111. Zukowski, E.E., Cool, T.A. and Gibson, E.G.: Experiments concerning non-equilibrium conductivity in a seeded plasma, *AIAA Jl.*, **3**, 1410 (1964)

112. Zukowski, E.E. and Cool, T.A.: Non-equilibrium electrical conductivity measurements in argon and helium seeded plasma, *AIAA Jl*, **3**, pp. 370 (1965)

113. Knoche, K.F.: Waermediagramme von Argon Plasma bis $10^5\,°K$ mit Anwendungsbeispiele, Ph.D. Dissertation, Tech. Univ., Braunschweig, Germany (1961)

List of Symbols

The more commonly used symbols are given below followed by units. The numbers denote the chapters in which the symbol is mainly used.

A surface area, m^2, 4

A' surface element area, m^2, 4

A_{mn} coefficient for probability of spontaneous emission, s^{-4}, 4,10

a a constant, 3

a_s isentropic sonic speed, ms^{-1}, 9

$\mathbf{B}$ magnetic induction, Vsm^{-2}, 7

B_p blowing parameter, 11

B_{mn} coefficient for probability of induced emission, $\text{m}^2\text{J}^{-1}\text{s}^{-1}$, 4

B_{nm} coefficient for probability of absorption, $\text{m}^2\text{J}^{-1}\text{s}^{-1}$, 4

B_λ spectral intensity of radiation per unit wavelength, Wm^{-3}, 4

B_ν spectral intensity of radiation per unit frequency, J.m^{-2}, 4

B^* total intensity of equilibrium radiation, Wm^{-2}, 4

B_λ^* spectral intensity of equilibrium radiation per unit wave length, Wm^{-3}, 4

B_ν^* spectral intensity of equilibrium radiation per unit frequency, Jm^{-2}, 4

b distance, m, 5

b_j mobility coefficient of j-th species, $\text{m}^2\text{V}^{-1}\text{s}^{-1}$, 7,11

C capacitance, AsV^{-1}, 9

C_1, C_2 constants, 4

C_p, C_v molar specific heat at constant pressure and constant volume, respectively, J(kmole.K)^{-1}, 3

c velocity of light in vacuum, ms^{-1}

c Alfven speed, ms^{-1}, 12

c circumference, m, 12

c_f friction coefficient, 12

c_p, c_v specific heat at constant pressure and constant volume, respectively, J(kg.K)^{-1}

$c_{p,eff}$ effective specific heat at constant pressure, J(kg.K)^{-1}

$\mathbf{D}$ electric displacement vector, Asm^{-2}

D diameter or hydraulic diameter, m, 12

D_j diffusion coefficient of j-th species, m^2s^{-1}, 7

D_{jk} binary diffusion coefficient between j-th and k-th species, m^2s^{-1}, 7

D_o characteristic diameter, m, 12
D_{amb} ambi-polar diffusion coefficient, $m^2 s^{-1}$, 7
d diameter of a rigid particle, m, 5
$\mathbf{E}$ (total) electric field, Vm^{-1}, 7
$\mathbf{E'}$ externally applied electric field, Vm^{-1}, 7
$\mathbf{E}_{amb}$ ambi-polar electric field, Vm^{-1}, 7
E energy, J, 2,3
E specific internal energy, Jkg^{-1}, 11
E_{ac} activation energy, J, 5
E_D dissociation energy, J, 2
E_J energy of a rotor, J, 2
E_{lim} cut-off energy, J, 3
E^o total specific internal energy, Jkg^{-1}, 11
E_{mi} ionization potential (energy) per unit mass ion, Jkg^{-1}, 8
E_{rot} total rotational energy, J, 5
E_v vibrational energy of a single oscillator, J, 2,5
E_{vib} total vibrational energy, J, 5
E emissive power, Wm^{-2}, 4
e elementary charge, As, 2
$\dot{e}_a$ absorbed radiant energy, Wm^{-3} or $Wm^{-3}.sterad.^{-1}$, 4
$\dot{e}_e$ emitted radiant energy, Wm^{-3}, 4
$\dot{e}_i$ radiant energy release by induced emission, $Jm^{-3}.sterad.^{-1}$, 4
$\dot{e}_s$ radiant energy release by spontaneous emission, $Jm^{-2}.sterad.^{-1}$,4
$\dot{e}_R$ volumetric radiation energy release, W or Wm^{-3} or $Wm^{-3}.sterad.^{-1}$, 4
$\dot{e}_\nu$ spectral volumetric radiative energy release, Jm^{-3} or $J/m^{-3}.sterad.^{-1}$, 4
$\mathbf{F}$ force, N, 5
$\mathbf{F}$ volumetric force, Nm^{-3}, 5
F a function of electron mole fraction, 6
F free energy, $J(kg.K)^{-1}$ or $J(kmole.K)^{-1}$, 3
$\mathbf{G}$ mass averaged speed, ms^{-1}, 5
G free enthalpy, $J(kg.K)^{-1}$ or $J(kmole.K)^{-1}$, 3
$\mathbf{g}$ earth gravitation vector, ms^{-2}
g statistical weight, 2
g number of boxes (energy levels), 2
g relative speed, ms^{-1}
g_j mass fraction of j-th species
H magnetic field, Am^{-1}
H molar enthalpy, $Jkmole^{-1}$, 3
h specific enthalpy, Jkg^{-1}, 9
h Planck's constant, Js, 2
I electric current, A, 7
I_i ionization potential, J, 3,8
I_λ angular spectral intensity of radiation with respect to wave length, $Wm^{-3}.sterad.^{-1}$, 4

I_ν angular spectral intensity of radiation with respect to frequency, $Jm^{-2}.sterad.^{-1}$, 4

I^* total intensity of angular equilibrium radiation, $Wm^{-2}.sterad.^{-1}$, 4

I_λ^* angular spectral intensity of equilibrium radiation with respect to wavelength, $Wm^{-3}.sterad.^{-1}$, 4

I_ν^* angular spectral intensity of equilibrium radiation with respect to frequency, $Jm^{-2}.sterad.^{-1}$, 4

I_L intensity of line radiation, $Wm^{-2}.sterad.^{-1}$, 4

i ionization state

i index for primary components, 6

i volumetric line intensity, $Wm^{-3}.sterad.^{-1}$, 10

J rotational quantum number, 2

j index for all components, 6

j emission coefficient, $Wm^{-3}.sterad.^{-1}$, 4

$\mathbf{j}_c$ conduction current density, Am^{-2}

$\mathbf{j}_j$ current density of the j-the species, Am^{-2}

j spectral mass emission coefficient, $Jm^{-3}.sterad.^{-1}$, 4

K_x equilibrium constant based on mole fraction, 6

K_n equilibrium condtant based on number density, 6

K_p equilibrium constant based on partial pressure, 6

$\mathbf{k}$ electric current unit vector, 12

$\mathbf{k}_c$ convection electric current vector, 12

k (total)thermal conductivity coefficient, $Wm^{-1}K^{-1}$, 7

k spring constant, kgs^{-2}, 2

k reaction rate constant, m^2s^{-1} or s^{-1}, 5

k_B Boltzmann constant, JK^{-1}

k_c thermal conductivity by pure conduction, $Wm^{-1}K^{-1}$, 7

k_D diffusive (reactive) thermal conductivity coefficient, $Wm^{-1}K^{-1}$, 7

k_R radiative thermal conductivity coefficient, $Wm^{-1}K^{-1}$, 4

k_{eff} effective thermal conductivity coefficient, $Wm^{-1}K^{-1}$, 7

k_{cj} contribution of j-th species to the total heat conductivity coefficient, $Wm^{-1}K^{-1}$, 7,8

L a characteristic length, m

L inductance, VsA^{-1}, 9

L latent heat of vaporization of electrons, $Jkmole^{-1}$, 8

L_o latent heat of vaporization of electrons at absolute zero temperature, $Jkmole^{-1}$, 8

Le Lewis number

l directional cosine, 4

l optical thickness of a radiating gas, m, 4

M Mach number, 12

M mass of a particle, kg, 2,3

M_A Alfven Mach number, 12

M_j mass of the j-th species, kg, 2

M_p magnetic blowing parameter, 12
m average mole mass
m_j mole mass of the j-th species, $kgkmole^{-1}$
m_R mass rate of production, $kgm^{-3}s^{-1}$, 11
$\dot{m}$ mass flow rate, kgs^{-1}, 12
N dimensionless number
N number of balls (particles), 2,5
N_A Avogadro number, $kmole^{-1}$
n principal quantum number, 2
n number density of the gas, m^{-3}, 3
n total number of components, 6
n_c charge density, Asm^{-3}, 7
n_j number density of the j-th species, m^{-3}
n_R number density of photons, m^{-3}, 4
$\dot{\mathbf{n}}$ flux of particles, $m^{-2}s^{-1}$
$\dot{\mathbf{n}}_f$ flux of photons, $m^{-2}s^{-1}$, 4
Pr Prandtl number, 7
p pressure, bar or Nm^{-2}
p particle momentum, $kgms^{-1}$, 2,5
p_ν spectral radiative pressure, $Jsm^{-3}.sterad.^{-1}$, 4
p^* total equilibrium radiative pressure, Nm^{-2}, 4
p_ν^* spectral radiative pressure for equilibrium radiation, $Jsm^{-3}.sterad.^{-1}$, 4
Q collision cross-section, m^2, 5
Q total or integrated radiation, W, 4
Q_{jf} energy production in electromagnetic field, Wm^{-3}, 11
$Q_{j,coll}$ energy production due to collision, Wm^{-3}, 11
$\mathbf{q}$ heat flux, Wm^{-2}
$\mathbf{q}_b$ heat flux at free-fall edge, Wm^{-2}, 8
$\mathbf{q}_w$ heat flux to the wall, Wm^{-2}, 8
$\mathbf{q}^R$ radiant heat flux, Wm^{-2}, 4
q electric charge, As
q production rate of electron-ion pair, $m^{-3}s^{-1}$, 9
q_k primary mole fraction ratio, 5
q_{eff} effective charge, As, 3
q_{eff} effective charge number, 6
R external resistance, VA^{-1}, 9
R_j reaction rate of j-th species, $m^{-3}s^{-1}$
R_j gas constant of j-th species, $Jkg^{-1}K^{-1}$
R^* universal gas constant, $Jkmole^{-1}K^{-1}$
R_H Rydberg constant, m^{-1}, 2
Re Reynolds number, 11
R_H magnetic pressure parameter, 12
R magnetic Reynolds number, 12
R_d radiation parameter, 12

R_h Hartmann number, 12
R_{cj} Larmor or cyclotron radius, m, 7
r radial distance, m
r distance between two particles, m, 3
r number of primary components
S molar entropy, Jkmole^{-1}K^{-1}, 3
S Saha function, 6
S magnetic force parameter, 12
Sc Schmidt number, 7
S source function, Wm^{-2}sterad.$^{-1}$, 4
S^* entropy of equilibrium radiation, 6
s line element, m, 2
s spin quantum, 2
T temperature, K, 9
T_j translational temperature of the j-th species, K
T_d characteristic dissociation temperature. K, 7
t time, s
$\mathbf{U}$ characteristic velocity, ms^{-1}, 12
$\mathbf{U}$ velocity vector, ms^{-1}, 12
U electric potential, V, 8,9
U_o externally applied potential, V, 9,10
u potential drop, V, 9
u gas velocity in x-coordinate direction, ms^{-1}, 11
u^* total equilibrium radiative energy, Jm^{-3}, 6
u spectral internal energy of radiation, Jsm^{-3}.sterad.$^{-1}$, 4
u_ν^* spectral equilibrium internal radiative energy, Jsm^{-3}sterad.$^{-1}$, 4
$\mathbf{V}$ fluid velocity, ms^{-1}
$\mathbf{V}_j$ velocity of the j-th species, ms^{-1}, 11
$\mathbf{V}_j'$ relative diffusive velocity of j-th species, ms^{-1}, 11
$\mathbf{V}_{fj}$ mass-average velocity of j-th species in the electro-magnetic field, ms^{-1}, 11
V^* molar volume, m^{3}kmole^{-1}, 3
v vibration quantum number
v kinetic speed of a single particle, ms^{-1}, 3,5
v reduced kinetic speed, ms^{-1}, 3
v',v'' kinetic speed of a single particle before and after collision, respectively, ms^{-1}, 5
W number of possibilities, 3
w_j velocity of a single particle of j-th species, ms^{-1}
x coordinate, m, 2
x a measure for anharmonicity, 2
x_j mole fraction of the j-th species
Z partition function, 3,10
z number of interferometric fringes, 10

α influence coefficient for vibration on characteristic rotational temperature, K, 2

α number of ionizing collision per unit distance, m^{-1}, 9

α number of ionizing collision per unit time, s^{-1}, 9

α inverse Alfven Mach number, 12

α_i degree of ionization, 6

β positive ion coefficient, m^{-1}, 9

Γ collision frequency, s^{-1}, 5

Γ' volumetric collision frequency, $m^{-3}s^{-1}$, 5

γ specific heat ratio, 1

Δh^o heat of reaction, $Jkmole^{-1}$, 7

$\Delta \dot{e}$ gain (or loss) of heat flux in free-fall region, Wm^{-2}, 8

δ length of the gas column, m, 10

δ_{rs} Kroneker delta, 11

Δ arc deflection, m, 12

ϵ emissivity coefficient, 4

ϵ dielectric of a medium, $AsV^{-1}m^{-1}$, 7

ϵ_o dielectric constant in vacuum, $AsV^{-1}m^{-1}$, 7

ϵ_R radiative energy gain (or loss), Wm^{-3}, 11

Θ_r characteristic rotational temperature, K, 3

Θ_v characteristic vibrational temperature, K, 3

θ temperature ratio, 6

$\bar{\kappa}$ average absorption coefficient, $m^{-1}s^{-1}$, 4

κ_ν spectral absorption coefficient, m^{-1}, 4

κ_R Rosseland mean absorption coefficient, m^{-1}, 4

λ wavelength, m, 2

λ Lagrange first undetermined multiplier, 3

λ mean free path, m, 5

λ_D Debye shielding distance, m, 6

μ reduced mass, kg, 2

μ Lagrange second undetermined multiplier, 3

μ permittivity of gas, 7

μ dynamic viscosity coefficient, $kgm^{-1}s^{-1}$, 7

μ mass-flux rate, $kgm^{-2}s^{-1}$, 9

μ refractive index, 10

μ_o magnetic permeability in vacuum, $VsA^{-1}m^{-1}$, 7

ν frequency of radiation, s^{-1}, 2

$\bar{\nu}$ wave number, m^{-1}, 2

ν a matrix of chemical valency, 6

ν_d frequency of discharge, s^{-1}, 9

ν_j chemical valency of the j-th component, 6

ν_p plasma frequency, s^{-1}, 10

ν_{orb} orbiting frequency, s^{-1}, 2

ξ radian cyclotron frequency to collision frequency ratio, 7

ξ_{jk} kinetic velocity function for binary collision, ms^{-1}, 5
ρ density, kgm^{-3}, 7,9
ρ_d characteristic density for a dissociated gas, kgm^{-3}, 7
σ Boltzmann constant of radiation, $Wm^{-2}.K^{-4}$, 4
σ electrical conductivity, $AV^{-1}m^{-1}$, 7,11
σ_o scalar electrical conductivity, $AV^{-1}m^{-1}$, 7,11
τ radiative momentum flux, Jsm^{-3}, 4
τ optical length, 4
τ relaxation time, s, 5
τ shear stress, Nm^{-2}, 11
ϕ potential (energy), J, 5
ϕ steric factor, 5
ϕ dissipation function, $Nm^{-2}.s^{-1}$, 11
ϕ heat conductivity potential, Wm^{-1}, 11
ϕ_o work function, V
ϕ_w wall potential, V
χ angle of deflection, 5
ψ collision angle, 5
ψ wave function, 2
ψ solid angle, steradian, 4
ω coefficient matrix, 6
ω velocity gradient or vorticity, s^{-1}
ω_o radian cyclotron frequency, $radians.s^{-1}$, 7

Subscripts

a atom
e electron
h heavy particles
i singly-charged ion
j j-th species
w wall condition

Index

Printing: Saladruck Berlin
Binding Lüderitz&Bauer, Berlin